LINEAR PROGRAMMING FOR OPERATIONS RESEARCH

LINEAR PROGRAMMING FOR OPERATIONS RESEARCH

DONALD M. SIMMONS

W. R. Grace & Co.
Cambridge, Massachusetts

HOLDEN-DAY, INC.
San Francisco, Cambridge, London, Amsterdam

500 Sansome Street, San Francisco, California 94111

Library of Congress Catalog Card Number: 70–188129
ISBN: 0–8162–7986–1

Printed in the United States of America

for Juanita

PREFACE

Of all the techniques of operations research, linear programming is by far the best known and most widely practiced. In this day and age, all sensible planners and decision-makers—even those who by nature are somewhat suspicious of any quantitative method more sophisticated than long division—will acknowledge that some linear optimization problems are worth the trouble and expense of solving. The solution methods for these problems have the virtues of being respectably old, easily understood, highly efficient, and readily available in computer-coded form. Also contributing heavily to the popularity and importance of linear programming are the great difficulties which are generally encountered in solving nonlinear optimization problems. Because of these difficulties, the operations research analyst is frequently tempted to apply the methods of linear programming to problems that have serious nonlinearities.

The pros and cons of using linear models in nonlinear situations constitute just one of the many practical questions discussed in this book. The book is principally intended for the future practitioner of operations research; its two major concerns are the art of recognizing and formulating linear programming problems and the science of solving them. The art of formulation is given heavy emphasis throughout the text; in fact, an entire chapter is devoted to the modeling of various real-world problems as linear programs, with close attention being paid to which sorts of relationships can be treated as linear and which cannot.

The development of solution algorithms is also consistent with the requirements of operations research: Coverage is broad, computational advantages and disadvantages are discussed, and the underlying mathematical theory is carefully established. Regarding this last point, it is important to provide a firm mathematical foundation, especially in duality theory, for those students who will want to do further work in nonlinear programming (a subject, incidentally, which is covered by this author in another volume, *Nonlinear Programming for Operations Research*, to be published in late 1972).

Apart from my concern for operations research, I had a more general purpose in writing this book, namely, to make linear programming more accessible to students of widely varied skills and experience. A basic axiom is that it is not necessary to be a genius to understand this material; at present, too many intelligent students with modest mathematical backgrounds are shying away from linear programming in the belief that it is more sophisticated and difficult than is actually the case. Accordingly, a great deal of care has been taken in this book to state theorems explicitly and to prove them in a clear and orderly progression, with the eventual destination (i.e., the establishment of some solution algorithm) being kept in sharp focus all along the way. Digressions from the main thread of discourse for historical background, computational clarification, and interpretive discussion are usually isolated and presented under their own section headings.

The book is intended to accompany the standard one-semester or one-trimester course that appears, with variations, in management, engineering, and economics curricula. The chief prerequisite is, of course, linear algebra, but because of the organization of the text the student need not have already taken it; the material required for linear programming is normally presented in the first month or so of a linear-algebra course. In order to allow students without this background to take linear algebra and linear programming concurrently, this book has been arranged so that Chapter 3 (Duality Theory) may be covered before Chapter 2 (Mathematical Background). The instructor wishing to reorder the material in this way can introduce Chapter 3 by merely defining matrices, vectors, and their transposes, and then showing how a set of simultaneous linear equations or inequalities can be written compactly in matrix form. On the other hand, instructors who prefer to make linear algebra a prerequisite for the course will find in Chapter 2 a compendium of all necessary definitions and results from that field. In both cases, it should be noted that no familiarity whatever with simultaneous equation systems or convex set theory is assumed; these topics are presented de novo in Chapter 2.

There is, however, one inescapable prerequisite for this text: a semester of differential calculus. Because of the fundamental role duality theory plays throughout mathematical programming, it is accorded in these pages the position of first importance and is in fact used to establish the optimality condition of the simplex method. The proof given for the crucial duality theorem is that of Dreyfus and Freimer, which requires a first-order Taylor expansion. Moreover, differential calculus is also used in the various interpretations of reduced costs and dual variables.

The precise topics included in the text and the degree of emphasis placed on each are described in the table of contents. Note that substantial coverage is given to postoptimality problems and sensitivity analysis, which are matters of major concern in the practice of operations research. Wherever possible, interpretations of mathematical results are presented within an economic or resource-allocation framework, both because this is the natural context of operations research in any professional discipline and because experience shows that students find these sorts of interpretations easiest to understand. Computational topics are included insofar as they would be of interest to operations research analysts: solution times for linear programs of various shapes and sizes, relative advantages and disadvantages of competing algorithms, computational considerations which might affect problem formulation, and so on. No descriptions of specific computer codes, however, will be found.

In closing, I would like to acknowledge and thank Dr. Felipe Ochoa and Professor Alan Martin, who introduced me to mathematical programming; Edward Millman and Professor Gilbert Howard, who provided suggestions and criticism; and to my typist, Connie Spargo, who read my handwriting.

Donald Simmons
Cambridge, Mass.
January 1972

CONTENTS

PREFACE vii

1 INTRODUCTION TO LINEAR PROGRAMMING 1

1.1 Optimization problems 1
1.2 A note on operations research 3
1.3 Deterministic and probabilistic models 5
1.4 Mathematical programming problems 6
1.5 Linear programs 8
1.6 Some sample formulations 10
1.7 The geometry of linear programming 14
1.8 Outline of the text 17
Exercises 18

2 MATHEMATICAL BACKGROUND 23

2.1 Prerequisites 23
2.2 Linear algebra assumed 24
2.3 Simultaneous linear equations 31
2.4 Hyperplanes and convex sets 37
2.5 Some additional definitions 41
Exercises 44

3 DUALITY THEORY 47

3.1 An example of dual linear programs 47
3.2 The role of duality 50
3.3 Canonical form of the linear program 50
3.4 Transformations to canonical form 51
3.5 Standard form and transformations to it 54
3.6 The dual problem 55
3.7 Some properties of primal and dual problems 62
3.8 The duality theorem 64
3.9 Economic interpretation of primal and dual 69
3.10 The existence theorem 75
3.11 Complementary slackness 77
Exercises 82

4 THE SIMPLEX METHOD 89

4.1 Introduction to the method 89
4.2 Extreme points and basic feasible solutions 90
4.3 Desirable properties of a solution method 94
4.4 Existence of a basic feasible solution 95
4.5 Definitions 99
4.6 Increasing the objective value: the optimality theorem 101
4.7 Pivoting to a new basic feasible solution 106
4.8 Computational procedures 110
4.9 Degeneracy and convergence of the simplex method 118
4.10 Uniqueness and multiple optima 120
4.11 Minimization problems 121
4.12 Economic interpretation of the simplex method 122
4.13 Two remarks 124
Exercises 125

5 THE TWO-PHASE ALGORITHM AND THE REVISED SIMPLEX METHOD 131

5.1 Desirable adjuncts to the simplex method 131
5.2 Phase 1: the initial basic feasible solution 133
5.3 The redundant constraints 137
5.4 Artificial variables in phase 2 139
5.5 Summary 141
5.6 Further remarks on the two-phase method 145
5.7 A one-phase method for solving linear programs 147
5.8 The revised simplex method 149
5.9 Summary 152
5.10 Revised versus standard simplex 155
Exercises 158

6 ILLUSTRATIVE CASE PROBLEMS 163
6.1 Linear problems in a nonlinear world 163
6.2 Resource allocation 164
6.3 The diet problem 168
6.4 Machine scheduling 169
6.5 The transportation problem 173
6.6 Job training 178
6.7 An investment problem 181
6.8 A minimax strategy 184
6.9 Production and inventory planning 185
Exercises 194

7 THE DUAL SIMPLEX METHOD AND POSTOPTIMALITY PROBLEMS 197
7.1 Introduction and motivation 197
7.2 The associated dual solution 199
7.3 The dual simplex method 203
7.4 Summary 209
7.5 Postoptimality problems 212
7.6 Problems involving discrete parameter changes 213
7.7 Problems involving continuous parameter changes 220
7.8 Problems involving structural changes 230
Exercises 233

8 THE TRANSPORTATION PROBLEM 239
8.1 The problem in its basic context 239
8.2 Alternate formats and variations 241
8.3 Mathematical properties of the transportation problem 244
8.4 The transportation tableau 248
8.5 Loops and linear dependence 251
8.6 The initial basic feasible solution 254
8.7 The transportation algorithm: the reduced costs 260
8.8 The transportation algorithm: pivoting to a new BFS 263
8.9 Summary 267
8.10 Computational remarks 271
8.11 Network flow theory 272
Exercises 273

REFERENCES 282

INDEX 284

LINEAR PROGRAMMING FOR OPERATIONS RESEARCH

1

INTRODUCTION TO LINEAR PROGRAMMING

1.1 OPTIMIZATION PROBLEMS

One hears a great deal these days about our new national pastime, known variously as "problem-solving," "optimization," and "analysis." Spokesmen for the Defense Department, for example, appear on Capitol Hill and in the evening news discussing the optimal mix of strategic weapons: so much offense, so much defense. Since resources are not unlimited, they usually advocate "maximum payoff for the defense dollar" (in an older, less sophisticated age it was simply "a bigger bang for a buck"). Later, on the same news show, a well-known commentator does not merely report on the situation in the Middle East; instead, he offers a two-minute "analysis" of the problems there. Shifting to the local scene, the Mayor's administrative assistant announces that the best solution to the city's financial problems has been found. "Think of it as an optimum budget adjustment," he suggests, "rather than just another raise in property taxes"—if possible, he is standing in front of a computer. Meanwhile the school board has solved its budget problem by switching to split sessions and night classes in order to obtain "maximum utilization of educational facilities." Lawyers debate the optimal assignment of penalties to crimes: In making kidnapping a capital offense, does society gain more from deterring potential crime than it loses when kidnappers, having nothing more to lose, kill their victims? Medical researchers estimate optimum drug dosage as a function of body weight. Politicians take positions on issues after analyzing voter preference polls, while businessmen figure optimal inventory levels and sportswriters analyze the pennant race. Even painters and film directors speak of their work in terms of posing and solving problems.

Indeed, everybody's doing it, and it's getting to be a little confusing. Should any unsettled state of affairs requiring someone to make a decision be called a "problem"? What does it mean to "solve a problem"—is it the same as finding an "optimum"? What, exactly, is an optimum in contexts such as strategic weapons systems, drug dosage, and budget allocation? Presumably the optimum is "best," but best in what way?

Inasmuch as this entire textbook will be devoted to finding optimal solutions for a very special type of mathematical problem, we ought to begin by assigning precise meanings to some of these terms; the definitions we shall state are generally accepted in the professions of engineering, economics, applied mathematics, and management science (there being considerably less unanimity among film directors and sportswriters). An *optimization problem* ultimately faced by a decision-maker is one of choosing from many alternatives the one that yields a maximum or minimum value of some numerically measurable criterion of performance. The maximizing or minimizing policy is known as the *optimum* or *optimal solution*, and the process of finding it is called *solving* the problem. Note the phrase "ultimately faced by a decision-maker": The necessity of coming to a decision and implementing it in the real world gives focus and meaning to the optimization problem, but the actual work of solving it often is performed by a specialist or team of specialists who then report to the decision-maker.

In general, there are a great many alternative ways to solve a problem. Some, however, may be impossible to implement because they violate laws of physics or established rules and regulations or because they require the expenditure of more resources than are available. The conditions and restrictions that determine which courses of action can be adopted and which cannot are called *constraints*; the courses of action that satisfy all constraints are said to be *feasible*. Suppose, for example, that a plastic surgeon wants to invest up to $20,000 in stocks ABC and XYZ, now selling at $20 and $100 per share, respectively. There are many different ways for him to allocate his money (including buying no stock at all), but the upper limit of $20,000 on his total outlay constitutes a constraint. To buy 500 shares of ABC and 50 of XYZ would be a feasible policy, assuming there are no other constraints, because 500($20) + 50($100) = $15,000 is not greater than $20,000; however, any plan to purchase more than 200 shares of XYZ would be *infeasible* because it would violate the constraint.

The performance criterion is the *measure of effectiveness* of any policy and is therefore the crucial component of the optimization problem. It must be explicitly stated and must take on a single numerical value (possibly infinity) for each feasible policy. This implies that the value or effectiveness of each element of every policy must be measurable on the same scale: If it is not possible to evaluate everything in terms of a single unit (as in military decisions in which both dollars and lives are involved), then our definition of an optimization problem is not satisfied. By far the most frequently used measure of effectiveness is total dollar cost, which serves as the principal—if not exclusive—yardstick of businessmen and engineers. Particularly in business, measurement in terms of dollars is quite natural and convenient: Most of the required cost data (including wages, raw-material prices, inventory carrying charges, etc.) are readily available, and those that are not (such as the cost of a customer's ill will when his order cannot be filled on time) usually can be estimated in some rational manner.

1.2 A NOTE ON OPERATIONS RESEARCH

The optimization problem as we have defined it sounds rather scientific and mathematical and, in fact, it is. Although the terminology is general enough to admit problems from a great many professional disciplines, certain specific structural elements must be present, including a numerical measure of effectiveness. These requirements are somewhat restrictive (they seem to eliminate film directors, for example, and probably foreign-policy analysts as well), but they serve an important purpose in guaranteeing that *any optimization problem can be expressed in terms of mathematical relationships and then solved by means of computational methods.*

The process of translating a real-world situation into mathematical language is called formulating or *modeling* the problem. Ordinary algebraic variables are used to represent a decision-maker's initially unknown policy choices (e.g., "let x_1 be the number of new employees to be hired in month 1") with the value of each variable corresponding to some specific choice. The analyst or model-builder then selects equations and inequalities that define permissible or feasible sets of values for all the variables, ruling out those sets of values that are prohibited by the constraints of the problem. Finally, the measure of effectiveness is expressed as a single mathematical function, which is to be maximized or minimized subject to the variables' remaining within their feasible ranges.

Having formulated a mathematical optimization problem, the analyst must proceed to solve it. The methods or *algorithms* for doing so range in difficulty from simple pencil-and-paper arithmetic to advanced calculus requiring sophisticated computer programming. In this text we shall examine one of these algorithms, the *simplex method*, which, like most of the others, is applicable only to a certain type of optimization problem.

The art of modeling optimization problems and the mathematical techniques for solving them are the twin concerns of *operations research*, a discipline that might be described generally as a scientific methodology for making decisions. Operations research as we know it today has its origins, or at least its early formative development, during World War II. Military planning at the general staff level had become so complex and the problems so gigantic that the planners began to feel the need for an orderly, formalized approach to decision-making; this was particularly true with regard to identifying different goals and objectives and measuring the relative desirability of each. At a lower level, certain tactical problems (such as how to search for an enemy submarine) were recognized to be essentially mathematical in nature, and mathematical solution procedures were devised for some of them. After the war, the general methodology of operations research spread throughout the business world and then into various sectors of the federal government, most notably the Defense Department and the Bureau of the Budget. Tremendous forward strides were made in the techniques of applied mathematics, as methods for solving many different types of optimization problems were discovered and perfected. By the early 1960's, the influence of operations research was being felt in the fields of economics and statistical analysis and in all the social sciences. More recently, the new methodology has had a widespread bureaucratic impact. Universities now offer degrees in the "decision sciences," and many state and local governments have established offices for "program-planning" and "systems analysis."[1]

Concurrent with the development of operations research, and contributing heavily to it, has been the development of the electronic computer, which has made it possible to come to grips with the huge problems arising in industry, economics, and government. As a practical matter, these problems, which often include thousands of variables and hundreds of constraints, simply cannot be approached without the computer, even if they are relatively straightforward, because the number of man-hours of hand calculation required is usually prohibitive. Moreover, the sophisticated mathematical algorithms that have been devised during the last three decades for solving the more complicated optimization problems also owe much to the computer. Without it there would have been far less interest in finding a computational procedure, say, for minimizing a quadratic function subject to linear constraints because such a method could not be applied by hand to problems having more than 10 or 12 variables—again, too much arithmetic would be involved. The computer has provided much of the motivation for recent advances in applied mathematics and computational techniques. We can appreciate the truth of this remark by reflecting on the fact that the great flowering of operations research after World War II has really contributed rather little to basic mathematical theory. Most of the theorems underlying the new methods have been known for decades or even centuries—the

[1] It is probably impossible to state definitions for these terms that would be acceptable to everyone who uses them. *Program-planning* usually refers specifically to the preparation of a budget, while *systems analysis* is roughly synonymous with operations research and connotes the decision-making methodology and approach to optimization problems that have been described in this section.

applications are new. It seems reasonable to suppose that many of these applications would have been developed 50 years earlier if the means for harnessing and exploiting them had then been available.

1.3 DETERMINISTIC AND PROBABILISTIC MODELS

As operations research has developed, optimization problems have tended to divide into two broad classes, the *deterministic* and the *probabilistic*. Deterministic problems are those in which uncertainty is either negligible or entirely absent, so that, for any policy that the decision-maker might choose, the resources required and the eventual outcome can be predicted with complete certainty. In the real world, truly deterministic problems are quite rare (one example might be the cutting of a number of strips of fabric or sheet metal of various lengths and widths ordered by customers from a single parent roll of constant width in such a way as to minimize the total length of parent stock required). In most problems that are modeled as deterministic, however, small errors and uncertainties are neglected or deliberately ignored: Manufacturing costs are usually estimated rather than known, measuring instruments with 99.5% accuracy are treated as perfect, completion dates for construction projects are established with "normal" allowance for bad weather, and so on. In such circumstances the use of a deterministic model is justified only if the deviations in practice are expected to be both infrequent and small.

Probabilistic or *stochastic* optimization problems, on the other hand, involve a degree of uncertainty that is simply too great to be ignored. For example, under certain assumptions about random arrivals of passengers at an airport ticket counter, the number of people waiting in line for service might have a long-run average value of 5, but it might rise as high as 50 once or twice a day. It would clearly be wrong in this case to calculate the optimal number of ticket counters on the basis of a deterministic model that used only the average value of 5 and ignored the occasional 50. In order to represent uncertainty of this sort, stochastic optimization models use *random* variables, whose values are described by probability distributions rather than by single numbers; as a result, stochastic problems are generally much more complicated and harder to solve than deterministic problems.

Two rather different styles of treatment and families of solution methods have evolved for these two classes of optimization problems. More and more, they are being taught in different courses and out of different textbooks, and in view of the great proliferation of new theorems and algorithms over the past quarter century, it may already be fair to say that an operations research analyst can be expert in one area or the other, but rarely in both. This textbook is no exception to the general trend: We shall henceforth be concerned exclusively with the simpler deterministic class of problems or, rather, with a certain important subclass of them.

1.4 MATHEMATICAL PROGRAMMING PROBLEMS

Having described a general methodology for solving real-world decision problems by constructing mathematical models to represent them, we now turn our attention to the basic model itself. A *mathematical programming problem* or *mathematical program* is the algebraic formulation of a deterministic optimization problem and has the general format

$$\text{Max or Min} \quad f(x_1, \ldots, x_n) \tag{1-1}$$

$$\text{subject to} \quad g_i(x_1, \ldots, x_n) \begin{Bmatrix} \leq \\ = \\ \geq \end{Bmatrix} b_i \qquad i = 1, \ldots, m \tag{1-2}$$

where the b_i are scalar constants and all the functions f and g_i are real-valued. When (1-1) and (1-2) are used as a model for a decision problem, the variables represent the decision-maker's possible policy choices, the relations (1-2) express the constraints, and the function f is the measure of effectiveness.

Let us now leave the real world and consider the problem from a strictly mathematical point of view. Any particular assignment of numerical values to the variables $x_1, \ldots, x_n$ is called a *solution* to the problem and may be thought of as a point in Euclidean n-space E^n. The equations and/or inequalities (1-2) are the *constraints* and a solution is said to be *feasible* or *infeasible* according to whether it does or does not satisfy every one of the constraints. The collection of all feasible solutions or points in E^n is known as the *feasible region.* Incidentally, when no constraints are present (that is, when $m = 0$) we still use the term "mathematical program"; the feasible region then consists of all of E^n and there are no infeasible solutions.

The function f to be optimized—that is, maximized or minimized—will be referred to in this text as the *objective* or *objective function.* A feasible solution is an *optimal solution* or an *optimum* if it yields a value of the objective function that is as good (i.e., as large or as small) as or better than the value yielded by every other feasible solution. Thus "optimal" implies "feasible." From the definition it is clear that the optimal solution to a given problem need not be unique, but that the optimal value of the objective function must be.

As an example, consider the mathematical program

$$\begin{array}{ll} \text{Max} & x_2 \\ \text{subject to} & x_1^2 + x_2^2 \leq 1 \\ \text{and} & x_1, x_2 \geq 0 \end{array}$$

The feasible region consists of the sector of the unit circle that lies in the first quadrant; it is graphed in Figure 1-1. The points (0,0), (1,0), (0,1), ($\frac{1}{2}$,$\frac{1}{2}$), and

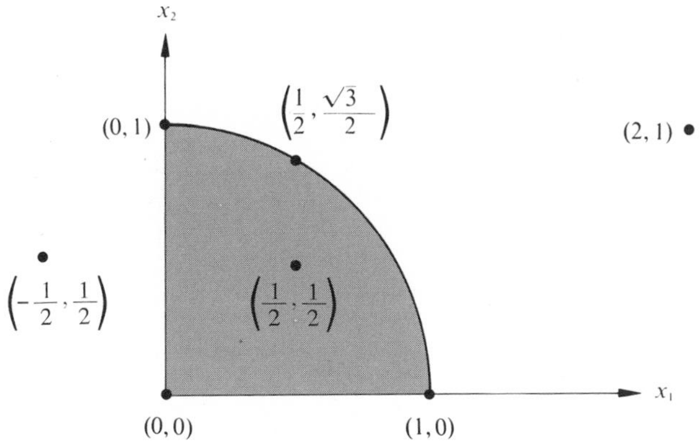

Figure 1-1

$(\frac{1}{2},\sqrt{3}/2)$ are feasible solutions, while (2,1) and $(-\frac{1}{2},\frac{1}{2})$ are infeasible. The optimal solution, which maximizes the objective function $f(x_1,x_2) = x_2$ subject to the given constraints, is (0,1) and is unique.

The goal in solving a mathematical programming problem is, of course, to identify an optimal solution (and its associated value of the objective); when there are multiple optima it may or may not be satisfactory to find just one of them. We should not assume, however, that an optimal solution always exists. It may happen in a given problem that there is no assignment of values to the variables that satisfies all the constraints. In such a case, when no feasible solution exists, the problem itself is said to be *infeasible*. Another possibility is that in a problem having feasible solutions the optimal value of the objective function is positively or negatively infinite. We say that such a problem is *unbounded* or has an unbounded optimal solution. When this occurs the feasible region itself is usually unbounded, here meaning "of infinite extent." A simple example would be the maximization of $f(x) = x^3$ subject to $x \geq 0$. But a nonlinear problem with a bounded (or finite) feasible region also can have an unbounded solution, as would be the case if $f(x) = x^{-1}$ is maximized subject to $-1 \leq x \leq +1$.

The branch of applied mathematics concerned with solving problems of the form (1-1) and (1-2) is known, appropriately, as *mathematical programming*. Various types of mathematical programs are distinguished according to the nature of the functions f and g_i. The simplest type, in which all the functions are linear, is called a *linear program* (or linear programming problem) and is, of course, the subject of this text. All other varieties may be referred to collectively as *nonlinear programs*. It is useful, however, to single out certain problem types within the general nonlinear class for which special solution methods have been devised. These include the *quadratic program*, which has linear constraints and a quadratic objective, and the *geometric program*, in which all the functions are polynomials of a certain form.

In addition, there are other kinds of programs. A *stochastic program* satisfies the format of (1-1) and (1-2), but takes its name from the fact that it is a deterministic formulation of what is actually a probabilistic decision problem (instead of the random variables themselves, the model uses the expected values of various functions of them). An *integer program* is a linear programming problem with the additional feature that some or all of the variables are required to take on integer values only. Finally, "dynamic programming" does not refer to a specific problem-type at all, but is the name given to a general strategy for solving optimization problems by proceeding one stage or one variable at a time; thus the term does not belong in the same semantic grouping with linear programming, mathematical programming, and the others.

1.5 LINEAR PROGRAMS

At last we are ready to begin discussion of our central topic. The great importance of the *linear programming problem* (LPP) is due to the coincidence of two factors:

(1) A wide variety of linear (or nearly linear) problems arise in engineering, management science, economics, and other areas.
(2) Several very efficient[1] techniques, or algorithms, are available for solving them.

Because nonlinear optimization problems, in general, are very difficult to solve, the second of these factors actually outweighs the first—so much so that problems with serious nonlinearities are frequently beaten and hammered into linear formulations in order that the LPP solution procedures can be applied. In fact, mild distortions of this sort are sometimes quite justifiable on practical grounds: For problems of moderate size or greater (upwards of, say, 50 constraints and 100 variables) nonlinear solution methods are intolerably time-consuming, even on the fastest computers.

We have defined the linear program as an optimization problem of the form (1-1) and (1-2), in which the objective and constraint functions are all linear. This definition emphasizes the fact that the LPP is simply a special type of mathematical programming problem. Most writers and researchers, however,

[1] An *efficient* method of solving a problem is one that requires relatively little time on an electronic computer or, equivalently, relatively few arithmetic operations. As a secondary meaning in another context, an efficient approach can also be one that requires relatively little computer storage space.

prefer that the definition of a linear program also include nonnegativity requirements for all the variables. They would define the LPP in the following mathematical terms:

$$\text{Max or Min } z = c_1x_1 + c_2x_2 + \cdots + c_nx_n \tag{1-3}$$

subject to the constraints

$$a_{i1}x_1 + a_{i2}x_2 + \cdots + a_{in}x_n \begin{Bmatrix} \leq \\ = \\ \geq \end{Bmatrix} b_i \qquad i = 1, \ldots, m \tag{1-4}$$

and to the *nonnegativity restrictions*

$$x_j \geq 0 \qquad j = 1, \ldots, n \tag{1-5}$$

As is customary, the scalar z represents the (variable) value of the objective function.

Because most problems that arise in the real world in fact require that the variables of interest take on only nonnegative values, the latter definition seems more natural to engineers and corporate planners. Surprisingly, although there seems to be a difference between these two definitions of the linear program, they are equivalent. Inasmuch as nonnegativity restrictions are simply linear constraints, any problem satisfying the formulation (1-3) through (1-5) also trivially satisfies our more general definition of an LPP [namely, a problem of type (1-1) and (1-2) with all functions linear]. Conversely, any problem that is an LPP under the general definition can be reformulated in terms of nonnegative variables by means of the following transformation:

$$\text{Let} \quad x_j = x_j' - x_j''$$
$$\text{where} \quad x_j' \geq 0 \quad \text{and} \quad x_j'' \geq 0$$

for $j = 1, 2, \ldots, n$. Observe that every real value of x_j can be expressed by nonnegative values of x_j' and x_j''. Thus an objective function and constraints that were formerly linear in n unrestricted variables are now linear in $2n$ nonnegative variables and the problem is essentially unchanged. It turns out, in fact, that a linear program can be represented in several distinct, but equivalent, formats; this topic is covered thoroughly in Chapter 3.

Actually, there is also an important mathematical reason for singling out the nonnegativity restrictions for special attention. In solving linear programming problems we shall treat nonnegativity differently from the way in which we treat the other constraints (1-4). The solution procedure we shall use for the LPP, called the *simplex method*, will specifically require that all variables be prohibited from assuming negative values. So efficient is this simplex method that even when we are given a linear program consisting only of (1-3) and (1-4), without the nonnegativity restrictions, we shall transform the variables as shown above in order to be able to make use of it.

1.6 SOME SAMPLE FORMULATIONS

In this section we present three examples of linear programming problems arising in various real-world contexts; others can be found in the exercises at the end of the chapter. The major purpose of these examples, aside from providing motivation for the study of linear programming, is to introduce the sometimes-tricky art of *modeling*, or mathematical formulation. In Chapter 6, after the student has become familiar with various properties of linear programs and methods for solving them, a number of case problems will be discussed in much greater detail; the pros and cons of treating certain dubious relationships as linear will be considered and a few modeling tricks will be revealed.

EXAMPLE 1.1. *Resource allocation.* The American Eagle Munitions Company manufactures three products: bullets, hand grenades, and artillery shells. Current company inventory includes 100 tons of gunpowder, 150 tons of lead, and 150 tons of steel. To manufacture a ton of bullets the company must use 0.7 ton of lead, 0.1 ton of gunpowder, and 0.2 ton of steel. One ton of hand grenades requires 0.5 ton each of steel and gunpowder, while a ton of artillery shells requires 0.5 ton of lead, 0.3 ton of gunpowder, and 0.2 ton of steel. The sole objective of American Eagle is, of course, to maximize profits, which are \$300, \$600, and \$900 per ton for bullets, grenades, and artillery shells, respectively. Using only the inventory on hand, how much of each should the company produce?

Let b, g, and a represent the numbers of tons of bullets, grenades, and artillery shells to be produced. Consider first the constraint imposed by the limited supply of gunpowder: Every ton of bullets manufactured must consume 0.1 ton of gunpowder, and so on. Since only 100 tons of it are available, we have the constraint

$$0.1b + 0.5g + 0.3a \leq 100$$

The inequality is needed because the company does not have to use all the available gunpowder. Similarly, the lead and steel constraints are

$$0.7b \qquad\quad + 0.5a \leq 150$$

and

$$0.2b + 0.5g + 0.2a \leq 150$$

Because negative production is impossible,

$$b, g, a \geq 0$$

Finally, the objective function must be

$$\text{Max } z = 300b + 600g + 900a$$

where z represents the total profit in dollars for any values of b, g, and a. We have thus formulated a linear program in three variables and six constraints (of which three are nonnegativity restrictions).

EXAMPLE 1.2. *The transportation problem.* A company has S_1 and S_2 tons of some commodity available at its two storage warehouses and wishes to transport various amounts of it to each of three retail stores, where the demands are for D_1, D_2, and D_3 tons, respectively. Let c_{ij} be the (nonnegative) cost per ton of shipping from the ith warehouse to the jth store. What shipping plan will satisfy the demands at a minimum cost?

The problem can be readily visualized with the aid of Figure 1-2. The major arrows in the diagram should be thought of as roads running from warehouses to stores; they are labeled with their per-unit shipping costs. We remark first that we must assume $S_1 + S_2 \geq D_1 + D_2 + D_3$; if more commodity were demanded than could be supplied, the problem would not have a feasible solution. Let x_{ij} be the number of tons shipped from the ith warehouse to the jth store. Then the objective function is

$$\text{Min } z = \sum_{i=1}^{2} \sum_{j=1}^{3} c_{ij}x_{ij}$$

a simple minimization of total cost. In order to insure that the total amount of

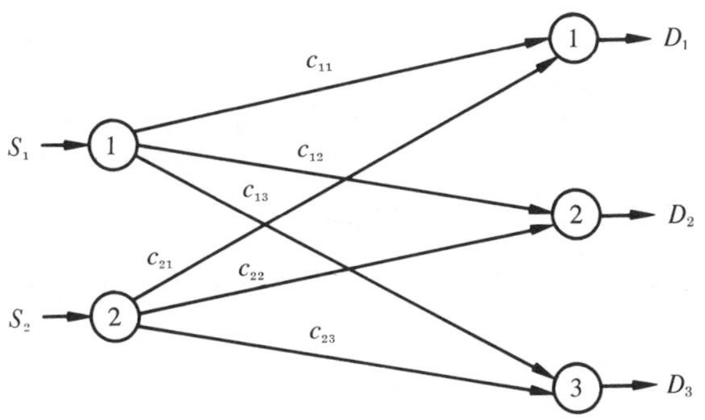

Figure 1-2

commodity arriving at each store satisfies the demand there, we must have the constraints

$$\sum_{i=1}^{2} x_{ij} = D_j \qquad j = 1, 2, 3$$

There is no need to allow for the possibility that the demand might be oversatisfied; because all transportation costs are nonnegative, it can never be optimal to ship more than is demanded. We must require also that the amount of commodity shipped out of each warehouse not exceed the amount originally stored there:

$$\sum_{j=1}^{3} x_{ij} \le S_i \qquad i = 1, 2$$

Here the inequality is appropriate because total supply may exceed total demand. Finally, negative amounts may not be shipped:

$$x_{ij} \ge 0 \qquad \text{for all } i \text{ and } j$$

This produces another linear program of the form (1-3) through (1-5). Incidentally, we shall see in Chapter 8 that the transportation problem is not just an ordinary linear program; its structure allows us to solve it by a special computational procedure that is even more efficient than the simplex method.

EXAMPLE 1.3. *Production scheduling.* A metal foundry employs 1000 workers. In one week each of them can either produce two swords from raw materials or convert one sword into a plowshare, but not both. At present the foundry has an ample supply of raw materials on hand, but no swords or plowshares. In three weeks all swords and plowshares will be sold for \$20 and \$75 apiece, respectively. How many of each item should be produced during each of the three weeks in order to maximize total sales? Assume for simplicity that a sword can be converted into a plowshare *only* in the week immediately after it has been produced.

Let s_i be the number of workers producing swords during the ith week. It is obviously not necessary to allow for the possibility that workers might be left idle because producing an extra couple of swords is always (at least) \$40 better than doing nothing. The number of men converting swords into plowshares in the ith week therefore must be $(1000 - s_i)$, and the manpower constraints are simply

$$s_i \ge 0$$

and

$$1000 - s_i \ge 0 \qquad i = 1, 2, 3$$

Because no swords are initially available for conversion during week 1, we must have $s_1 = 1000$, and the constraints reduce to

$$0 \leq s_2 \leq 1000 \tag{1-6}$$

and

$$0 \leq s_3 \leq 1000 \tag{1-7}$$

with $s_1 = 1000$ being treated as constant. Of course (1-6) and (1-7) are each equivalent to two simple linear constraints.

We also need to insure that in the ith week we do not assign more workers to the profitable conversion of swords into plowshares than there were swords produced in week $i - 1$; that is,

$$(1000 - s_i) \leq 2s_{i-1} \qquad \text{for } i = 2, 3$$

Furthermore, for $i = 2$ this restriction,

$$(1000 - s_2) \leq 2s_1 = 2000$$

is automatically satisfied by the fact that s_2 cannot be negative. Thus the only production-linking constraint that must be explicitly stated is

$$1000 - s_3 \leq 2s_2$$

or

$$2s_2 + s_3 \geq 1000 \tag{1-8}$$

The required constraints for this problem are therefore (1-6), (1-7), and (1-8).

As for the objective function, a man working on swords produces 2($20) = $40 worth of goods, while a plowshare worker "destroys" a sword to create a plowshare, thereby adding a value of $75 − $20 = $55. The objective function, therefore, is

$$\text{Max } z = 40(1000 + s_2 + s_3) + 55(0 + 1000 - s_2 + 1000 - s_3)$$

or

$$\text{Max } z = 150{,}000 - 15s_2 - 15s_3$$

The latter is equivalent to

$$\text{Max } z = -s_2 - s_3$$

as will be shown in Chapter 3. Again, we have formulated a linear program in nonnegative variables.

1.7 THE GEOMETRY OF LINEAR PROGRAMMING

It is worthwhile to learn to visualize linear programming problems geometrically. This is not to say that we shall offer graphical proofs of important theorems; on the contrary, the theoretical structure of this text is entirely algebraic. Nevertheless, the LPP and the simplex method have direct geometrical interpretations that provide a clear and accurate picture of what is going on—clarity that hardly can be obtained from the mathematical equations alone. Of course, we shall not be able to diagram problems of more than two variables, but the two-dimensional geometry we shall discuss is analogous in a straightforward way to that of linear programs in any number of variables.

Consider the example

$$\text{Max } z = x_1 + 3x_2$$
$$\text{subject to} \quad x_1 + x_2 \leq 1$$
$$\text{and} \quad x_1, x_2 \geq 0$$

For this problem the feasible region is the shaded right triangle shown in Figure 1-3. Any point (x_1,x_2) satisfying the first constraint must lie on or below the line $x_1 + x_2 = 1$, and any point satisfying the nonnegativity restrictions must lie in the first quadrant. Therefore, all points corresponding to feasible solutions of the linear program must lie in the *intersection* of these two regions, which is the triangle of Figure 1-3.

The three parallel lines shown represent the sets of points at which the objective function equals 1, 2, and 3, respectively. To economists these lines are known as *indifference curves* because all points on any one of them are equally desirable: If a decision-maker were given any line and asked to choose the best

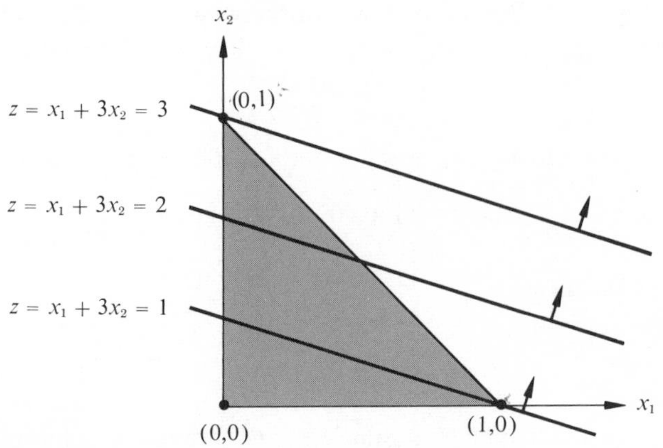

Figure 1-3

point on it he would be indifferent. For example, every point within the triangle on the line $z = 1$ is a feasible solution to the given LPP with an objective value of 1. Can this be the greatest attainable value? Obviously not, because a segment of the line $z = 2$ lies within the feasible triangle, as does a point on the line $z = 3$. Figure 1-3 demonstrates that this last is, in fact, the maximum value that the objective can assume at a feasible point. The same result could have been obtained by plotting any one of the indifference curves, which all have the form $z = x_1 + 3x_2 = k$ for some constant k, and then translating or moving this line parallel to itself in the direction of increasing z, as shown by the tiny attached arrows in Figure 1-3. Translation continues, with z increasing, so long as some part of the line touches the feasible region and stops finally when the "end" of the feasible region is reached. The last value of the objective is optimal. The whole process can be simulated by placing a pencil at the proper angle and rolling it across the diagram.

Although this example, like most LPPs, has a unique optimal point, it is quite possible for an infinite number of alternate optima to exist, all having the same objective value. An example is

$$\begin{aligned} &\text{Max } z = x_1 + x_2 \\ &\text{subject to} \quad x_1 + x_2 \leq 1 \\ &\text{and} \qquad\qquad x_1, x_2 \geq 0 \end{aligned}$$

Here the feasible region is again the right triangle of Figure 1-3, but now the indifference curves are parallel to the hypotenuse and the best feasible indifference curve is the hypotenuse itself. Thus any one of its points must be an optimal solution to the problem. Note that if the objective were to *minimize* $z = x_1 + x_2$, the optimum (minimum) would occur at the origin and would be unique.

Consider now a different example, whose feasible region is diagrammed in Figure 1-4:

$$\begin{aligned} &\text{Max } z = x_1 + x_2 \\ &\text{subject to} \quad x_1 - 2x_2 \leq 0 \\ &\qquad\qquad\quad 2x_1 - x_2 \leq 3 \\ &\text{and} \qquad\qquad x_1, x_2 \geq 0 \end{aligned}$$

In this case the feasible region is unbounded (the student should check that its borders match the given constraints.) Translation of an indifference curve in the direction of increasing z can go on indefinitely without ever exhausting the feasible region; some feasible point or points will lie on the line $x_1 + x_2 = k$

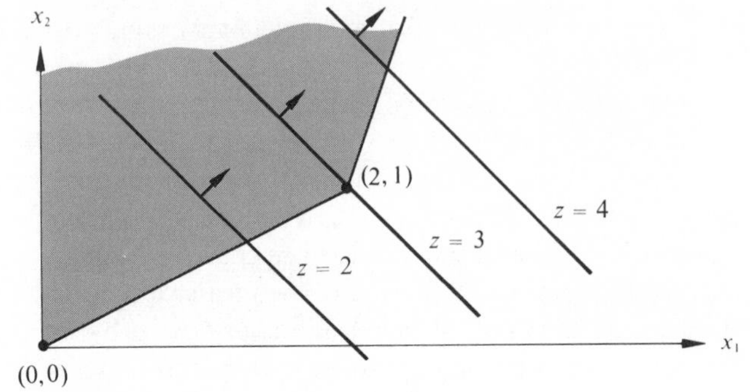

Figure 1-4

for every nonnegative k, no matter how large. Thus the objective function may become infinitely large, and we say that both the problem and the objective function z are unbounded. One optimal solution to the example would be $x_1 = 0$, $x_2 \to \infty$.

Recall that the feasible region may be unbounded though the problem is not. This would occur if the objective were to minimize $z = x_1 + x_2$ over the feasible region shown in Figure 1-4. Another possibility is that the problem is bounded but some infinitely distant points are optimal. For example, try minimizing x_1 in Figure 1-4.

A third major class of linear programs, in addition to those having finite and infinite optima, is exemplified by

$$\begin{aligned} \text{Max } z &= x_1 + x_2 \\ \text{subject to } \quad x_1 \phantom{{}-x_2} &\le 1 \\ x_1 - x_2 &\ge 2 \\ \text{and} \quad x_1, x_2 &\ge 0 \end{aligned}$$

This is a simple example of an infeasible problem (see Figure 1-5). The intersection of the four sets of points satisfying each of the four constraints is an empty set, so the given LPP has no feasible solutions.

The student should now have a clear mental picture of linear programming problems in two variables. For three variables the objective and constraint functions will be of the form $ax_1 + bx_2 + cx_3$, which is simply a plane in euclidean three-dimensional space E^3. The feasible region, then, must be a solid block bounded by planes, while the objective function's indifference surfaces are parallel planes slicing through it. As in the two-dimensional case, we can think of the objective plane being translated through the solid feasible region in the direction of improving z, with the last feasible point touched being the optimum.

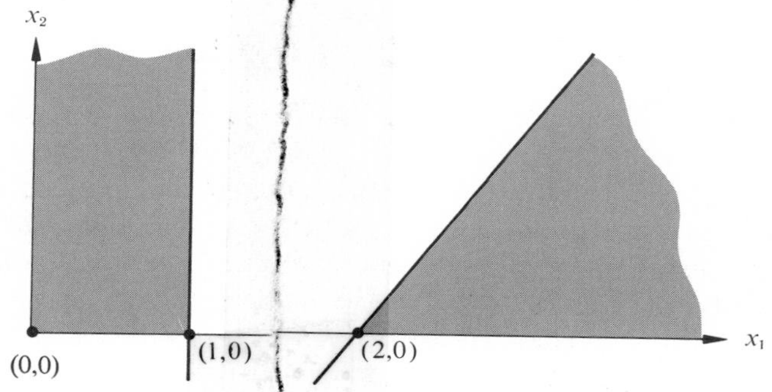

Figure 1-5

What about higher dimensions? Spatial visualization is impossible, of course, but everything that can happen in E^n can be deduced and understood by extrapolating from the analogous situations in two and three dimensions. Keep these simple pictures in mind and the occasional references to the geometry of linear programming will shed much light on the algebraic development.

Incidentally, we could also draw similar diagrams of *nonlinear* problems in two variables to illustrate feasible regions, unbounded optima, and so on. The parallel indifference curves would be produced by plotting the function $f(x_1,x_2) = k$ for different values of the constant k. Of all the indifference curves touching the feasible region, the optimal point would lie on the one whose value of k is best (i.e., largest in a maximization problem).

1.8 OUTLINE OF THE TEXT

Before rolling up our sleeves and coming to grips with the simplex method for solving linear programs, we shall turn our attention first to its theoretical basis. The most important property of the linear program is that it coexists with a second, or *dual*, LPP, which is derived from the first by a simple algebraic manipulation. The two are associated in an intricate and symmetric way; for example, it turns out that when we solve a given LPP via the simplex method we are also solving its dual simultaneously! This relationship is of fundamental importance throughout optimization theory. Lagrange multipliers, for example, can be interpreted as the variables in a sort of dual problem, and nonlinear programming is founded largely upon a generalization of the same idea. We shall develop "duality theory" first and then use it to prove a crucial theorem on which the simplex method depends. As a result, the student will know a fair amount about linear programming problems before he learns (Chapter 4) how actually to solve one. From this perspective, the simplex method will appear, as is fitting, to be a computational black box, a cookbook subroutine that an analyst can call upon after identifying his decision problem and formulating it as a linear program.

After covering linear programming in some detail, we shall devote an entire chapter to a collection of basic case problems. Several variations of each will be discussed, with emphasis on whether certain commonly occurring real-world relationships can be modeled as linear. Then, in Chapter 7, we shall consider some slightly more advanced topics, including various procedures for obtaining additional information about a linear program after its optimal solution has been found. The final chapter of our text will be concerned with a very important subspecies of linear program called the *transportation problem* and will show how the highly specialized structure of its constraints allows us to solve it by methods even more efficient than the simplex algorithm.

Here and there throughout the book we shall rely upon the student's basic mathematical background. Although none of the material we shall need—vectors and matrices, simultaneous linear equations, and convex set theory—is at all advanced (some of it harks back to high school), experience has shown that most students have either forgotten or never learned various parts of it. Therefore, we have collected this basic background material in Chapter 2, and we strongly advise the student to read it carefully. In any case, the various theorems stated or proved in this preliminary chapter will be used frequently.

As a final note, a few of the leading texts in linear programming are listed in the references. That of Dantzig [5] deserves special mention inasmuch as he can be considered the father of the field. The student who seeks clarification or amplification of a topic in our present text, however, should probably turn first to Hadley [15], with whom we have much notation and terminology in common.

EXERCISES

Section 1.5

1-1. Nonmathematicians often make statements such as "If I am elected, I promise to deliver the best possible government at the lowest possible cost to the taxpayers." What, in general, is wrong with this sort of objective?

1-2. Reformulate the following linear programming problem in terms of six non-negative variables:

$$\text{Max } z = 5x_2 - x_3$$

$$\text{subject to} \quad x_1 + 2x_2 + 3x_3 \le 20$$

$$2x_2 - x_3 = 0$$

$$\text{and} \quad x_1 - x_2 + x_3 \ge 10$$

$$\text{with} \quad x_1, x_2, \text{ and } x_3 \text{ unrestricted}$$

Now reformulate it using only four nonnegative variables. How many nonnegative variables would have been required if the linear program had had N unrestricted variables?

1-3. Show that the optimal solution to the linear programming problem

$$\begin{aligned} &\text{Max } z = 2x_1 + x_2 \\ &\text{subject to} \quad x_1 + x_2 = 2 \\ &\text{and} \qquad\qquad x_1, x_2 \geq 0 \end{aligned}$$

is *not* the same as the optimal solution to the problem which is obtained by solving the constraint for x_2 in terms of x_1 and substituting into the objective:

$$\begin{aligned} &\text{Max } z = x_1 + 2 \\ &\text{subject to} \quad x_1 \geq 0 \end{aligned}$$

Why not?

Section 1.6

Formulate Exercises 1-4 through 1-7 as linear programming problems, taking the conservative approach of including all constraints that might conceivably be relevant. Then discuss which of the constraints (if any) could immediately be deleted without affecting the optimal solution.

1-4. (a) The E-Z Gro Fertilizer Company has decided to manufacture a large supply of various plant foods to be sold during the upcoming spring planting season. The company can invest up to \$25,000 in the three basic ingredients, which are nitrates, costing \$800 per ton; phosphates, at \$400 per ton; and potash, at \$1000 per ton. Three standard grades of plant food will be produced from these ingredients: E-Z Regular, in which the nitrates, phosphates, and potash are combined, respectively, in a 3:6:1 ratio by weight and which can be sold for \$750 per ton; E-Z Xtra, a 4:4:1 mixture selling for \$800 per ton; and E-Z Super, a 6:4:3 mixture selling for \$900 per ton. If E-Z Gro's objective is simply to maximize profits (i.e., total sales minus total expenditures for ingredients) and if its production capacity permits it to manufacture no more than 40 tons of plant food overall, how much of each ingredient should it buy and how much of each grade of plant food should it produce?

(b) Repeat part (a) subject to the additional proviso that E-Z Gro can somehow earn an immediate 10% on all capital *not* invested in nitrates, phosphates, and potash; that is, we are now supposing that the company can make an extra profit of \$1 for every \$10 not spent on the three ingredients.

1-5. Let us consider Example 1.2 one stage earlier in the distribution process. Suppose D tons of the commodity, where $D = D_1 + D_2 + D_3$, are available at the factory for shipping to the two warehouses, where it will be stored for a while before being shipped to the retail outlets. Storage costs at the two warehouses are a_1 and a_2 dollars per ton, and their capacities are k_1 and k_2 tons. Ignoring the cost

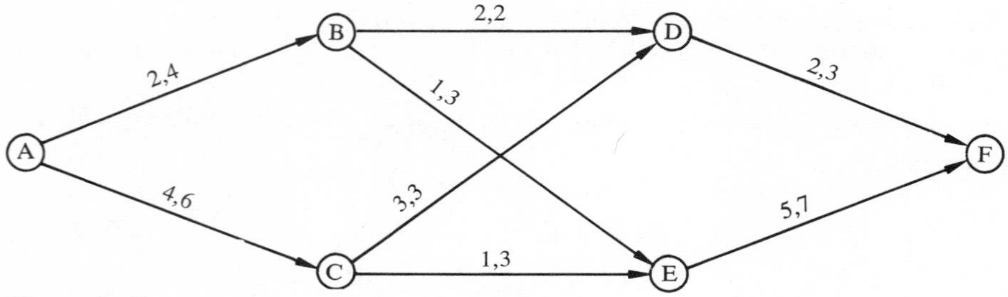

Figure 1–6

of sending the commodity from the factory to the warehouses (Can this cost actually be absorbed into the a_i?), what overall shipping plan will satisfy the retail-store demands at minimum total cost?

1-6. Nine units of some commodity must be shipped from city A to city F over the roads diagramed in Figure 1-6. Each arrow represents a road and is labeled with its per-unit shipping cost and capacity, in that order; for example, a maximum of 3 units may be shipped from city B to city E at a cost of \$1 per unit. How should the nine units be shipped from A to F in order to minimize the total shipping cost?

1-7 (a) Antoine's Meat Market sells two different kinds of hamburger, both of which are ground fresh every day. The regular grade contains 20% ground beef and 80% oatmeal and is priced at p_1 cents per pound, while the deluxe grade is 30% beef and 70% oatmeal and sells for p_2 cents per pound. Antoine can buy up to L pounds of beef per day at c_1 cents per pound, but he must pay c_2 cents for each additional pound, where $c_2 > c_1$. Oatmeal costs c_3 cents per pound, regardless of quantity. If Antoine can spend a total of D dollars per day on these two ingredients, how much of each should he purchase daily and how much regular and deluxe hamburger should he make? We are assuming that Antoine's customers will buy all the hamburger he can grind.

(b) Now suppose $c_2 < c_1$. How does this change affect the model? Is it still possible to formulate a linear program?

Section 1.7

For Exercises 1-8 through 1-18 graph or sketch the feasible region and a few indifference curves of the objective function, including the optimal one (if any). Indicate which are linear programs and comment on matters such as feasibility, existence and uniqueness of the optimum, and boundedness of feasible region and objective function.

1-8. Max $z = 2x_1 + x_2$

$$\begin{aligned} \text{subject to} \quad 2x_1 - 3x_2 &\le 2 \\ -x_1 + x_2 &\le 3 \\ x_1 + x_2 &\le 6 \\ x_2 &\le 4 \\ \text{and} \quad x_1, x_2 &\ge 0 \end{aligned}$$

1-9. Max $z = x_1 + x_2$, subject to the constraints of Exercise 1-8.

1-10. Min $z = 2x_1 + x_2$

$$\begin{aligned} \text{subject to} \quad x_1 - x_2 &\geq 0 \\ x_1 + x_2 &\leq -1 \\ \text{and} \quad x_2 &\geq 0 \end{aligned}$$

1-11. Min $z = x_1 + 2x_2$

$$\begin{aligned} \text{subject to} \quad x_1 + x_2 &\leq 2 \\ x_1 \quad &\geq 1 \\ x_1 - x_2 &\leq 0 \end{aligned}$$

1-12. Max $z = x_1 + x_2$

$$\text{subject to} \quad x_1^2 + x_2^2 \leq 2$$

1-13. Min $z = (x_1 - 2)^2 + (x_2 - 2)^2$

$$\begin{aligned} \text{subject to} \quad x_1 + x_2 &\leq 2 \\ \text{and} \quad x_1, x_2 &\geq 0 \end{aligned}$$

1-14. Min $z = x_1 + x_2$

$$\begin{aligned} \text{subject to} \quad x_1 - x_2 &\geq -1 \\ \text{and} \quad x_1, x_2 &\geq 0 \end{aligned}$$

1-15. Max $z = x_1 - x_2$

$$\begin{aligned} \text{subject to} \quad x_1 - x_2 &\geq -1 \\ \text{and} \quad x_1, x_2 &\geq 0 \end{aligned}$$

1-16. Min $z = x_1 - x_2$

$$\begin{aligned} \text{subject to} \quad x_1 - x_2 &\geq -1 \\ \text{and} \quad x_1, x_2 &\geq 0 \end{aligned}$$

1-17. Max $z = x_2$

$$\begin{aligned} \text{subject to} \quad x_1^2 - x_2 &\geq 0 \\ x_1 \quad &\leq 2 \\ x_1 \quad &\geq -2 \end{aligned}$$

1-18. Max $z = (x_1^2 + x_2^2)^{-1}$

$$\begin{aligned} \text{subject to} \quad x_1 + x_2 &\geq -2 \\ x_1 \quad &\leq 3 \\ x_2 &\leq 4 \end{aligned}$$

General

1-19. Given a set of constraints $g_i(x_1, \ldots, x_n) \leq b_i$, $i = 1, \ldots, m$, suppose you wish to find the maximum feasible value of *either* $f_1(x_1, \ldots, x_n)$ *or* $f_2(x_1, \ldots, x_n)$, whichever is greater; that is, you are trying to choose that set of feasible values of the variables $x_1, \ldots, x_n$ that maximizes the quantity

$$\text{Max}\ \{f_1, f_2\}$$

Can you formulate a single mathematical program of format (1-1) and (1-2) to accomplish this? If not, can you manage it with more than one mathematical program?

1-20. Why, intuitively, does an optimal solution of a linear program with a bounded feasible region always occur at one of the "corners"? (This will be proved rigorously in Chapter 4; it is not true for the general nonlinear problem.)

1-21. It was mentioned that the optimal solution of a linear program in two variables could be found by rolling a pencil across its diagram. Can you describe an algebraic algorithm that would solve any LPP by, in effect, pushing an indifference curve (or plane, or whatever it is in higher dimensions) across the feasible region? What difficulties would be encountered? How complicated would the computations be?

1-22. If the strategy of Exercise 1-21 seems undesirable, might it be reasonable to find the LPP optimum by searching over the boundaries? Or by searching over the corners? What sorts of computations would be required? (These questions are really designed only to provoke thought; they will be answered in later chapters.)

2

MATHEMATICAL BACKGROUND

2.1 PREREQUISITES

We shall proceed from the assumption that the student is familiar with elementary differential calculus and has had a very modest introduction to linear algebra; no further background is required. Strictly speaking, the calculus is not essential to the development of linear programming, and in fact, the subject has customarily been taught with no reference to it whatsoever. We shall amend this procedure by introducing calculus exactly once in this textbook, namely, in the proof of the duality theorem in Chapter 3. Although other proofs using only linear algebra are available, our approach provides valuable insights into certain properties of linear programming problems, properties whose generalized versions lie at the heart of mathematical programming theory.

What linear programming does require, however, is some acquaintance with linear algebra. It is not necessary to have taken an entire one-semester course in the subject: Chapters 2 and 3 of Hadley's text [14], for example, *with the material on determinants omitted*, constitute a background that is just sufficient for the study of linear programming. For the convenience of those who have been fully exposed to matrices and vectors and merely wish to refresh their memories, a checklist of important definitions and properties is provided in Section 2.2. The reader is hereby advised not to proceed further without a sound understanding of everything mentioned in that section. All other topics covered in this chapter will be presented de novo, with no prior introduction assumed. Much of the material will be quite familiar—hence its compilation under the heading "Mathematical Background." Nevertheless, this chapter should be treated as required reading, not only because one's memory is sometimes deceiving, but also because throughout this book we shall refer repeatedly to specific results derived in the following sections.

2.2 LINEAR ALGEBRA ASSUMED

An *$m \times n$ matrix* is a rectangular array of numbers having m rows and n columns. It is written as follows:

$$\mathbf{A} = \begin{bmatrix} a_{11} & a_{12} & \cdots & a_{1n} \\ a_{21} & a_{22} & \cdots & a_{2n} \\ \vdots & \vdots & & \vdots \\ a_{m1} & a_{m2} & \cdots & a_{mn} \end{bmatrix}$$

where the (i,j)th element a_{ij} is the entry in the ith row and jth column of $\mathbf{A}$. We shall use a boldface capital letter to denote a matrix. An $m \times 1$ matrix is called a *column vector*, while a $1 \times n$ matrix is a *row vector*. A column vector $\mathbf{u}$ can be written as

$$\mathbf{u} = \begin{bmatrix} u_1 \\ u_2 \\ \vdots \\ u_n \end{bmatrix}$$

and a row vector as

$$\mathbf{u} = [u_1, u_2, \ldots, u_n]$$

where u_i denotes the ith element or component. Throughout this text, *all vectors will be understood to be of the column variety* unless otherwise specified. When we need to represent a $1 \times n$ array of numbers in our theoretical development, we shall frequently use a transposed column vector rather than a row vector. A boldface small letter will represent a vector, as will a row or column of numbers enclosed in parentheses or brackets. Finally, a 1×1 matrix is simply a *scalar*, that is, an ordinary real number.

Although an n-component vector is defined as a matrix with n rows and 1

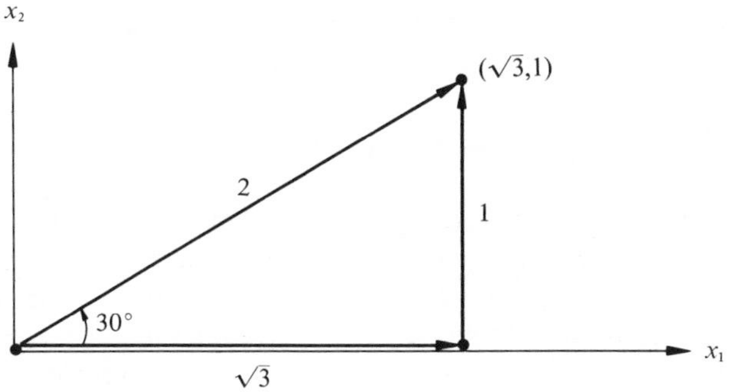

Figure 2-1

column (or with 1 row and n columns), it also may be interpreted geometrically as a unique point in euclidean n space E^n. For this reason an n-component vector is also referred to as an *n-dimensional* vector. As an example, if the coordinate axes in E^2 are labeled x_1 and x_2, the vector $(\sqrt{3},1)$ identifies the point that is reached by proceeding $\sqrt{3}$ units away from the origin in the positive x_1 direction, and then 1 unit in the positive x_2 direction (see Figure 2-1). Notice that this point could have been reached also by proceeding directly from the origin for a distance of 2 units along a straight line oriented at 30° from the x_1 axis. Thus, a vector, in addition to specifying a point, also specifies a directed straight-line segment from the origin to that point. It follows that a vector can be thought of as having direction and magnitude (or length), an interpretation that is very useful for describing a force in physics. In this textbook we shall have occasion, at one time or another, to use all three interpretations of vectors: as simple arrays of numbers, as points in E^n, and as directions or directed line segments.

From here on, the reader will be expected to know and understand each of the following definitions from linear algebra. It should be noted that our compendium does *not* include symmetric matrices, the determinant, or the rank of a matrix, none of which is required in the development of linear programming.

Let $\mathbf{A}$ be an $m \times n$ matrix whose (i,j)th element is a_{ij}. Then the *negative* of $\mathbf{A}$ is the $m \times n$ matrix $-\mathbf{A}$ whose (i,j)th element is $-a_{ij}$, for all i and j, and the *scalar multiple* of $\mathbf{A}$ by some scalar number λ is the $m \times n$ matrix $\lambda\mathbf{A}$ whose (i,j)th element is λa_{ij}. In addition, the *transpose* of $\mathbf{A}$ is the $n \times m$ matrix $\mathbf{A}^T$ whose (j,i)th element is a_{ij}; that is, the ith row (jth column) of $\mathbf{A}$ becomes the ith column (jth row) of $\mathbf{A}^T$. Because a vector is a type of matrix, the negative, scalar multiple, and transpose of a vector are defined exactly as above; note, in particular, that the transpose of a row vector is a column vector and vice versa.

Let $\mathbf{A}$ and $\mathbf{B}$ be two $m \times n$ matrices whose (i,j)th elements are a_{ij} and b_{ij}, respectively. Then $\mathbf{A}$ and $\mathbf{B}$ are said to be *equal* if and only if $a_{ij} = b_{ij}$ for all i

and j; in that case we write $\mathbf{A} = \mathbf{B}$. The *sum* of $\mathbf{A}$ and $\mathbf{B}$ is defined to be an $m \times n$ matrix $\mathbf{C}$ whose (i,j)th element is

$$c_{ij} = a_{ij} + b_{ij} \qquad \text{for all } i \text{ and } j$$

In that case we write $\mathbf{C} = \mathbf{A} + \mathbf{B}$. By extension, any number of $m \times n$ matrices are equal if their respective elements are all equal, and any number of $m \times n$ matrices can be added together. But note that two or more matrices cannot be equal nor can they be added unless they all have the same number of rows and the same number of columns. Again, the above definitions apply to vectors as well as matrices.

Vector inequalities. The n-component vector $\mathbf{u}$ is said to be less than or equal to the n-component vector $\mathbf{v}$, written $\mathbf{u} \leq \mathbf{v}$, if and only if $u_i \leq v_i$ for all i, $i = 1, \ldots, n$; that is, vector inequality implies inequality in all components. Other types of vector inequality relationships are defined analogously.

A *null vector* is any vector whose components are all zero. An n-component or n-dimensional null vector may be written as $\mathbf{0}_n$ or simply as $\mathbf{0}$. A *unit vector* is any vector whose components are all 0 except for a single 1; a unit vector having a 1 in the kth position is usually written as $\mathbf{e}_k$, regardless of dimensionality (which must be otherwise specified). Thus, depending on context, the row vector $\mathbf{e}_1$ could be $[1, 0, 0, 0]$ or $[1, 0]$ or even $[1]$.

Given a set of n-component vectors $\mathbf{u}_1, \mathbf{u}_2, \ldots, \mathbf{u}_m$, the n-component vector

$$\mathbf{u} = \lambda_1 \mathbf{u}_1 + \lambda_2 \mathbf{u}_2 + \cdots + \lambda_m \mathbf{u}_m = \sum_{i=1}^{m} \lambda_i \mathbf{u}_i$$

where $\lambda_1, \ldots, \lambda_m$ is any set of scalars, is called a *linear combination* of $\mathbf{u}_1, \ldots, \mathbf{u}_m$. Notice that, because addition is not defined for vectors whose dimensionality is different, all vectors participating in a linear combination must have the same number of components.

A set of n-component vectors $\mathbf{u}_1, \ldots, \mathbf{u}_m$ is said to be *linearly dependent* if and only if some linear combination of them, in which at least one of the scalars λ_i differs from 0, adds up to the null vector; that is, $\mathbf{u}_1, \ldots, \mathbf{u}_m$ are linearly dependent if and only if there exist scalars $\lambda_1, \ldots, \lambda_m$ *not all zero* such that

$$\sum_{i=1}^{m} \lambda_i \mathbf{u}_i = \mathbf{0}_n$$

If the only linear combination of $\mathbf{u}_1, \ldots, \mathbf{u}_m$ that adds to the null vector has all $\lambda_i = 0$, however, then the set of vectors is said to be *linearly independent*.

A set of n-component vectors is said to *span* E^n if every vector in E^n can be written as a linear combination of them. A set of linearly independent vectors that spans E^n constitutes a *basis* for E^n.

The *dot product* or *scalar product* of two n-component vectors $\mathbf{u}$ and $\mathbf{v}$, written $\mathbf{u}\cdot\mathbf{v}$, is the scalar quantity

$$z = \mathbf{u}\cdot\mathbf{v} = \sum_{i=1}^{n} u_i v_i$$

The dot product is not defined for vectors of different dimensionality.

Matrix multiplication. Given an $m \times r$ matrix $\mathbf{A}$ and an $r \times n$ matrix $\mathbf{B}$, the ordered product $\mathbf{AB}$, sometimes called the *matrix product*, is the $m \times n$ matrix $\mathbf{C}$ whose (i,j)th element c_{ij} is the dot product of the ith row of $\mathbf{A}$ and the jth column of $\mathbf{B}$; that is,

$$c_{ij} = \sum_{k=1}^{r} a_{ik} b_{kj} \qquad \text{for all } i = 1, \ldots, m \text{ and } j = 1, \ldots, n$$

In forming the product $\mathbf{AB}$, we say that $\mathbf{B}$ is *premultiplied* by $\mathbf{A}$ and that $\mathbf{A}$ is *postmultiplied* by $\mathbf{B}$. The product $\mathbf{AB}$ is defined if and only if the number of columns of $\mathbf{A}$ equals the number of rows of $\mathbf{B}$, in which case $\mathbf{A}$ and $\mathbf{B}$ are said to be *conformable* for multiplication in that order (that is, $\mathbf{A}$ first). The fact that $\mathbf{AB}$ is defined does not imply that $\mathbf{BA}$ is defined, and if both are defined, it is generally not true that $\mathbf{AB} = \mathbf{BA}$, even when the two product matrices have the same number of rows and columns. Note that, because a vector is a type of matrix, the dot product of two n-component column vectors $\mathbf{u}$ and $\mathbf{v}$ also can be written as the matrix product $\mathbf{u}^T\mathbf{v}$ or $\mathbf{v}^T\mathbf{u}$, that is, as the product of a $1 \times n$ and an $n \times 1$ matrix.

Any matrix having exactly as many rows as it has columns is called a *square matrix*. The *main diagonal* of an $n \times n$ square matrix $\mathbf{A}$ runs from the upper left corner to the lower right corner and includes the elements $a_{11}, a_{22}, \ldots$, and a_{nn}. An *identity matrix of order n*, written $\mathbf{I}_n$, is an $n \times n$ matrix having 1's on the main diagonal and 0's everywhere else; thus, the identity matrix of order 4 would be

$$\mathbf{I}_4 = \begin{bmatrix} 1 & 0 & 0 & 0 \\ 0 & 1 & 0 & 0 \\ 0 & 0 & 1 & 0 \\ 0 & 0 & 0 & 1 \end{bmatrix}$$

Given an $n \times n$ matrix **A**, if there exists an $n \times n$ matrix **B** such that

$$\mathbf{AB} = \mathbf{BA} = \mathbf{I}_n$$

then **B** is called the *inverse* of **A**; the symbol for the inverse of **A** is $\mathbf{A}^{-1}$. The square matrix **A** is said to be *nonsingular* if $\mathbf{A}^{-1}$ exists and *singular* if it does not. Note that only a square matrix can have an inverse.

Let any r rows and any s columns of an $m \times n$ matrix **A** be deleted. If the remaining $m - r$ rows and $n - s$ columns are pressed together to fill the open spaces, the resulting rectangular array of numbers is called a *submatrix* of **A**.

A *partitioning* or *partition* of a matrix **A** is produced when **A** is divided by imaginary horizontal and vertical cuts into a rectangular array of submatrices $\mathbf{A}_{ij}$ arranged in p rows and q columns as follows:

$$\mathbf{A} = \begin{bmatrix} \mathbf{A}_{11} & \mathbf{A}_{12} & \cdots & \mathbf{A}_{1q} \\ \mathbf{A}_{21} & \mathbf{A}_{22} & \cdots & \mathbf{A}_{2q} \\ \vdots & \vdots & & \vdots \\ \mathbf{A}_{p1} & \mathbf{A}_{p2} & \cdots & \mathbf{A}_{pq} \end{bmatrix}$$

Each cut goes all the way through **A**, so that for any given i, $i = 1, \ldots, p$, every submatrix in the ith row of submatrices has the same number of rows, and for any given j, $j = 1, \ldots, q$, every submatrix in the jth column has the same number of columns. Note that in a partitioning of **A** either p or q may equal 1.

Let the two $m \times n$ matrices **A** and **B** be partitioned as follows:

$$\mathbf{A} = \begin{bmatrix} \mathbf{A}_{11} & \mathbf{A}_{12} & \cdots & \mathbf{A}_{1q} \\ \mathbf{A}_{21} & \mathbf{A}_{22} & \cdots & \mathbf{A}_{2q} \\ \vdots & \vdots & & \vdots \\ \mathbf{A}_{p1} & \mathbf{A}_{p2} & \cdots & \mathbf{A}_{pq} \end{bmatrix} \quad \text{and} \quad \mathbf{B} = \begin{bmatrix} \mathbf{B}_{11} & \mathbf{B}_{12} & \cdots & \mathbf{B}_{1q} \\ \mathbf{B}_{21} & \mathbf{B}_{22} & \cdots & \mathbf{B}_{2q} \\ \vdots & \vdots & & \vdots \\ \mathbf{B}_{p1} & \mathbf{B}_{p2} & \cdots & \mathbf{B}_{pq} \end{bmatrix}$$

If, given any i, $i = 1, \ldots, p$, and any j, $j = 1, \ldots, q$, the submatrices $\mathbf{A}_{ij}$ and $\mathbf{B}_{ij}$ have the same number of rows and the same number of columns, then **A** and **B** are said to be *partitioned conformably for addition* and may be added "blockwise" to produce the partitioned matrix

$$\mathbf{C} = \begin{bmatrix} \mathbf{C}_{11} & \mathbf{C}_{12} & \cdots & \mathbf{C}_{1q} \\ \mathbf{C}_{21} & \mathbf{C}_{22} & \cdots & \mathbf{C}_{2q} \\ \vdots & \vdots & & \vdots \\ \mathbf{C}_{p1} & \mathbf{C}_{p2} & \cdots & \mathbf{C}_{pq} \end{bmatrix}$$

where $\mathbf{C} = \mathbf{A} + \mathbf{B}$ and $\mathbf{C}_{ij} = \mathbf{A}_{ij} + \mathbf{B}_{ij}$ for all i and j.

Let the $m \times r$ matrix $\mathbf{A}$ and the $r \times n$ matrix $\mathbf{B}$ be partitioned as follows:

$$\mathbf{A} = \begin{bmatrix} \mathbf{A}_{11} & \mathbf{A}_{12} & \cdots & \mathbf{A}_{1s} \\ \mathbf{A}_{21} & \mathbf{A}_{22} & \cdots & \mathbf{A}_{2s} \\ \vdots & \vdots & & \vdots \\ \mathbf{A}_{p1} & \mathbf{A}_{p2} & \cdots & \mathbf{A}_{ps} \end{bmatrix} \quad \text{and} \quad \mathbf{B} = \begin{bmatrix} \mathbf{B}_{11} & \mathbf{B}_{12} & \cdots & \mathbf{B}_{1q} \\ \mathbf{B}_{21} & \mathbf{B}_{22} & \cdots & \mathbf{B}_{2q} \\ \vdots & \vdots & & \vdots \\ \mathbf{B}_{s1} & \mathbf{B}_{s2} & \cdots & \mathbf{B}_{sq} \end{bmatrix}$$

If the columns of $\mathbf{A}$ are partitioned at the same intervals as are the rows of $\mathbf{B}$—that is, if for any k, $k = 1, \ldots, s$, the number of columns in the submatrices $\mathbf{A}_{ik}$ equals the number of rows in the submatrices $\mathbf{B}_{kj}$—then $\mathbf{A}$ and $\mathbf{B}$ are said to be *partitioned conformably for multiplication* and may be multiplied "blockwise" in that order to produce the partitioned matrix

$$\mathbf{C} = \begin{bmatrix} \mathbf{C}_{11} & \mathbf{C}_{12} & \cdots & \mathbf{C}_{1q} \\ \mathbf{C}_{21} & \mathbf{C}_{22} & \cdots & \mathbf{C}_{2q} \\ \vdots & \vdots & & \vdots \\ \mathbf{C}_{p1} & \mathbf{C}_{p2} & \cdots & \mathbf{C}_{pq} \end{bmatrix}$$

where $\mathbf{C} = \mathbf{AB}$ and

$$\mathbf{C}_{ij} = \sum_{k=1}^{s} \mathbf{A}_{ik}\mathbf{B}_{kj}$$

for all i and j.

A great many elementary properties of matrices and vectors—far too many for us to mention here—are developed in every linear algebra textbook: Matrix addition is associative and commutative; $\mathbf{A} = \mathbf{C}$ and $\mathbf{B} = \mathbf{C}$ together imply $\mathbf{A} = \mathbf{B}$; if $\mathbf{AB}$ is defined, then $(\mathbf{AB})^T = \mathbf{B}^T\mathbf{A}^T$; the inverse of any given matrix, if it exists, is unique; and so on. We shall assume that the student is thoroughly familiar with these and also with the following intermediate-level properties, which are singled out here so that they may be referred to specifically later on.

(1) If $\mathbf{B}$ is an $m \times n$ matrix and $\mathbf{y} = (y_1, y_2, \ldots, y_n)$ is an n-component column vector, then

$$\mathbf{By} = y_1\mathbf{b}_1 + y_2\mathbf{b}_2 + \cdots + y_n\mathbf{b}_n \tag{2-1}$$

where the vector $\mathbf{b}_j$ is the jth column of $\mathbf{B}$, $j = 1, \ldots, n$.

(2) A set of vectors is linearly dependent if and only if some one of them is a linear combination of all the others.

(3) The set of n-component unit vectors $\mathbf{e}_1, \mathbf{e}_2, \ldots, \mathbf{e}_n$ constitutes a basis for euclidean n space E^n.
(4) The representation of any vector as a linear combination of a given set of basis vectors is unique.
(5) Any basis for E^n contains exactly n vectors, and any n linearly independent vectors in E^n form a basis. No more than n vectors in E^n can be linearly independent, and no fewer than n can span E^n.
(6) If either the columns or the rows of the square matrix $\mathbf{A}$ form a linearly independent set of vectors, then $\mathbf{A}$ has an inverse. If $\mathbf{A}$ has an inverse, then both the rows of $\mathbf{A}$ and the columns of $\mathbf{A}$ are linearly independent.

We have already mentioned one reference for this material, Hadley [14]; others are Birkhoff and MacLane [1], Hildebrand [16], and Noble [20].

It is also worthwhile in this section to present a detailed proof of a certain theorem from linear algebra that will have important applications later. Inasmuch as many readers will have seen it before, this result could perhaps be considered background, but it is rather more specialized and less well known than the results listed above.

THEOREM 2.1. Let $\mathbf{a}_1, \mathbf{a}_2, \ldots, \mathbf{a}_m$ constitute a basis for E^m and let

$$\mathbf{b} = \sum_{i=1}^{m} \lambda_i \mathbf{a}_i \tag{2-2}$$

be the unique representation of the vector $\mathbf{b}$ in E^m. If any vector $\mathbf{a}_j$ for which λ_j is nonzero is removed from the basic set and $\mathbf{b}$ is substituted for it, then the resulting set is also a basis for E^m.

Since this theorem provides part of the foundation for property 5 above, we must prove it by showing that the new set of vectors is linearly independent *and* spans E^m, thereby satisfying the definition of a basis. Suppose the original basis vectors were numbered so that $\lambda_m \neq 0$. Our first step is to establish that the new set $\mathbf{a}_1, \ldots, \mathbf{a}_{m-1}, \mathbf{b}$ is linearly independent; accordingly, assume that

$$\sum_{i=1}^{m-1} \mu_i \mathbf{a}_i + \mu_m \mathbf{b} = \mathbf{0} \tag{2-3}$$

for some set of scalars μ_i, $i = 1, \ldots, m$. Substituting (2-2) into (2-3) yields

$$\sum_{i=1}^{m-1} \mu_i \mathbf{a}_i + \sum_{i=1}^{m} \mu_m \lambda_i \mathbf{a}_i = \sum_{i=1}^{m-1} (\mu_i + \mu_m \lambda_i) \mathbf{a}_i + \mu_m \lambda_m \mathbf{a}_m = \mathbf{0}$$

Because the basis vectors $a_1, \ldots, a_m$ are linearly independent, all scalar coefficients in the above linear combination must be zero. In particular $\mu_m \lambda_m = 0$, which implies $\mu_m = 0$. But this reduces the left side of (2-3) to a linear combination of the $m - 1$ linearly independent vectors $\mathbf{a}_1, \mathbf{a}_2, \ldots, \mathbf{a}_{m-1}$; hence $\mu_1 = \ldots = \mu_{m-1} = 0$. Thus (2-3) can hold only if $\mu_i = 0$, $i = 1, \ldots, m$, and the new set of vectors must be linearly independent.

If we can show that the new set also spans E^m, then it must be a basis. We know that any vector in E^m can be represented as a linear combination of the basis $\mathbf{a}_1, \ldots, \mathbf{a}_m$. Because $\lambda_m \neq 0$, we may divide (2-2) by λ_m to produce

$$\mathbf{a}_m = \frac{1}{\lambda_m}\mathbf{b} - \sum_{i=1}^{m-1} \frac{\lambda_i}{\lambda_m}\mathbf{a}_i$$

Thus any vector in E^m also can be expressed as a linear combination of $\mathbf{a}_1, \ldots, \mathbf{a}_{m-1}$, and

$$\frac{1}{\lambda_m}\mathbf{b} - \sum_{i=1}^{m-1} \frac{\lambda_i}{\lambda_m}\mathbf{a}_i$$

which is to say, as a linear combination of the vectors in the new set. This completes the proof of Theorem 2.1.

As an example, we know that the unit vectors $\mathbf{e}_1$, $\mathbf{e}_2$, and $\mathbf{e}_3$ form a basis for E^3. The vector $\mathbf{a} = (3, 2, -4)$ may be represented as

$$\mathbf{a} = 3\mathbf{e}_1 + 2\mathbf{e}_2 - 4\mathbf{e}_3$$

By Theorem 2.1 we could substitute $\mathbf{a}$ for any one of the unit vectors and still have a basis. Suppose we let it replace $\mathbf{e}_3$, so that the new basis consists of $\mathbf{e}_1$, $\mathbf{e}_2$, and $\mathbf{a}$. Now the vector $\mathbf{b} = (3, -4, 8)$ can be represented as

$$\mathbf{b} = 9\mathbf{e}_1 + 0\mathbf{e}_2 - 2\mathbf{a}$$

Therefore $\mathbf{b}$, $\mathbf{e}_2$, and $\mathbf{a}$ form a basis, as do $\mathbf{e}_1$, $\mathbf{e}_2$, and $\mathbf{b}$; however, $\mathbf{e}_1$, $\mathbf{b}$, and $\mathbf{a}$ do not.

2.3 SIMULTANEOUS LINEAR EQUATIONS

The remainder of this chapter will make no further demands on the reader's mathematical background. In this section we shall present some definitions and establish a few properties of systems of linear equations that will be of use in our development of linear programming.

Consider a set or *system* of m simultaneous linear equations in the unknowns $x_1, x_2, \ldots, x_n$:

$$\begin{aligned} a_{11}x_1 + a_{12}x_2 + \cdots + a_{1n}x_n &= b_1 \\ a_{21}x_1 + a_{22}x_2 + \cdots + a_{2n}x_n &= b_2 \\ \vdots \\ a_{m1}x_1 + a_{m2}x_2 + \cdots + a_{mn}x_n &= b_m \end{aligned} \tag{2-4}$$

where the a_{ij} and b_i are given constants. If we define

$$\mathbf{A} \equiv \begin{bmatrix} a_{11} & a_{12} & \cdots & a_{1n} \\ a_{21} & a_{22} & \cdots & a_{2n} \\ \vdots & \vdots & & \vdots \\ a_{m1} & a_{m2} & \cdots & a_{mn} \end{bmatrix} \qquad \mathbf{b} \equiv \begin{bmatrix} b_1 \\ b_2 \\ \vdots \\ b_m \end{bmatrix} \qquad \text{and} \qquad \mathbf{x} \equiv \begin{bmatrix} x_1 \\ x_2 \\ \vdots \\ x_n \end{bmatrix}$$

then the system (2-4) may be represented by the matrix equation

$$\mathbf{Ax} = \mathbf{b}$$

Because $\mathbf{Ax}$ and $\mathbf{b}$ are $m \times 1$ column vectors, this single equation implies all of the m equality relationships of (2-4). Any assignment of values to the variables $x_1, x_2, \ldots, x_n$ that satisfies all the equations is called a *solution* to the system. If no solution exists, the equations are said to be *inconsistent*, but if more than one exists, the system is *indeterminate*.

Consider now the row vectors

$$\mathbf{r}_i \equiv [a_{i1}, a_{i2}, \ldots, a_{in}]$$

and

$$\mathbf{s}_i \equiv [a_{i1}, a_{i2}, \ldots, a_{in}, b_i]$$

defined for each equation in (2-4). The kth of these equations is said to be *redundant* if $\mathbf{s}_k$ is exactly equal to a linear combination of all the other $\mathbf{s}_i$, $i = 1, \ldots, k-1, k+1, \ldots, m$. This means that the kth equation could have been derived by adding together multiples of some of the others; thus, it need not have been listed in (2-4) and is redundant in the usual (nonmathematical) sense. By means of some obvious algebraic manipulation, any equation whose $\mathbf{s}_i$ was included in the linear combination adding up to $\mathbf{s}_k$ can be seen to be redundant as well. We also say that the system (2-4) is redundant if any of its members are redundant.

Suppose that $\mathbf{r}_k$ is a linear combination of the other $\mathbf{r}_i$:

$$\mathbf{r}_k = \sum_{\substack{i=1 \\ i \neq k}}^{m} \lambda_i \mathbf{r}_i \qquad \text{for some } k,\ 1 \leq k \leq m \tag{2-5}$$

but suppose, on the other hand, that

$$\mathbf{s}_k \neq \sum_{\substack{i=1\\i\neq k}}^{m} \lambda_i \mathbf{s}_i \tag{2-6}$$

where k and the multipliers λ_i are the same as in (2-5). This pair of conditions implies an internal contradiction in the system of simultaneous equations. To see why, let us assume that the set of values $\mathbf{x}_0$ constitutes a solution to (2-4). Taking the dot products of the column vector $\mathbf{x}_0$ with the various row vectors of (2-5), we have

$$\mathbf{r}_k\mathbf{x}_0 = \sum_{i\neq k} \lambda_i \mathbf{r}_i \mathbf{x}_0$$

Because $\mathbf{x}_0$ is assumed to be a solution to (2-4), this becomes

$$b_k = \sum_{i\neq k} \lambda_i b_i$$

However, (2-5) and (2-6) together imply

$$b_k \neq \sum_{i\neq k} \lambda_i b_i$$

Therefore, no set of x values can satisfy (2-4) and the equations must be inconsistent. Notice that *if (2-5) holds, then the system of equations (2-4) is redundant, or it is inconsistent*, or both.[1] The converse is also true.

Let us return to $\mathbf{Ax} = \mathbf{b}$, the matrix equivalent of (2-4), and consider the relationship between the number of equations m and the number of variables n. If $m > n$, then (2-5) must hold for some k because, by property 5 of the preceding section, no more than n vectors $\mathbf{r}_i$ can be linearly independent. It follows that, *when $m > n$, the system of equations (2-4) is inconsistent or redundant.*

If $m = n$, we ask whether (2-5) holds for any k. If not, then the rows of $\mathbf{A}$ are linearly independent, and from property 6 we know that $\mathbf{A}$ has an inverse. From $\mathbf{Ax} = \mathbf{b}$ it follows that $\mathbf{x} = \mathbf{A}^{-1}\mathbf{b}$; the inverse is unique, so this provides a unique solution to (2-4). If (2-5) does hold, then the system may be inconsistent or redundant. Provided that the equations are consistent, the removal of redundant constraints is desirable because it speeds up whatever computational procedure is being used to find solutions. As soon as even one such constraint is removed we enter the domain of the third case, where $m < n$.

Suppose $m < n$ and the system of equations is not inconsistent. Let us eliminate redundant constraints until the rows $\mathbf{r}_i$ of the $\mathbf{A}$ matrix are linearly

[1] As an example of this last case, consider the three equations $x_1 + x_2 = 1$, $2x_1 + 2x_2 = 2$, and $2x_1 + 2x_2 = 3$.

independent. The elimination procedure is not obvious and not simple; fortunately, the simplex method for solving linear programming problems will provide a means for identifying and ignoring redundant constraints (this will be discussed in Chapter 5). In the meantime, let us assume that this job has been done. We are left with a system of linear equations

$$\mathbf{Ax} = \mathbf{b} \tag{2-7}$$

in which the rows of **A** are linearly independent, that is, in which there are no redundant or inconsistent constraints. Using the familiar notation, we take **A** to be $m \times n$; in order to have greater generality, let $m \leq n$. We now prove

THEOREM 2.2. If **A** is an $m \times n$ matrix with $m \leq n$ and the rows of **A** are linearly independent, then there exist m linearly independent columns of **A**.

If $m = n$, property 6 implies that **A** is invertible and, from this, that the m columns of **A** are linearly independent, Q.E.D.

If $m < n$, we shall eliminate a column of **A** in such a way that the shortened rows of **A** remain linearly independent. We know from property 5 that at most m columns $\mathbf{a}_j$ of **A** can be linearly independent, so at the moment the columns of **A** are linearly dependent. Order them so that the nth is a linear combination of the others. Then

$$\mathbf{a}_n = \sum_{j=1}^{n-1} \lambda_j \mathbf{a}_j$$

and for each element, in particular,

$$a_{in} = \sum_{j=1}^{n-1} \lambda_j a_{ij} \qquad \text{for } i = 1, 2, \ldots, m \tag{2-8}$$

Let us now define

$$\mathbf{r}_i^k \equiv [a_{i1}, a_{i2}, \ldots, a_{ik}] \qquad \text{for } i = 1, \ldots, m \text{ and } k = 1, \ldots, n$$

For example, $\mathbf{r}_i^n$ is the ith row of **A**. The $\mathbf{r}_i^n$ are, by hypothesis, linearly independent and we wish to show that the $\mathbf{r}_i^{n-1}$ are also. Therefore, *assume* that scalars $\mu_1, \ldots, \mu_m$ exist such that

$$\sum_{i=1}^{m} \mu_i \mathbf{r}_i^{n-1} = \mathbf{0}$$

here using **0** as a row of zeros. This implies

$$\sum_{i=1}^{m} \mu_i a_{ij} = 0 \qquad \text{for } j = 1, 2, \ldots, n - 1 \tag{2-9}$$

It follows that

$$\begin{aligned} \sum_{i=1}^{m} \mu_i a_{in} &= \sum_{i=1}^{m} \mu_i \sum_{j=1}^{n-1} \lambda_j a_{ij} && \text{from (2-8)} \\ &= \sum_{j=1}^{n-1} \lambda_j \sum_{i=1}^{m} \mu_i a_{ij} = 0 && \text{from (2-9)} \end{aligned}$$

This combines with (2-9) to guarantee that

$$\sum_{i=1}^{m} \mu_i \mathbf{r}_i^n = \mathbf{0}$$

The $\mathbf{r}_i^n$ are linearly independent, so all the μ_i must be 0, and the $\mathbf{r}_i^{n-1}$ are linearly independent as well.

Accordingly, let the column $\mathbf{a}_n$ be deleted from **A**. If $m = n - 1$, the rows of **A** (which are now the vectors $\mathbf{r}_i^{n-1}$) form a square matrix, which must be invertible by property 6. Therefore, the remaining $n - 1$ columns of **A** are linearly independent, and the theorem is proved. If $m < n - 1$, we repeat this process, reordering the columns (if necessary) to drop the $(n - 1)$th column and proving that the resulting row vectors $\mathbf{r}_i^{n-2}$ are still linearly independent. Eventually the matrix will shrink to m linearly independent columns, as required.

We now return to the consistent, nonredundant system of equations

$$\mathbf{Ax} = \mathbf{b} \tag{2-7}$$

which, according to the identity (2-1), also may be represented as

$$x_1\mathbf{a}_1 + x_2\mathbf{a}_2 + \cdots + x_n\mathbf{a}_n = \mathbf{b}$$

Theorem 2.2 tells us that there exists a set of m linearly independent columns of **A** (although it does not tell us how to find them; again the simplex method will provide this capability). Suppose we have found them and reordered the columns of **A** so that $\mathbf{a}_1, \mathbf{a}_2, \ldots, \mathbf{a}_m$ are linearly independent; by property 5 they form a basis. It follows that some unique linear combination of them, with scalar coefficients x_j, must add up to $\mathbf{b}$. Thus, even if we set $x_{m+1}, x_{m+2}, \ldots, x_n$ (the variables associated with $\mathbf{a}_{m+1}, \ldots, \mathbf{a}_n$) equal to zero, leaving

$$x_1\mathbf{a}_1 + x_2\mathbf{a}_2 + \cdots + x_m\mathbf{a}_m = \mathbf{b} \tag{2-10}$$

we can always find unique values of $x_1, \ldots, x_m$ that satisfy (2-10) and solve the original system of equations, regardless of $\mathbf{b}$. These values, supplemented by setting $x_{m+1}, \ldots, x_n$ equal to zero, constitute what is known as a *basic solution*. It is characterized by having at most m, and perhaps fewer, variables with nonzero values.[1]

This notion should be clarified by the following rigorous definition. Suppose m linearly independent columns have been selected from the $\mathbf{A}$ matrix of our familiar system of equations (2-7). If we set the variables *not* associated with these columns equal to zero, then the solution to the resulting system of m equations in m unknowns is called a *basic solution*. The variables that may differ from zero are the *basic variables*. Introducing some notation that will be used in later chapters, let us form the m linearly independent columns into a nonsingular $m \times m$ submatrix $\mathbf{B}$, whose columns are denoted $\mathbf{b}_1, \ldots, \mathbf{b}_m$. Note that the columns of $\mathbf{A}$ were not first reordered in any special way; in general, the first *basic column* $\mathbf{b}_1$ might be $\mathbf{a}_6$, and so on. If we now collect the m basic variables into a vector $\mathbf{x}_\mathbf{B}$, ordering them in the same way as their associated columns are ordered in $\mathbf{B}$, the resulting set of m equations in m basic variables then can be written

$$x_{\mathbf{B}1}\mathbf{b}_1 + x_{\mathbf{B}2}\mathbf{b}_2 + \cdots + x_{\mathbf{B}m}\mathbf{b}_m = \mathbf{B}\mathbf{x}_\mathbf{B} = \mathbf{b}$$

where the $x_{\mathbf{B}i}$ are the (ordered) components of $\mathbf{x}_\mathbf{B}$. The unique solution to this is

$$\mathbf{x}_\mathbf{B} = \mathbf{B}^{-1}\mathbf{b} \tag{2-11}$$

and, by definition, the basic solution to (2-7) associated with the basis $\mathbf{B}$ is given by (2-11) with the additional stipulation that all nonbasic variables are assigned values of zero. Of course, this is, in general, neither the only solution nor the only basic solution to the original system (2-7). There may be other basic solutions associated with other basic submatrices.

It should be emphasized that inherent in this definition of a basic solution is the invertibility of $\mathbf{B}$. If the set of equations $\mathbf{Ax} = \mathbf{b}$ were inconsistent or redundant, then, in view of (2-5), some row of $\mathbf{B}$ would be a linear combination of the other rows, no matter what columns were chosen from $\mathbf{A}$. Then property 6 would guarantee that $\mathbf{B}$ was singular: *A redundant or inconsistent set of equations cannot have a basic solution.* In addition, even when all the rows of

[1] The basic solution will include fewer than m variables with nonzero values whenever one or more of the $x_1, \ldots, x_m$ must take on zero values in order to satisfy (2-10).

A are linearly independent, it still may be possible to find m linearly dependent columns that form some singular **B**. We then say that no basic solution exists for that particular **B**.

For the system (2-7) with $m < n$ and no redundant or inconsistent equations, there are as many possible different basic solutions as there are ways of selecting m different columns from a total of n. This number is $n!/m!\,(n-m)!$. Usually, the number of basic solutions will be less than this theoretical maximum, because several different sets of m column vectors $\mathbf{a}_i$ will turn out to be linearly dependent.

One additional definition ought to be supplied here: A basic solution to $\mathbf{Ax} = \mathbf{b}$ is said to be *degenerate* if one or more of its basic variables equals zero. This is an event of special interest in linear programming.

We close this section with an example consisting of two equations:

$$x_1 - x_2 + x_3 + x_4 = 6$$

and

$$2x_2 + x_3 + x_4 = 9$$

There could be as many as $4!/2!\,(4-2)! = 6$ basic solutions. We find one of them by setting $x_3 = x_4 = 0$. The resulting basic solution is $x_1 = 10.5$, $x_2 = 4.5$. If we set $x_1 = x_2 = 0$ we find that the remaining columns are $\begin{bmatrix}1\\1\end{bmatrix}$ and $\begin{bmatrix}1\\1\end{bmatrix}$, obviously linearly dependent. Hence there is no basic solution involving x_3 and x_4.

2.4 HYPERPLANES AND CONVEX SETS

Having established some algebraic properties of simultaneous equation systems, we turn now to an examination of linear equations and inequalities from a *geometric* point of view. Whereas the preceding algebraic results will be exploited directly in the development of the simplex method for solving linear programs, the geometric material is intended rather to enrich the reader's understanding by providing an alternate way of looking at things. Convex sets and hyperplanes are, strictly speaking, algebraic notions, but all of the definitions and theorems that follow have direct and obvious spatial interpretations that offer the reader an easily visualized frame of reference. From time to time, the theory of convex sets will be called upon to illustrate various properties of linear programs; in particular, it will allow us to "watch" what happens when an LPP is solved by the simplex method.

We begin with the familiar concept of a straight line, widely celebrated as "the shortest distance between two points." Its definition in vector notation, which we shall use, is likely to be rather less familiar. Let **a** and **b** be two points in euclidean n space E^n. Then the *straight line* passing through $\mathbf{a}$ and $\mathbf{b}$ is the set of points $\mathbf{x}$ in E^n satisfying

$$\mathbf{x} = \lambda\mathbf{a} + (1 - \lambda)\mathbf{b}$$

for any real value of λ, where $\mathbf{a}$, $\mathbf{b}$, and $\mathbf{x}$ are taken to be column vectors. In mathematical shorthand this set of points is represented by

$$\{\mathbf{x} \mid \mathbf{x} = \lambda\mathbf{a} + (1 - \lambda)\mathbf{b}, \quad \lambda \text{ real}\} \tag{2-12}$$

It should be emphasized that (2-12) is merely a definition: It states an algebraic rule for identifying certain sets of points that we shall call "straight lines." The choice of this particular name was, of course, not arbitrary: The set of points defined by (2-12) happens to coincide with what we think of geometrically as the shortest path between two points (in euclidean space, anyway). The student may convince himself that this is so by considering the case of E^2. Suppose $\mathbf{a} = (a_1,a_2)$ and $\mathbf{b} = (b_1,b_2)$. Then (2-12) implies two linear equations:

$$x_1 = \lambda a_1 + (1 - \lambda)b_1$$

and

$$x_2 = \lambda a_2 + (1 - \lambda)b_2$$

where (x_1,x_2) are the coordinates of the general point $\mathbf{x}$. Solving one of these for λ and substituting into the other produces

$$\frac{a_2 - b_2}{a_1 - b_1} = \frac{x_2 - b_2}{x_1 - b_1}$$

which will be recognized (from elementary analytic geometry) as the straight line between $\mathbf{a}$ and $\mathbf{b}$.

One advantage of our vector notation is that it allows us to distinguish between a straight line, which is infinitely long, and a line segment, which is limited by two end points. The *line segment* connecting two given points $\mathbf{a}$ and $\mathbf{b}$ is given by

$$\{\mathbf{x} \mid \mathbf{x} = \lambda\mathbf{a} + (1 - \lambda)\mathbf{b}, \quad 0 \le \lambda \le 1\} \tag{2-13}$$

This concept is crucial to the development of convex set theory. The algebraic definition (2-13) has a simple geometric interpretation. When λ is near zero, the point $\mathbf{x}$ is close to $\mathbf{b}$; as λ increases, $\mathbf{x}$ moves along the line segment toward $\mathbf{a}$, arriving there when $\lambda = 1$.

We now can introduce an important generalized definition. A *hyperplane* in E^n is the set of all points $(x_1, x_2, \ldots, x_n)$ satisfying

$$c_1x_1 + c_2x_2 + \cdots + c_nx_n = z$$

where z and $c_1, \ldots, c_n$ are given constants with at least one of the c_i differing from zero. Using vector notation, a hyperplane in E^n is given by

$$\{\mathbf{x} \mid \mathbf{c}^T\mathbf{x} = z\} \tag{2-14}$$

for some given column vector $\mathbf{c}$ and scalar z. [Note that the transpose of $\mathbf{c}$ is required in (2-14) in order that $\mathbf{c}^T\mathbf{x}$ will be a scalar expression, because both $\mathbf{c}$ and $\mathbf{x}$ are column vectors.] In E^2 a hyperplane is just a straight line, $c_1x_1 + c_2x_2 = z$; in E^3 it corresponds to our ordinary geometrical notion of a plane, $c_1x_1 + c_2x_2 + c_3x_3 = z$. In Exercise 2-12 the student will be asked to prove that two hyperplanes in E^3 intersect to form a straight line, that is, a set of points satisfying our definition (2-12).

What if various inequality signs are substituted for the equality in (2-14)? We define a *closed half-space* in E^n to be the set

$$\{\mathbf{x} \mid \mathbf{c}^T\mathbf{x} \leq z\} \tag{2-15}$$

Geometrically, a closed half-space includes a hyperplane plus all points in E^n lying to one side of it. This definition obviously includes the set $\{\mathbf{x} \mid \mathbf{c}^T\mathbf{x} \geq z\}$ because we could reverse the inequality and obtain the format of (2-15) merely by multiplying through by -1. Incidentally, although we shall not use the concept in this book, an *open* half-space in E^n is a set $\{\mathbf{x} \mid \mathbf{c}^T\mathbf{x} < z\}$.

We shall see that all these sets of points have in common the property of convexity. *A set of points is convex if the straight-line segment between any two of its members lies entirely in the set.* In mathematical terms, a set in E^n is convex if, for any $\mathbf{x}_1$ and $\mathbf{x}_2$ in the set, every point

$$\mathbf{x} = \lambda\mathbf{x}_1 + (1 - \lambda)\mathbf{x}_2 \qquad \text{where } 0 \leq \lambda \leq 1 \tag{2-16}$$

also belongs to the set. Thus, a sphere, a triangle, the plane E^2, a straight line, and a single point are convex sets, but a doughnut, the letter "L," and the surface of a sphere (which is a set of points in E^3) are not.

An *extreme point* of a convex set X is a point $\mathbf{x}$ in X that *cannot* be expressed as $\lambda\mathbf{x}_1 + (1 - \lambda)\mathbf{x}_2$, with $\mathbf{x}_1$ and $\mathbf{x}_2$ being two different points in X and $0 < \lambda < 1$. This implies geometrically that an extreme point of a convex set does not lie on a straight-line segment between any two other points in the set. For example, taking as our convex set a solid sphere in E^3, any point on the surface of the sphere would be an extreme point; so would any of the three vertices of a triangle, any of the four corners of a square, and so on. A point on one side (not a corner) of a square, however, is not an extreme point because it lies between two corner points.

It should be mentioned also that, while a straight line

$$\mathbf{x} = \lambda\mathbf{x}_1 + (1 - \lambda)\mathbf{x}_2$$

is represented by a linear combination of two given points, where λ can take on any real value, a line segment (2-13) is represented by a *convex combination* of two given points, where λ is restricted to being no less than zero and no greater than one. Thus, a set is convex if and only if every convex combination of any two of its points is also in the set. More generally, given any set of points $\mathbf{x}_1, \ldots, \mathbf{x}_m$ in E^n, any summation of the form

$$\sum_{i=1}^{m} \lambda_i \mathbf{x}_i \qquad \text{where all } \lambda_i \geq 0 \text{ and } \sum_{i=1}^{m} \lambda_i = 1$$

is called a convex combination of those points. A convex combination is a special kind of linear combination.

Now we can establish a series of lemmas that will culminate in a very important theorem about linear programming problems. First, let us prove that *a hyperplane is a convex set.* Suppose $\mathbf{x}_1$ and $\mathbf{x}_2$ lie on the hyperplane $\mathbf{c}^T\mathbf{x} = z$. Then any point on the straight-line segment between them is given by $\hat{\mathbf{x}} = \lambda\mathbf{x}_1 + (1 - \lambda)\mathbf{x}_2$ for $0 \leq \lambda \leq 1$. But

$$\begin{aligned} \mathbf{c}^T\hat{\mathbf{x}} &= \mathbf{c}^T[\lambda\mathbf{x}_1 + (1 - \lambda)\mathbf{x}_2] = \lambda\mathbf{c}^T\mathbf{x}_1 + (1 - \lambda)\mathbf{c}^T\mathbf{x}_2 \\ &= \lambda z + (1 - \lambda)z = z \end{aligned}$$

Therefore, $\hat{\mathbf{x}}$ lies on the hyperplane.

By substituting inequalities as required, it can be shown also that *open and closed half-spaces are convex sets.* This means that each of the constraint equations that can appear in a linear programming problem represents a convex set because it is either a hyperplane of the form

$$a_{i1}x_1 + a_{i2}x_2 + \cdots + a_{in}x_n = b_i$$

or a closed half-space of one of the following types:

$$a_{i1}x_1 + a_{i2}x_2 + \cdots + a_{in}x_n \leq b_i$$

$$a_{i1}x_1 + a_{i2}x_2 + \cdots + a_{in}x_n \geq b_i$$

or

$$x_j \geq 0$$

We can show also that *the intersection of two convex sets is a convex set.* Let X be the intersection[1] of two convex sets X_1 and X_2 and let $\mathbf{x}_1$ and $\mathbf{x}_2$ be any two points in X. Because $\mathbf{x}_1$ and $\mathbf{x}_2$ are in X_1, so is the line segment between them. It is also in X_2, hence it is in their intersection X. This proves X is convex. Simple extensions can be used to prove that the intersection of any finite number of convex sets is convex. In particular, the intersection of any finite number of hyperplanes and closed half-spaces is convex. This immediately yields

THEOREM 2.3. The set of feasible solutions to a linear programming problem is a convex set.

This statement actually includes infeasible LPPs because an empty set trivially satisfies the definition of convexity. The reader should take note that convexity has nothing to do with boundedness. The convex set of solutions to a linear program may be bounded, as in Figure 1-3, or unbounded, as in Figure 1-4.

2.5 SOME ADDITIONAL DEFINITIONS

In addition to the straight line and the line segment, another continuous collinear set of points can be defined. A *ray* is a line anchored by a given point at one end and extending away toward infinity. More rigorously, for any two given points $\mathbf{a}$ and $\mathbf{b}$ in E^n, where $\mathbf{b}$ is not the null vector, we shall define a *ray* to be the set of points

$$\{\mathbf{x} \mid \mathbf{x} = \mathbf{a} + \lambda\mathbf{b}, \quad \lambda \geq 0\} \tag{2-17}$$

For $\lambda = 0$, $\mathbf{x}$ is at the anchor point $\mathbf{a}$. As λ increases, $\mathbf{x}$ moves away from $\mathbf{a}$ along a straight-line path whose direction is determined by the components of $\mathbf{b}$, thereby generating the ray. To verify this geometric interpretation for the two-dimensional case, we can write the two scalar equations implied by the definition (2-17):

$$x_1 = a_1 + \lambda b_1 \qquad \text{and} \qquad x_2 = a_2 + \lambda b_2$$

Solving the first for λ yields

$$\lambda = \frac{x_1 - a_1}{b_1}$$

[1] The *intersection* of two sets of points A and B is just the set of all points that belong to both A and B.

Assume, arbitrarily, that b_1 is positive; then, because λ is required to be nonnegative, we must stipulate $x_1 \geq a_1$. Substitution now yields

$$x_2 = \frac{b_2}{b_1} x_1 + \left(a_2 - \frac{a_1 b_2}{b_1}\right)$$

which is the equation of a straight line with slope b_2/b_1; the line starts at $x_1 = a_1$ and extends without end in the direction of increasing x_1. Note that the slope of the line (direction of the ray) depends only on **b**.

It can also be shown rigorously, using our definitions, that a ray (2-17) in any number of dimensions would simply be a straight line if λ were permitted to be negative. Algebraic manipulation of (2-17) produces

$$\{\mathbf{x} \mid \mathbf{x} = \lambda \mathbf{c} + (1 - \lambda)\mathbf{a}\}$$

where $\mathbf{c} \equiv \mathbf{a} + \mathbf{b}$ is a point in E^n. If λ can assume any real value, this is a straight line passing through **c** and **a**.

We need the notion of a ray in order to visualize geometrically a linear programming problem whose feasible region is unbounded. When this occurs, some of the boundaries of the feasible region will be defined by infinitely long rays, instead of by edges of finite length. To develop this idea and to introduce others, let us begin with the following definition: The hyperplane $z_0 = \mathbf{c}^T\mathbf{x}$ is called a *bounding* or *supporting hyperplane* of the convex set X if it passes through at least one point $\hat{\mathbf{x}}$ in X (that is, if $z_0 = \mathbf{c}^T\hat{\mathbf{x}}$) and if all of X lies in one of the closed half-spaces defined by the hyperplane. Clearly the bounding hyperplane cannot pass through any interior points[1] of the convex set X because if it did there would be points of X on both sides of it. This implies that $\hat{\mathbf{x}}$ is not an interior point.

Six bounding hyperplanes are shown for the convex set of Figure 2-2, namely, A, B, C, D, and the two axes. As another example, any straight line tangent to the circle $x_1^2 + x_2^2 = 1$ in E^2 would be a bounding hyperplane of the convex set $\{(x_1,x_2) \mid x_1^2 + x_2^2 \leq 1\}$. Finally, the plane $x_1 + x_2 = 2$ is a bounding hyperplane of the unit cube[2] in E^3: It passes through the vertex (1, 1, 1), and every point of the unit cube lies in the closed half-space $x_1 + x_2 \leq 2$.

[1] A point **y** in E^n is an *interior point* of a set S in E^n if there exists some positive number ϵ such that all points in E^n within a distance ϵ of **y** are in S. The reader may think in terms of the familiar, everyday concept of distance; more precisely, the distance between two points **a** and **b** in E^n is defined to be

$$\left[\sum_{i=1}^{n} (a_i - b_i)^2\right]^{1/2}$$

Any point in S that is not an interior point is a *boundary point* of S. This implies that if **y** is a boundary point of S, then for any $\epsilon > 0$ there exists within a distance ϵ of **y** at least one point that is not in S.

[2] The unit cube, or unit hypercube in E^3, lies in the nonnegative orthant of E^3 and has length, width, and height all equal to one unit. Its vertices are at (0, 0, 0), (1, 0, 0), (0, 1, 0). (1, 1, 0), (0, 0, 1), (1, 0, 1), (0, 1, 1), and (1, 1, 1).

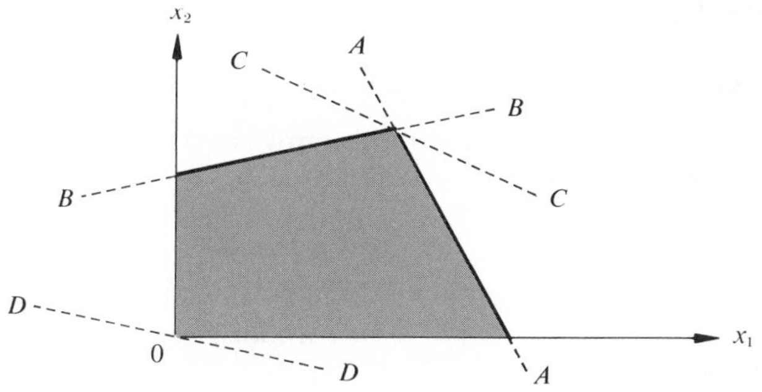

Figure 2-2

Now recall our definition of an extreme point as a point in a convex set that does not lie between any two other points in the set. A line segment between two different extreme points of a convex set X is called an *edge* of the set if it is the intersection of X with some bounding hyperplane; the two extreme points connected by the edge are said to be *adjacent*. This, of course, corresponds to our intuitive notion of an edge (the word was used two paragraphs above). For example, consider again the unit cube in E^3. One of its faces lies entirely in the $x_1 - x_2$ plane; it is the intersection of the cube with the bounding hyperplane $x_3 = 0$. Within that face is a line segment between the extreme points (0, 0, 0) and (1, 1, 0), but this line is not an edge because it is not the (complete) intersection of the cube with any bounding hyperplane. On the other hand, the line segment between (0, 0, 0) and (1, 0, 0) may not be the intersection of the cube with $x_3 = 0$, but it *is* the intersection of the cube with the bounding hyperplane $x_2 + x_3 = 0$. Therefore, it is an edge.

As will be seen in Chapter 4, the edges of a convex set play an important role in our geometric picture of what is going on when a linear programming problem is solved via the simplex method. In order to visualize this process in the case of an unbounded feasible region we shall need one final definition, which combines the notions of a ray and an edge. A ray lying entirely in a convex set and originating at one of its extreme points is an *extreme ray* if it is the complete intersection of the convex set and some bounding hyperplane. For example, the nonnegative orthant of E^3, which contains the unit cube, consists (in some sense) of one-eighth of all of E^3 and is bounded by the planes $x_1 = 0$, $x_2 = 0$, and $x_3 = 0$. It is a convex set of infinite extent and, as the reader can verify with a simple diagram, its three extreme rays are the three coordinate axes. Although the ray extending from the origin through the point (1, 1, 0) and onward lies entirely within the intersection of the nonnegative orthant with the bounding hyperplane $x_3 = 0$, it is not an extreme ray because it does not itself constitute the (complete) intersection.

EXERCISES

Section 2.2

2-1. By writing out the individual elements, prove the identity

$$\mathbf{By} \equiv y_1\mathbf{b}_1 + y_2\mathbf{b}_2 + \cdots + y_n\mathbf{b}_n$$

where $\mathbf{B}$ is an $m \times n$ matrix, $\mathbf{y} \equiv (y_1, \ldots, y_n)$, and $\mathbf{b}_i$ is the ith column of $\mathbf{B}$.

2-2. Suppose that the matrix $\mathbf{A}$ has been partitioned as indicated by the dotted lines in the following:

$$\mathbf{A} = \left[\begin{array}{cc:c} 2 & -3 & 1 \\ 1 & 0 & 4 \\ \hdashline 2 & 0 & 5 \end{array}\right] = \begin{bmatrix} \mathbf{A}_{11} & \mathbf{A}_{12} \\ \mathbf{A}_{21} & \mathbf{A}_{22} \end{bmatrix}$$

where $\mathbf{A}_{11}$ is a 2×2 submatrix, and so on. Using only a single cut, partition the matrix

$$\mathbf{B} = \begin{bmatrix} 2 & 0 \\ 0 & 1 \\ 3 & -1 \end{bmatrix}$$

so that it is conformable for premultiplication by $\mathbf{A}$. Now calculate the product $\mathbf{AB}$ both by block-wise multiplication and by ordinary matrix multiplication (ignoring the partitioning).

2-3. Prove that the inverse of any square matrix $\mathbf{A}$, if it exists, is unique.

2-4. What is the inverse of the partitioned matrix

$$\mathbf{A} = \begin{bmatrix} \mathbf{X} & \mathbf{Y} \\ \mathbf{0} & \mathbf{I} \end{bmatrix}$$

where $\mathbf{X}$ is square and invertible, $\mathbf{I}$ is an identity submatrix, and $\mathbf{0}$ is a submatrix of zeros?

2-5. Prove that if a set of vectors is linearly independent, so is any subset of it.

2-6. It is convenient, and somewhat elegant, to develop linear programming without using the concept of the determinant. The most difficult step in doing so is the proof of a lemma: The columns of a square matrix $\mathbf{A}$ are linearly independent if and only if $\mathbf{A}$ has an inverse. This proposition forms part of property 6 in Section 2.2. Prove it without using the determinant.

2-7. Which of the following sets of vectors form bases?

(a) $\mathbf{b}_1 = (1, 0, 0)$, $\mathbf{b}_2 = (1, 1, 0)$, $\mathbf{b}_3 = (1, 1, 1)$

(b) $\mathbf{b}_1 = (3, -4, -1)$, $\mathbf{b}_2 = (3, -4, -1)$, $\mathbf{b}_3 = (7, 2, -4)$

(c) $\mathbf{b}_1 = (3, 8, 0)$, $\mathbf{b}_2 = (-4, 1, 3)$, $\mathbf{b}_3 = (-5, 10, 6)$

(d) $\mathbf{b}_1 = (-1, 4)$, $\mathbf{b}_2 = (2, 3)$, $\mathbf{b}_3 = (5, -7)$

(e) $\mathbf{b}_1 = (1, 1, 1, 1)$, $\mathbf{b}_2 = (1, 1, 1, 2)$, $\mathbf{b}_3 = (1, 1, 1, 3)$

2-8. The following vectors form a basis in E^4: $\mathbf{a}_1 = (1, 0, 0, 0)$, $\mathbf{a}_2 = (0, 3, 4, 2)$, $\mathbf{a}_3 = (-1, -2, 0, 0)$, and $\mathbf{a}_4 = (0, 0, 0, 1)$. For which of these vectors $\mathbf{a}_i$ could each of the following vectors be substituted so as to preserve the basis?

(a) $\mathbf{b}_1 = (0, 1, 4, -3)$

(b) $\mathbf{b}_2 = (0, 3, 4, 2)$

(c) $\mathbf{b}_3 = (-1, 4, 8, 5)$

(d) $\mathbf{b}_4 = (0, 0, 0, 0)$

(e) $\mathbf{b}_5 = (-2, -1, 4, 2)$

2-9. Suppose that $\mathbf{a}_1, \mathbf{a}_2, \ldots, \mathbf{a}_n$ form a basis for E_n, with $n \geq 3$, and suppose that, if either $\mathbf{b}_1$ or $\mathbf{b}_2$ were substituted for any one of the $\mathbf{a}_i$, another basis would result. Would it then be true that a basis would necessarily result if both $\mathbf{b}_1$ and $\mathbf{b}_2$ were substituted for any *two* of the $\mathbf{a}_i$? Prove or give a counterexample.

Section 2.3

2-10. Tell whether each of the following sets of simultaneous linear equations is inconsistent, indeterminate, and/or redundant. How many different solutions and how many different basic solutions exist for each?

(a)
$$\begin{aligned} 2x_1 + x_2 + 3x_3 &= 11 \\ -x_1 + x_2 \qquad\quad &= 4 \end{aligned}$$

(b)
$$\begin{aligned} x_1 + x_2 \qquad\quad &= 10 \\ 2x_1 \qquad - x_3 &= 1 \\ x_1 + x_2 + x_3 &= 19 \end{aligned}$$

(c)
$$\begin{aligned} x_1 - x_2 + 2x_3 &= 2 \\ -x_1 + 3x_2 - 2x_3 &= -2 \\ 3x_1 - x_2 + 6x_3 &= 16 \end{aligned}$$

(d)
$$\begin{aligned} x_1 + x_2 \qquad\quad &= 0 \\ 2x_1 - x_2 + 3x_3 &= 0 \\ -x_1 \qquad + x_3 &= 0 \end{aligned}$$

(e)
$$\begin{aligned} x_1 + x_2 + x_3 &= 0 \\ 2x_1 - x_2 \qquad &= 0 \\ x_1 - 2x_2 - x_3 &= 0 \end{aligned}$$

(f)
$$\begin{aligned} x_1 + 2x_2 \qquad &= 1 \\ x_1 \qquad + x_3 &= -1 \\ 2x_2 + x_3 &= 10 \\ x_1 - 2x_2 - x_3 &= -4 \end{aligned}$$

(g)
$$\begin{aligned} x_1 - 2x_2 - x_3 &= 4 \\ 2x_1 + x_2 - 5x_3 &= 7 \\ 4x_1 - 3x_2 - 7x_3 &= 15 \\ x_1 + 3x_2 - 4x_3 &= 3 \end{aligned}$$

2-11. Find all basic solutions of the following system of equations:

$$\begin{aligned} 2x_1 - 3x_2 - x_3 + x_4 &= 2 \\ 6x_2 \qquad - 2x_4 &= 9 \end{aligned}$$

Section 2.4

2-12. Using our algebraic definitions, prove that two hyperplanes in E^3 intersect to form a straight line (if they intersect at all). Comment on how this proposition would be generalized for n-dimensional space.

2-13. Which of the following sets of points are convex?

(a) $\{(x_1,x_2) \mid x_1x_2 \leq 1, \quad x_1 \geq 0, \quad x_2 \geq 0\}$

(b) $\{(x_1, x_2, x_3) \mid x_3 = 8\}$

(c) $\{(x_1,x_2) \mid x_1^2 + x_2 \geq 3, \quad x_1 \geq 0, \quad x_2 \geq 0\}$
(d) $\{(x_1,x_2) \mid x_1^2 + x_2^2 = 1\}$
(e) $\{\mathbf{x} \mid x_1 \geq x_2 \geq \cdots \geq x_n\}$
(f) $\{(x_1,x_2) \mid x_1 \geq 3 \quad or \quad x_2 \geq 2\}$
(g) $\{(x_1, x_2, x_3) \mid x_1 \geq x_2 \quad or \quad x_2 \geq x_3\}$
(h) $\{x_1 \mid x_1$ is an integer$\}$

2-14. Given a set of points $\mathbf{x}_1, \ldots, \mathbf{x}_m$ in the convex set X in E_n, show that any convex combination of them is also in X. (*Hint:* This proposition, which is true by definition for $m = 2$, can be proved via induction.)

2-15. (a) Let S be a set of points $\mathbf{x}$ in E_n. Suppose that, for any two points $\mathbf{x}_1$ and $\mathbf{x}_2$ in S, there exists some λ, where $0 < \lambda < 1$, such that the point

$$\mathbf{x}_3 = \lambda\mathbf{x}_1 + (1 - \lambda)\mathbf{x}_2$$

is also in S. Show that S is not necessarily a convex set.

(b) Let S be a set of points $\mathbf{x}$ in E_n. Suppose that there exists some particular λ_0, where $0 < \lambda_0 < 1$, such that, for any two points $\mathbf{x}_1$ and $\mathbf{x}_2$ in S, the point

$$\mathbf{x}_3 = \lambda_0\mathbf{x}_1 + (1 - \lambda_0)\mathbf{x}_2$$

is also in S. Show that S is not necessarily a convex set.

Section 2.5

2-16. Is the set of points $\{(x_1,x_2) \mid x_1^2 + x_2^2 < 1\}$ a convex set? Does it have any bounding hyperplanes?

2-17. Consider the proposition that it is possible for an unbounded convex set in E_n to have exactly k extreme rays. Tell whether this proposition is true or false in each of the following cases:

(a) $n = 2, k = 0$
(b) $n = 2, k = 1$
(c) $n = 2, k = 2$
(d) $n = 2, k \geq 3$
(e) $n = 3, k = 0$
(f) $n = 3, k = 1$
(g) $n = 3, k = 2$
(h) $n = 3, k = 3$
(i) $n = 3, k = 8$

2-18. Repeat Exercise 2-17 for the following proposition: It is possible for an unbounded convex set *in the nonnegative orthant* (i.e., where $\mathbf{x} \geq \mathbf{0}$) of E^n to have exactly k extreme rays.

2-19. A *cone* C in E^n is a set of points having the property that, if $\mathbf{x}$ is in the set, so is $\lambda\mathbf{x}$ for all $\lambda \geq 0$. Prove or disprove (possibly by finding an example or a counter-example) each of the following statements:

(a) The origin $\mathbf{0}$ is included in any cone.
(b) Every cone is a convex set.
(c) An extreme ray must form part of the boundary of any cone in E^n that does not include all of E^n.
(d) An extreme ray anchored at the origin in any space E^n is a cone.

3

DUALITY THEORY

3.1 AN EXAMPLE OF DUAL LINEAR PROGRAMS

Let us begin by considering an ordinary resource-allocation problem of the type that is frequently faced by plant managers and corporate planners. A certain factory can produce n different *outputs*, or products, using a total of m different raw materials as *inputs*. Each unit of the jth product can be sold for c_j dollars, $j = 1, \ldots, n$, and requires for its manufacture the amounts a_{ij} of the various inputs, $i = 1, \ldots, m$. The total amount of the ith input on hand is b_i. Using only the available resources, how much of each output should the manager of the factory plan to produce in order to maximize overall receipts from sales?

The reader has already been introduced to this problem in resource allocation, or *input-output analysis*—recall Example 1-1 of Chapter 1—so we shall not linger over the formulation. Letting x_j be the amount of the jth output to be produced, $j = 1, \ldots, n$, the objective function is simply

$$\text{Max } z = c_1x_1 + c_2x_2 + \cdots + c_nx_n$$

The limited availability of the raw materials imposes a resource constraint

$$a_{i1}x_1 + a_{i2}x_2 + \cdots + a_{in}x_n \leq b_i$$

for each input, $i = 1, \ldots, m$. Finally, negative production levels are prohibited:

$$x_j \geq 0 \qquad j = 1, \ldots, n$$

This completes the formulation of a linear programming problem. It may be represented in matrix/vector form as follows:

$$\begin{aligned} &\text{Max } z = \mathbf{c}^T\mathbf{x} \\ &\text{subject to} \quad \mathbf{Ax} \leq \mathbf{b} \\ &\text{and} \qquad\qquad \mathbf{x} \geq \mathbf{0} \end{aligned}$$

where $\mathbf{x}$ and $\mathbf{c}$ are n-component column vectors, $\mathbf{b}$ is an m-component column vector, and $\mathbf{A}$ is an $m \times n$ matrix.

Suppose we consider a different problem faced by the factory's manager, or perhaps by its vice president in charge of finance. In order to protect the dollar income that will be realized when the finished products have been sold, the raw materials on hand should be insured against fire, theft, and so on. Insurance policies cost money, and it will be desirable to place small per-unit valuations on the raw materials; on the other hand, the valuations should be large enough to compensate the factory fully—that is, to replace the income that would otherwise accrue from sales—in the event of accident or disaster. Subject to these considerations, what is the optimal insurance scheme?

Let u_i be the valuation placed on each unit of the ith input, $i = 1, \ldots, m$, so that the overall amount of the ith input currently on hand is given a total valuation of $b_i u_i$. No raw material can have a negative value because, at worst, it could simply be discarded; hence

$$u_i \geq 0 \qquad i = 1, \ldots, m$$

If we take the cost of insurance to be a simple multiple of the total valuation of all items insured, then the objective of minimizing insurance cost is

$$\text{Min } z' = b_1 u_1 + b_2 u_2 + \cdots + b_m u_m$$

How can our vice president decide how large his valuations must be in order to guarantee full compensation for any loss? Not having solved the linear program we formulated above, he has no way of knowing how the various raw materials will be used; that is, if one unit of the sixth input were destroyed by fire, he doesn't know exactly how much less dollar income the factory would realize, and that income differential is precisely the insurance valuation u_6 that he wants to place on the sixth input.

If the vice president is imaginative, however, he will reflect that the value of a unit of any raw material exists only insofar as it can be converted into a salable product. Whatever the individual values of the various inputs may be, it is clear that the combined values of all inputs required to produce exactly one unit of the first product (that is, a_{11} units of the first input, a_{21} units of the second, and so on) must be at least c_1; or, in mathematical terms,

$$a_{11}u_1 + a_{21}u_2 + \cdots + a_{m1}u_m \geq c_1$$

This is so because, *at the very least*, that set of inputs could be assembled into one unit of product 1 and sold for c_1 dollars—and it is also possible that they could be used cleverly in other ways (perhaps requiring that production levels for certain other outputs be modified) in order to further increase total sales income. But raw materials can be made to produce income in only n different ways, that is, via the production of only n different products. Therefore, to assure adequate valuations for insurance purposes, it is necessary only to guarantee that the total valuation of all inputs required to produce a unit of any particular output be at least equal to the income obtainable from that unit of output:

$$a_{1j}u_1 + a_{2j}u_2 + \cdots + a_{mj}u_m \geq c_j$$

for $j = 1, \ldots, n$. The correct valuations are obtained then by solving the linear program

$$\begin{aligned} &\text{Min } z' = b_1u_1 + b_2u_2 + \cdots + b_mu_m \\ &\text{subject to} \quad a_{1j}u_1 + a_{2j}u_2 + \cdots + a_{mj}u_m \geq c_j \qquad j = 1, \ldots, n \\ &\text{and} \qquad\qquad\qquad\qquad\qquad\qquad\quad u_i \geq 0 \qquad i = 1, \ldots, m \end{aligned}$$

Let us place the matrix formulation of this linear program side by side with the first:

Production problem	*Insurance problem*
Max $z = \mathbf{c}^T\mathbf{x}$	Min $z' = \mathbf{b}^T\mathbf{u}$
subject to $\mathbf{Ax} \leq \mathbf{b}$	subject to $\mathbf{A}^T\mathbf{u} \geq \mathbf{c}$
and $\mathbf{x} \geq \mathbf{0}$	and $\mathbf{u} \geq \mathbf{0}$

These two problems are remarkably symmetric: The cost vector in the objective function of each is just the right-hand-side vector in the other's set of constraints, and the constraint matrices are simply transposes of each other. Because of these relationships, the two problems are known as *dual* linear programs—the definition will be stated precisely later on. In addition to the fact that they are derived from the same real-world situation, another and more startling connection between the two problems is that the optimal values of their objective functions are exactly the same! This reflects the fact that the total insurance valuation of all raw materials should be precisely equal to the maximum dollar income that can be extracted from them.

3.2 THE ROLE OF DUALITY

In addition to serving as an introduction, the foregoing example was presented for motivational purposes. Before describing a computational algorithm for solving either the production or the insurance problem, we shall prove several theorems about the complex and involuted relationships between them; we hope, therefore, that the previous section has whetted the student's curiosity. Actually, we must confess that in the rather glib discussion of the insurance problem certain difficulties were glossed over. For example, suppose that 100 units of the fourth input are on hand initially and the optimal production scheme consumes only 90 of them. Because the maximum dollar income would not be affected at all if two or three units of this input were destroyed, it would appear that, for insurance purposes, a zero valuation should be assigned: $u_4 = 0$. On the other hand, if all 100 units were destroyed, the proceeds from sales would be seriously reduced, perhaps to zero. Does this mean, then, that the correct insurance valuation equals the total reduction of sales income divided by 100? For the moment, let us postpone this and other questions; we shall have much more to say in Section 3.10 about the interpretation of dual linear programs.

In the meantime the reader is advised that the present chapter is extremely important. The notion of duality plays a central role in optimization theory, not only in linear programming, but throughout all of mathematical programming as well (and even in the classical procedure of optimization via Lagrange multipliers). We shall make frequent use of the various theorems about duality, first to help establish the simplex method, then to develop a variant of it for solving linear programs under certain special conditions, and finally to derive a highly efficient solution algorithm for the transportation problem. Incidentally, it will turn out that, in solving a linear program by the simplex method, we shall simultaneously be solving its associated dual problem as well.

3.3 CANONICAL FORM OF THE LINEAR PROGRAM

Before we begin proving all these theorems, however, our first order of business must be to straighten out exactly what a linear programming problem is, or rather, to specify a particular format in which it is to be expressed. It will be recalled from Section 1.5 that the constraints of an LPP may be equations or inequalities, while the objective may call for maximization or minimization. We also remarked that a problem with unrestricted variables can be formulated as a true LPP, which formally requires nonnegative variables. In the face of such flexible structure it would be difficult to derive theorems about linear programs; the proofs would have to cover too many different cases. It is more economical to identify a particular format for the LPP and to prove that any linear program can be represented within it.

A linear programming problem is said to be in *canonical form* if it is presented as follows: Find values of $x_1, \ldots, x_n$ in order to

$$\text{Max } z = c_1x_1 + c_2x_2 + \cdots + c_nx_n \tag{3-1}$$

$$\text{subject to} \quad a_{11}x_1 + a_{12}x_2 + \cdots + a_{1n}x_n \leq b_1 \tag{3-2}$$

$$\vdots \qquad\qquad\qquad \vdots$$

$$a_{m1}x_1 + a_{m2}x_2 + \cdots + a_{mn}x_n \leq b_m \tag{3-3}$$

$$\text{and} \qquad x_j \geq 0 \qquad j = 1, \ldots, n \tag{3-4}$$

The problem has m constraints and n nonnegative variables; notice that we do *not* include the nonnegativity restrictions in counting the number of constraints. Vector notation also may be used to represent an LPP in canonical form:

$$\text{Max } z = \mathbf{c}^T\mathbf{x} \tag{3-5}$$

$$\text{subject to} \quad \mathbf{Ax} \leq \mathbf{b} \tag{3-6}$$

$$\text{and} \qquad \mathbf{x} \geq \mathbf{0} \tag{3-7}$$

where $\mathbf{A}$ is a given $m \times n$ coefficient matrix, $\mathbf{b}$ is a given m-component column vector, $\mathbf{c}$ is a given n-component column vector, $\mathbf{x}$ is an $n \times 1$ vector of unknowns, and $\mathbf{0}$ is an n-component null vector. Recall that the vector inequality sign implies inequality in each component.

The reader should be aware that the various components of the linear program have been rather haphazardly named. In the literature, the constraint matrix $\mathbf{A}$ is referred to usually as the *activity matrix* and occasionally as the *technology matrix*. The right-hand-side vector $\mathbf{b}$ is the *requirements vector*, and $\mathbf{c}$ is the *cost vector* or, less often, the *price vector*. The individual elements a_{ij} and c_j are called *activity coefficients* and *cost coefficients*, respectively. Obviously, this nomenclature is more descriptive of minimization problems.

3.4 TRANSFORMATIONS TO CANONICAL FORM

It is not difficult to see that any linear objective function can be put into the form (3-5). If the given objective is

$$\text{Min } z = \mathbf{c}^T\mathbf{x} \tag{3-8}$$

we simply define $\mathbf{d} = -\mathbf{c}$. Then (3-8) is equivalent to minimizing $z' = -\mathbf{d}^T\mathbf{x}$ or maximizing $z'' = \mathbf{d}^T\mathbf{x}$, which satisfies the format (3-5). This is easily proved: If $\mathbf{x}^*$ minimizes z', then

$$-\mathbf{d}^T\mathbf{x}^* \leq -\mathbf{d}^T\mathbf{x}$$

for all other feasible $\mathbf{x}$. This implies that

$$\mathbf{d}^T\mathbf{x}^* \geq \mathbf{d}^T\mathbf{x}$$

proving that $\mathbf{x}^*$ maximizes z'', Q.E.D.

Incidentally, it is not quite correct to say that the two objectives

$$\text{Min } z = \mathbf{c}^T\mathbf{x} \qquad \text{and} \qquad \text{Max } z'' = -\mathbf{c}^T\mathbf{x}$$

are equivalent. After all, the optimal values of the objective functions in the two cases will be the negatives of each other. What is meant by equivalence is that the optimal values $\mathbf{x}^*$ of the variables would be the same regardless of which of the two objectives is applied to any given set of feasible solutions $\mathbf{x}$.

What about linear objectives of the form

$$\text{Max } z = \mathbf{c}^T\mathbf{x} + c_0 \tag{3-9}$$

where c_0 is a constant? This is a completely general representation of a linear objective, whereas (3-5) is not. The c_0 is not needed, however. The $\mathbf{x}^*$ that maximizes (3-9) also maximizes $z' = \mathbf{c}^T\mathbf{x}$. Proof of this assertion is trivial and follows the model of the last paragraph.

The next step is to show that any linear constraint can be expressed as

$$a_{i1}x_1 + a_{i2}x_2 + \cdots + a_{in}x_n \leq b_i \tag{3-10}$$

where all minus signs in the sum are carried as negative a_{ij}. If the given linear constraint has the opposite inequality, simple multiplication of both sides by -1 produces the form of (3-10). A linear equality

$$a_{i1}x_1 + a_{i2}x_2 + \cdots + a_{in}x_n = b_i \tag{3-11}$$

can be transformed by noting that (3-11) implies both

$$a_{i1}x_1 + a_{i2}x_2 + \cdots + a_{in}x_n \leq b_i \tag{3-12}$$

and

$$a_{i1}x_1 + a_{i2}x_2 + \cdots + a_{in}x_n \geq b_i \tag{3-13}$$

Multiplication of (3-13) by -1 now produces

$$-a_{i1}x_1 - a_{i2}x_2 - \cdots - a_{in}x_n \leq -b_i \tag{3-14}$$

Both (3-12) and (3-14) are of the desired form and together are equivalent to (3-11).

Finally, nonnegativity requirements always can be manufactured for a problem with unrestricted variables. Suppose we are given the linear program

$$\text{Max } z = c_1x_1 + \cdots + c_nx_n \tag{3-15}$$

$$\text{subject to} \quad a_{11}x_1 + \cdots + a_{1n}x_n \leq b_1 \tag{3-16}$$

$$\vdots$$

$$\text{and} \quad a_{m1}x_1 + \cdots + a_{mn}x_n \leq b_m \tag{3-17}$$

where the variables are permitted to take on any real values. For each x_j define two new nonnegative variables x_j' and x_j'' such that

$$x_j \equiv x_j' - x_j'' \tag{3-18}$$

where $\quad x_j', x_j'' \geq 0$

Despite their nonnegativity, these new variables can assume values that allow x_j to be positive, negative, or zero. Substitution of (3-18) into the linear program then yields

$$\text{Max } z = c_1(x_1' - x_1'') + \cdots + c_n(x_n' - x_n'')$$

$$\text{subject to} \quad a_{11}(x_1' - x_1'') + \cdots + a_{1n}(x_n' - x_n'') \leq b_1$$

$$\vdots$$

$$a_{m1}(x_1' - x_1'') + \cdots + a_{mn}(x_n' - x_n'') \leq b_m$$

$$\text{and} \quad x_j', x_j'' \geq 0 \qquad j = 1, \ldots, n$$

[when the original LPP has both nonnegative and unrestricted variables, the transformations (3-18) are required only for the latter]. Although we have twice as many variables as before, this is clearly in canonical form, with the variables x_j'' having activity coefficients $-a_{ij}$ and cost coefficients $-c_j$. Note that any x_j value can be expressed by nonnegative values of x_j' and x_j'' in an infinite number of different ways; that is, if $x_j = -3.5$, the values of x_j' and x_j'' might be 0 and 3.5, or 0.5 and 4.0, and so on. Therefore, the new linear program has an infinite number of solutions for every solution to the original LPP and, in particular, an infinite number of optimal solutions.

Actually, a far more economical set of transformations is available for the problem (3-15) through (3-17). For every $x_j, j = 1, \ldots, n$, define

$$x_j = x_j' - x_0 \tag{3-19}$$

where the new variables x_0 and $x'_1, \ldots, x'_n$ are restricted to be nonnegative. It is then possible to represent any set of real values $x_1, \ldots, x_n$ by means of nonnegative values of the new variables; this requires only that x_0 be at least as large as the absolute value of the most negative of the x_j. Substitution of (3-19) into the given LPP produces the canonical form, containing $n + 1$ nonnegative variables instead of n unrestricted ones.

In any case, we have seen that the objective function of any linear program can be presented as a maximization, that its constraints can always be converted into the form of (3-6) and that its variables can be transformed to comply with the nonnegativity restriction. Therefore, *any linear programming problem can be expressed in canonical form.*

By now the reader may suspect that our identification of (3-5) through (3-7) as the canonical form of the linear program was a somewhat arbitrary maneuver. He would be right. In other texts, the canonical LPP is sometimes a minimization, and sometimes the constraints (3-6) are replaced by $\mathbf{Ax} \geq \mathbf{b}$. All writers are consistent, however, in including *nonnegative variables* and *inequality constraints,* which are implicit in any definition of the canonical format.

3.5 STANDARD FORM AND TRANSFORMATIONS TO IT

The following linear programming problem is said to be in *standard form*:

$$\text{Max } z = \mathbf{c}^T\mathbf{x} \tag{3-20}$$

$$\text{subject to} \quad \mathbf{Ax} = \mathbf{b} \tag{3-21}$$

$$\text{and} \quad \mathbf{x} \geq \mathbf{0} \tag{3-22}$$

This is a less arbitrary definition; all writers agree that the standard form includes equality constraints and nonnegative variables. We are merely selecting maximization over minimization.

As we have seen, any LPP can be presented in canonical form, so we can show that any LPP can be presented also in standard form by demonstrating the transformations from canonical to standard. This requires only a modification of the constraints (3-6). Accordingly, given

$$a_{i1}x_1 + a_{i2}x_2 + \cdots + a_{in}x_n \leq b_i \tag{3-23}$$

we introduce a new variable s_i and write

$$a_{i1}x_1 + a_{i2}x_2 + \cdots + a_{in}x_n + s_i = b_i \tag{3-24}$$

Requiring that the inequality (3-23) be observed is equivalent to requiring that s_i be nonnegative in (3-24); that is, any set of values of the x_j that violates (3-23) would force s_i to take on a negative value in (3-24). Thus the constraints of an LPP in canonical form may be expressed as equalities by adding to each of them a new nonnegative variable. Because each new variable is added to only one constraint and not to the objective, it is free to take on any positive value without affecting the objective or acting on the other constraints. Note finally that it satisfies (3-22), as required. Therefore, *every canonical LPP has an equivalent standard form*, which, in general, requires more variables but no more constraints.

This new variable s_i is called a *slack variable*, and it plays a very important role in the solution of linear programming problems. The simplex method operates only upon the standard form of the LPP; thus, in order to solve a problem with constraints of type (3-23), it is necessary to introduce slack variables. Similarly, it may be necessary to transform a constraint of the form

$$a_{i1}x_1 + a_{i2}x_2 + \cdots + a_{in}x_n \geq b_i \tag{3-25}$$

directly into an equality by subtracting from the left-hand side a *surplus variable*. As before, if we write

$$a_{i1}x_1 + a_{i2}x_2 + \cdots + a_{in}x_n - s_i = b_i \tag{3-26}$$

then the inequality (3-25) implies and is implied by the nonnegativity of s_i in (3-26).

These auxiliary variables for transforming inequalities into equations are appropriately named. The slack variable measures the amount by which the left-hand side of (3-23) falls short of the right. It must be added to take up the slack and create equality. Similarly, the surplus variable measures the surplus, or excess, of the left-hand side of (3-25) over the right.

3.6 THE DUAL PROBLEM

The standard form of the linear program will be used for solution via the simplex method, but the canonical form was defined for a different and theoretical reason. We now introduce a very important definition: Given a linear programming problem P in canonical form,

$$\text{Max } z = \mathbf{c}^T\mathbf{x} \tag{3-27}$$

$$\text{subject to} \quad \mathbf{Ax} \leq \mathbf{b} \tag{3-28}$$

$$\text{and} \quad \mathbf{x} \geq \mathbf{0} \tag{3-29}$$

where $\mathbf{A}$ is $m \times n$, $\mathbf{c}$ and $\mathbf{x}$ are $n \times 1$, and $\mathbf{b}$ is $m \times 1$, the following linear program is called its *dual* or *dual problem D*:

$$\text{Min } z' = \mathbf{b}^T\mathbf{u} \tag{3-30}$$

$$\text{subject to} \quad \mathbf{A}^T\mathbf{u} \geq \mathbf{c} \tag{3-31}$$

$$\text{and} \quad \mathbf{u} \geq \mathbf{0} \tag{3-32}$$

where $\mathbf{u}$ is an $m \times 1$ vector of unknowns and all other vectors and matrices are taken from the given linear program P. When speaking of P in this context of having a dual, we shall call it the *primal* problem, or more simply, the *primal*. The $\mathbf{u}$ are the *dual variables*, and (3-31) are the *dual constraints*; their primal counterparts are similarly named. Notice that we have begun to refer to an entire LPP by a single letter, such as P or D. This is much easier than the awkward phrase "(3-27) through (3-29)" and will be our practice throughout this book.

Observe that if there are n primal variables, there must be n dual constraints; similarly, the number of dual variables equals the number of primal constraints. To each primal variable x_j is associated a column $\mathbf{a}_j$ of $\mathbf{A}$ that contains the scalar coefficients of x_j. This is demonstrated by writing $\mathbf{Ax} \leq \mathbf{b}$ as

$$x_1\mathbf{a}_1 + x_2\mathbf{a}_2 + \cdots + x_n\mathbf{a}_n \leq \mathbf{b}$$

following the scheme of (2-1). Each of these columns becomes a row of $\mathbf{A}^T$ and provides the coefficients for one dual constraint. Furthermore, the cost coefficient c_j of the jth primal variable becomes the right-hand side of the jth dual constraint. Thus each primal variable is *associated with*, or corresponds to, a dual constraint in the above senses. Similar relationships associate primal constraints and dual variables. These relationships can be illustrated in a tableau credited to A. W. Tucker:

$$\begin{array}{c|cccc|c}
 & x_1 & x_2 & \cdots & x_n & \\
\hline
u_1 & a_{11} & a_{12} & \cdots & a_{1n} & \leq b_1 \\
u_2 & a_{21} & a_{22} & \cdots & a_{2n} & \leq b_2 \\
\vdots & \vdots & \vdots & & \vdots & \vdots \\
u_m & a_{m1} & a_{m2} & \cdots & a_{mn} & \leq b_m \\
\hline
 & \geq c_1 & \geq c_2 & \cdots & \geq c_n &
\end{array}$$

Primal constraints are read across the rows, multiplying each a_{ij} by its respective

x_j; dual constraints are read vertically. Note that each row (column) contains a dual (primal) variable and cost coefficient along with the associated primal (dual) constraint coefficients.

Before proceeding further we offer an example.

EXAMPLE 3.1. If the primal problem is

$$\begin{aligned} &\text{Max } z = x_1 + 3x_2 - 10x_3 \\ &\text{subject to} \quad x_1 + 2x_2 + 3x_3 \leq 8 \\ &\qquad\qquad\quad 4x_1 \qquad\quad - 5x_3 \leq 7 \\ &\text{and} \qquad\qquad\quad x_1, x_2, x_3 \geq 0 \end{aligned}$$

then the dual problem is

$$\begin{aligned} &\text{Min } z' = 8u_1 + 7u_2 \\ &\text{subject to} \quad u_1 + 4u_2 \geq 1 \\ &\qquad\qquad\quad 2u_1 \qquad \geq 3 \\ &\qquad\qquad\quad 3u_1 - 5u_2 \geq -10 \\ &\text{and} \qquad\qquad u_1, u_2 \geq 0 \end{aligned}$$

Inasmuch as any linear programming problem can be converted into canonical form, it must be true that *every LPP has a dual problem*, namely, the dual of its canonical equivalent. The route by which the dual of a noncanonical linear program can be found is illustrated in the proof of the following result.

THEOREM 3.1. The dual of the dual is the primal.

In order to find the dual of the problem (3-30) through (3-32), we convert it to canonical form, using our now-familiar algebraic procedures. The objective (3-30) becomes

$$\text{Max } z'' = (-\mathbf{b}^T)\mathbf{u} \tag{3-33}$$

and the constraints are equivalent to

$$-\mathbf{A}^T\mathbf{u} \leq -\mathbf{c} \tag{3-34}$$

Thus, the original dual problem of the definition is equivalent to (3-33), (3-34), and $\mathbf{u} \geq \mathbf{0}$, which is in canonical form. Applying the definition, the dual of the original dual is

$$\text{Min } z''' = -\mathbf{c}^T\mathbf{x} \tag{3-35}$$

$$\text{subject to} \quad -\mathbf{A}\mathbf{x} \geq -\mathbf{b} \tag{3-36}$$

$$\text{and} \quad \mathbf{x} \geq \mathbf{0} \tag{3-37}$$

where $\mathbf{x}$ must have the same number of components as $\mathbf{c}$, namely n, and where we have used the fact that the transpose of $\mathbf{A}^T$ is $\mathbf{A}$. Finally, we observe that (3-35) is equivalent to (3-27) and (3-36) is equivalent to (3-28). We have shown that the dual of the dual is the primal.

This theme of recasting a linear program into canonical form, applying the definition of the dual, and modifying the resulting problem appears again in the derivation of our next result.

THEOREM 3.2. The linear program in standard form (3-20), (3-21), (3-22) has the dual problem

$$\text{Min } z' = \mathbf{b}^T\mathbf{w} \tag{3-38}$$

$$\text{subject to} \quad \mathbf{A}^T\mathbf{w} \geq \mathbf{c} \tag{3-39}$$

where $\mathbf{w}$ is an m-component vector of unrestricted variables.

To prove this, we first convert the standard LPP into canonical form:

$$\text{Max } z = \mathbf{c}^T\mathbf{x} \tag{3-40}$$

$$\text{subject to} \quad \mathbf{A}\mathbf{x} \leq \mathbf{b} \tag{3-41}$$

$$-\mathbf{A}\mathbf{x} \leq -\mathbf{b} \tag{3-42}$$

$$\text{and} \quad \mathbf{x} \geq \mathbf{0} \tag{3-43}$$

where the constraints (3-41) and (3-42) are equivalent to (3-21). Now the definition of the dual problem is applied, with dual variables $\mathbf{u}$ and $\mathbf{v}$ being "associated with" (3-41) and (3-42), respectively. The dual problem becomes

$$\text{Min } z' = \mathbf{b}^T\mathbf{u} - \mathbf{b}^T\mathbf{v} \tag{3-44}$$

$$\text{subject to} \quad \mathbf{A}^T\mathbf{u} - \mathbf{A}^T\mathbf{v} \geq \mathbf{c} \tag{3-45}$$

$$\text{and} \quad \mathbf{u}, \mathbf{v} \geq \mathbf{0} \tag{3-46}$$

The reader may want to derive this rigorously from the definition by rewriting (3-41) and (3-42) as

$$\begin{bmatrix} \mathbf{A} \\ -\mathbf{A} \end{bmatrix} \mathbf{x} \le \begin{bmatrix} \mathbf{b} \\ -\mathbf{b} \end{bmatrix} \tag{3-47}$$

This compresses the constraints into a single matrix equation, imposing upon the primal the exact canonical form specified in (3-27) through (3-29). By using a partitioned vector $(\mathbf{u} \quad \mathbf{v})$ of dual variables instead of the $\mathbf{u}$ of the definition, the dual constraints are seen to be

$$[\mathbf{A}^T \quad -\mathbf{A}^T] \begin{bmatrix} \mathbf{u} \\ \mathbf{v} \end{bmatrix} \ge \mathbf{c} \tag{3-48}$$

which are precisely in the form of (3-31). Block-wise multiplication of (3-48) then yields (3-45).

We have not yet finished. The dual problem (3-44) through (3-46) can be portrayed more compactly via the transformation

$$\mathbf{w} \equiv \mathbf{u} - \mathbf{v} \tag{3-49}$$

Substitution yields

$$\text{Min } z' = \mathbf{b}^T\mathbf{w}$$
$$\text{subject to} \quad \mathbf{A}^T\mathbf{w} \ge \mathbf{c}$$

The variables $\mathbf{w}$ must be unrestricted since $\mathbf{u}$ and $\mathbf{v}$ can assume any nonnegative values. This completes the proof.

If equality constraints in the primal lead to unrestricted dual variables, Thereom 3.1 suggests that unrestricted primal variables should yield equality constraints in the dual. By now it should be easy for the student to show this: Let the primal problem P be

$$\text{Max } z = \mathbf{c}^T\mathbf{x}$$
$$\text{subject to} \quad \mathbf{A}\mathbf{x} \le \mathbf{b}$$

We begin by transforming the variables

$$\mathbf{x} \equiv \mathbf{x}' - \mathbf{x}'' \tag{3-50}$$

where $\mathbf{x}'$ and $\mathbf{x}''$ are required to be nonnegative, just as in Section 3.4. This substitution converts P into canonical form:

$$\text{Max } z = \mathbf{c}^T\mathbf{x}' - \mathbf{c}^T\mathbf{x}''$$
$$\text{subject to} \quad \mathbf{A}\mathbf{x}' - \mathbf{A}\mathbf{x}'' \le \mathbf{b}$$
$$\text{and} \quad \mathbf{x}', \mathbf{x}'' \ge \mathbf{0}$$

We omit the intermediate steps of expressing the objective and constraints as single matrix equations and immediately write the dual of P:

$$\begin{aligned} \text{Min } z' &= \mathbf{b}^T\mathbf{u} \\ \text{subject to}\quad \mathbf{A}^T\mathbf{u} &\geq \mathbf{c} \\ -\mathbf{A}^T\mathbf{u} &\geq -\mathbf{c} \\ \text{and}\quad \mathbf{u} &\geq \mathbf{0} \end{aligned}$$

The two constraint equations here are equivalent to $\mathbf{A}^T\mathbf{u} = \mathbf{c}$, and our assertion is proved.

If both of these noncanonical features, equality constraints and unrestricted variables, are present in the primal problem, then, by the arguments we used before, the corresponding dual will have both unrestricted variables and equality constraints. That is, the dual of

$$\text{Max } z = \mathbf{c}^T\mathbf{x} \qquad \text{subject to } \mathbf{A}\mathbf{x} = \mathbf{b}$$

is

$$\text{Min } z' = \mathbf{b}^T\mathbf{u} \qquad \text{subject to } \mathbf{A}^T\mathbf{u} = \mathbf{c}$$

Moreover, if only *some* of the primal constraints are equalities, it can be shown that only those dual variables associated with the primal equality constraints are unrestricted, but those associated with inequalities must be nonnegative. Finally, by virtue of the reflexive Theorem 3.1, all these properties have their parallels in the opposite direction.

At this point, let us collect these various small results into a theorem, most of which has been proved in the last few paragraphs.

THEOREM 3.3. If the primal linear programming problem P is

$$\begin{aligned} &\text{Max } z = c_1x_1 + c_2x_2 + \cdots + c_nx_n \\ &\text{subject to}\quad a_{11}x_1 + a_{12}x_2 + \cdots + a_{1n}x_n \leq b_1 \\ &\qquad\vdots \\ &\qquad a_{r1}x_1 + a_{r2}x_2 + \cdots + a_{rn}x_n \leq b_r \\ &\qquad a_{r+1,1}x_1 + a_{r+1,2}x_2 + \cdots + a_{r+1,n}x_n = b_{r+1} \\ &\qquad\vdots \\ &\qquad a_{m1}x_1 + a_{m2}x_2 + \cdots + a_{mn}x_n = b_m \\ &\text{and}\qquad x_j \geq 0 \quad j = 1, \ldots, s \end{aligned}$$

with the variables $x_{s+1}, \ldots, x_n$ unrestricted, then the corresponding dual problem D is

$$\begin{aligned}
\text{Min } z' = b_1u_1 + b_2u_2 + \cdots + b_mu_m \\
\text{subject to } \quad & a_{11}u_1 + a_{21}u_2 + \cdots + a_{m1}u_m \geq c_1 \\
& \vdots \\
& a_{1s}u_1 + a_{2s}u_2 + \cdots + a_{ms}u_m \geq c_s \\
& a_{1,s+1}u_1 + a_{2,s+1}u_2 + \cdots + a_{m,s+1}u_m = c_{s+1} \\
& \vdots \\
& a_{1n}u_1 + a_{2n}u_2 + \cdots + a_{mn}u_m = c_n \\
\text{and} \quad & u_i \geq 0 \qquad i = 1, \ldots, r
\end{aligned}$$

with the variables $u_{r+1}, \ldots, u_m$ unrestricted. Moreover, the dual of D is P.

Provided that he can keep track of all the subscripts, the reader will find that the proof of Theorem 3.3 is straightforward (the proof involves converting the problem P into canonical form and applying the definition of the dual, then reconverting to produce the problem D). Notice that this result completely subsumes Theorem 3.2; there we wanted to single out the dual of the standard form of the LPP because the standard form is very important—the simplex method operates on it.

Let us summarize the content of Theorem 3.3 in plain English. The dual of a linear programming *maximization* problem may be immediately identified and written provided that none of its constraints is of the form[1]

$$a_{i1}x_1 + \cdots + a_{in}x_n \geq b_i \tag{3-51}$$

The dual will have a nonnegative variable associated with each primal inequality ($\leq$) constraint and an unrestricted variable associated with each primal equality constraint; it will have an inequality ($\geq$) constraint associated with each nonnegative primal variable and an equality constraint associated with each unrestricted primal variable. Similarly, the dual of a linear programming *minimization* problem may be immediately written provided that none of its constraints is of the form

$$a_{i1}x_1 + \cdots + a_{in}x_n \leq b_i \tag{3-52}$$

[1] This is not quite an accurate statement. It is perfectly legitimate to write a dual problem directly from any primal, even one that includes constraints of the type (3-51). Such constraints (in a maximization problem) would give rise to *nonpositive* dual variables—the proof of this is required by Exercise 3-7. Nonpositive variables seem physically unrealistic, however, and are, in any case, of no particular theoretical interest.

Again, unrestricted variables are associated with equality constraints and non-negative variables with inequalities. In either of these cases, if a "forbidden" constraint appears, it may simply be multiplied on both sides by -1; then the dual may be written directly.

As a final gesture, we summarize the summary in the following mnemonic table:

Primal	*Dual*
Maximize	Minimize
constraint $\leq$	variable $u_i \geq 0$
constraint $=$	variable unrestricted
variable $x_j \geq 0$	constraint $\geq$
variable unrestricted	constraint $=$

The headings "Primal" and "Dual" are arbitrary, of course, and could have been reversed. This table is quite easy to use. For example, if the primal problem is

$$\begin{aligned} &\text{Max } z = x_1 + 3x_2 \\ &\text{subject to} \quad x_1 + 2x_2 + 3x_3 \leq 9 \\ &\qquad\qquad\quad x_1 \qquad\quad - x_3 = -5 \\ &\text{and} \qquad\qquad\qquad x_2, x_3 \geq 0 \end{aligned}$$

the dual can be written from it immediately, as follows:

$$\begin{aligned} &\text{Min } z' = 9u_1 - 5u_2 \\ &\text{subject to} \quad u_1 + u_2 = 1 \\ &\qquad\qquad\quad 2u_1 \qquad \geq 3 \\ &\qquad\qquad\quad 3u_1 - u_2 \geq 0 \\ &\text{and} \qquad\qquad\quad u_1 \geq 0 \end{aligned}$$

3.7 SOME PROPERTIES OF PRIMAL AND DUAL PROBLEMS

For the rest of this chapter we shall be referring to the two problems P and D that were used in the definition of the dual. We shall repeat them for convenience: The primal P is

$$\text{Max } z = \mathbf{c}^T\mathbf{x} \tag{3-27}$$

$$\text{subject to} \quad \mathbf{Ax} \le \mathbf{b} \tag{3-28}$$

$$\text{and} \quad \mathbf{x} \ge \mathbf{0} \tag{3-29}$$

and the dual D is

$$\text{Min } z' = \mathbf{b}^T\mathbf{u} \tag{3-30}$$

$$\text{subject to} \quad \mathbf{A}^T\mathbf{u} \ge \mathbf{c} \tag{3-31}$$

$$\text{and} \quad \mathbf{u} \ge \mathbf{0} \tag{3-32}$$

Properties of the solutions to primal and dual problems will be developed in the next few theorems, whose proofs will use P and D. Nevertheless, the results will apply to *all* pairs of primal and dual problems because they can be shown by the methods of Section 3.4 to be equivalent to P and D.

At this point, it would be well to recall some definitions from Section 1.4. In a linear programming problem—as in any mathematical program—a set of values for the variables that satisfies all constraints, including nonnegativity if required, is said to be a *feasible solution.* The collection of all feasible solutions is known as the *feasible region.* A feasible solution whose associated objective value is at least as good as that of any other feasible solution is called an *optimal solution.*

THEOREM 3.4 (Duality Lemma 1). If $\mathbf{x}_0$ is a feasible solution to the primal P and $\mathbf{u}_0$ is a feasible solution to the dual D, then $\mathbf{c}^T\mathbf{x}_0 \le \mathbf{b}^T\mathbf{u}_0$.

The proof is extremely simple. Since $\mathbf{x}_0$ is a feasible solution to P, $\mathbf{Ax}_0 \le \mathbf{b}$. The nonnegativity of $\mathbf{u}_0$ allows us to write

$$\mathbf{u}_0{}^T\mathbf{Ax}_0 \le \mathbf{u}_0{}^T\mathbf{b} = \mathbf{b}^T\mathbf{u}_0 \tag{3-53}$$

where we can freely transpose $\mathbf{u}_0{}^T\mathbf{b}$ because it is a scalar. Similarly, we are given $\mathbf{A}^T\mathbf{u}_0 \ge \mathbf{c}$, which may be transposed to $\mathbf{u}_0{}^T\mathbf{A} \ge \mathbf{c}^T$. This implies

$$\mathbf{u}_0{}^T\mathbf{Ax}_0 \ge \mathbf{c}^T\mathbf{x}_0 \tag{3-54}$$

Taken together, (3-53) and (3-54) imply $\mathbf{c}^T\mathbf{x}_0 \le \mathbf{b}^T\mathbf{u}_0$, Q.E.D.

THEOREM 3.5 (Duality Lemma 2). If $\mathbf{x}_0$ and $\mathbf{u}_0$ are feasible solutions to P and D and $\mathbf{c}^T\mathbf{x}_0 = \mathbf{b}^T\mathbf{u}_0$, then both $\mathbf{x}_0$ and $\mathbf{u}_0$ are optimal solutions.

If $\hat{\mathbf{x}}$ is any other feasible solution to P, we know from Theorem 3.4 that

$$\mathbf{c}^T\hat{\mathbf{x}} \le \mathbf{b}^T\mathbf{u}_0 = \mathbf{c}^T\mathbf{x}_0 \tag{3-55}$$

Because P is a maximization problem, $\mathbf{x}_0$ must be an optimal solution. A parallel argument shows that $\mathbf{u}_0$ must be an optimal solution to D.

These two lemmas may be attributed to Gale, Kuhn, and Tucker [11].

3.8 THE DUALITY THEOREM

The foundation now has been laid for the derivation of the single most important theoretical result in the field of linear programming. It is probably the most important result in all of mathematical programming as well. We shall begin simply by stating it.

THEOREM 3.6 (Duality Theorem). A feasible solution $\mathbf{x}_0$ to the primal problem P is optimal if and only if there exists a feasible solution $\mathbf{u}_0$ to the dual problem D such that $\mathbf{c}^T\mathbf{x}_0 = \mathbf{b}^T\mathbf{u}_0$.

This theorem was originally stated by John von Neumann, and a proof first appeared in 1956 in an article by Goldman and Tucker [13]. Their proof made use of an old result in linear algebra known as *Farkas' lemma* or the Minkowski–Farkas theorem. For the historically minded, that lemma reads as follows:

FARKAS' LEMMA. Given any matrix $\mathbf{A}$ and any vector $\mathbf{b}$, one and only one of the following statements is true:

(1) The system $\mathbf{A}\mathbf{x} = \mathbf{b}$ has a nonnegative solution $\mathbf{x}_0$.

(2) The system $\begin{Bmatrix} \mathbf{u}^T\mathbf{A} \geq \mathbf{0} \\ \mathbf{u}^T\mathbf{b} < \mathbf{0} \end{Bmatrix}$ has a solution $\mathbf{u}_0$.

By taking a corollary to this lemma and applying more linear algebra, it was possible to prove the duality theorem. In 1962, an entirely new derivation by Dreyfus and Freimer appeared in a text on dynamic programming [9]. This is the proof that we now present.

Assume $\mathbf{u}_0$ is a feasible solution to D and $\mathbf{c}^T\mathbf{x}_0 = \mathbf{b}^T\mathbf{u}_0$. By hypothesis, $\mathbf{x}_0$ is a feasible solution to P; therefore, from Theorem 3.5, $\mathbf{x}_0$ is optimal. This, of course, is trivial. It is the other direction, the "only if," that is of interest to us.

Recall that our primal problem P is

$$\text{Max } z = \mathbf{c}^T\mathbf{x} \tag{3-27}$$

$$\text{subject to} \quad \mathbf{A}\mathbf{x} \leq \mathbf{b} \tag{3-28}$$

$$\text{and} \quad \mathbf{x} \geq \mathbf{0} \tag{3-29}$$

If the vector $\mathbf{b}$ were algebraically increased, more and more solutions would be added to the feasible set. The objective function either would continue to have the same maximum value or would be able to attain a greater one. Define the scalar function $f(\boldsymbol{\beta})$ to be the maximum possible value of the objective when the right-hand side of (3-28) takes on the value $\boldsymbol{\beta}$, where $\mathbf{A}$ and $\mathbf{c}$ are held fixed. By hypothesis the optimal solution to P is $\mathbf{x}_0$, so $f(\mathbf{b}) = \mathbf{c}^T\mathbf{x}_0$.

Now perturb this optimal solution by increasing its jth component x_{0j} by an infinitesimal positive amount ϵ.[1] In order to be sure that this new solution satisfies the constraints of P, we add $\epsilon\mathbf{a}_j$ to the right-hand side $\mathbf{b}$, where $\mathbf{a}_j$ is the jth column of $\mathbf{A}$. We have ensured thereby that

$$\hat{\mathbf{x}} = \mathbf{x}_0 + \epsilon\mathbf{e}_j \tag{3-56}$$

where $\mathbf{e}_j$ is the unit vector with a one in the jth position, is a feasible solution to the linear program

$$\begin{aligned} &\text{Max } z = \mathbf{c}^T\mathbf{x} \\ &\text{subject to} \quad \mathbf{Ax} \le \mathbf{b} + \epsilon\mathbf{a}_j \\ &\text{and} \qquad\qquad \mathbf{x} \ge \mathbf{0} \end{aligned} \tag{3-57}$$

This feasible solution (3-56) yields an objective value of

$$\mathbf{c}^T(\mathbf{x}_0 + \epsilon\mathbf{e}_j) = f(\mathbf{b}) + \epsilon c_j \tag{3-58}$$

It is not necessarily true, however, that (3-56) is the optimal solution to the new problem (3-57). All we can say is that

$$f(\mathbf{b} + \epsilon\mathbf{a}_j) \ge f(\mathbf{b}) + \epsilon c_j \tag{3-59}$$

Because f is a continuous function (the reader will have to let this slip by without proof), we can perform a Taylor expansion of the left side of (3-59) to obtain

$$f(\mathbf{b}) + \epsilon \sum_{i=1}^{m} \frac{\delta f}{\delta b_i} a_{ij} + (\text{terms in } \epsilon^2 \text{ and higher}) \ge f(\mathbf{b}) + \epsilon c_j \tag{3-60}$$

where we have used the more convenient notation $\delta f/\delta b_i$ in place of $\delta f(\mathbf{b})/\delta\beta_i$. Note that in (3-60) the argument $\mathbf{b}$ of the function f is undergoing a change ϵa_{ij} in *each* of its components b_i. For sufficiently small perturbations ϵ, division of (3-60) by ϵ yields

$$\sum_{i=1}^{m} \frac{\delta f}{\delta b_i} a_{ij} \ge c_j \qquad \textit{whenever } \epsilon \textit{ is positive} \tag{3-61}$$

[1] This will be the only use of calculus in the entire development of linear programming.

Now suppose that the value x_{0j} of the jth variable in the optimal solution to the original problem P is greater than zero. We could then permit ϵ to be a *negative* infinitesimal perturbation and (3-56) would still be a feasible solution to (3-57). Repeating the above derivation, we arrive again at (3-60). Now division by ϵ reverses the inequality and we get

$$\sum_{i=1}^{m} \frac{\delta f}{\delta b_i} a_{ij} \leq c_j \qquad \textit{whenever } \epsilon \textit{ is negative} \tag{3-62}$$

Because perturbation in both directions is possible when x_{0j} is greater than zero, we must conclude from (3-61) and (3-62) that

$$\sum_{i=1}^{m} \frac{\delta f}{\delta b_i} a_{ij} = c_j \qquad \text{if } x_{0j} > 0 \tag{3-63}$$

whereas

$$\sum_{i=1}^{m} \frac{\delta f}{\delta b_i} a_{ij} \geq c_j \qquad \text{if } x_{0j} = 0 \tag{3-64}$$

Now recall that the maximum value of the objective function of an LPP cannot be decreased by an algebraic increase in **b** or by an isolated increase in any one of its components. Such increases exclude no solutions **x** from the initially feasible set and may admit new ones. This implies that

$$\frac{\delta f}{\delta b_i} \geq 0 \qquad \text{for all } i = 1, \ldots, m \tag{3-65}$$

We now make the surprising observation that the vector $\mathbf{u}_0$ whose ith component is

$$u_{0i} \equiv \frac{\delta f}{\delta b_i} \tag{3-66}$$

is a *feasible solution to the dual problem D*! This follows immediately from (3-63), (3-64), and (3-65).

We still must show that the values of the primal and dual objectives are equal. The dual constraints (3-31) must hold for the feasible solution $\mathbf{u}_0$. Transposing and multiplying by the nonnegative vector $\mathbf{x}_0$ gives

$$\mathbf{u}_0{}^T\mathbf{A}\mathbf{x}_0 \geq \mathbf{c}^T\mathbf{x}_0 \tag{3-67}$$

Component-wise, this is the same as

$$\sum_{j=1}^{n} (\mathbf{u}_0{}^T\mathbf{A})_j x_{0j} \geq \sum_{j=1}^{n} c_j x_{0j} \tag{3-68}$$

where $(\mathbf{u}_0{}^T\mathbf{A})_j$ is the jth component of the row vector $\mathbf{u}_0{}^T\mathbf{A}$,

$$(\mathbf{u}_0{}^T\mathbf{A})_j \equiv \sum_{i=1}^{m} u_{0i}a_{ij} \tag{3-69}$$

From (3-63) and (3-64), either $x_{0j} = 0$ or, if not, $c_j = (\mathbf{u}_0{}^T\mathbf{A})_j$. In either case, equality obtains term by term in (3-68), so

$$\mathbf{u}_0{}^T\mathbf{A}\mathbf{x}_0 = \mathbf{c}^T\mathbf{x}_0 \tag{3-70}$$

One final step remains. We have assumed that $\mathbf{x}_0$ is a feasible and optimal solution to P, so the primal constraints (3-28) and the nonnegativity (3-65) of $\mathbf{u}_0$ allow us to write

$$\mathbf{u}_0{}^T\mathbf{A}\mathbf{x}_0 \leq \mathbf{u}_0{}^T\mathbf{b} \tag{3-71}$$

Representing this as a sum of terms,

$$\sum_{i=1}^{m} u_{0i}(\mathbf{A}\mathbf{x}_0)_i \leq \sum_{i=1}^{m} u_{0i}b_i \tag{3-72}$$

Now suppose that $\mathbf{x}_0$ satisfies the ith primal constraint as an *inequality*:

$$a_{i1}x_{01} + a_{i2}x_{02} + \cdots + a_{in}x_{0n} < b_i \tag{3-73}$$

Whenever this is true—that is, whenever $(\mathbf{A}\mathbf{x}_0)_i < b_i$—we argue from economics and common sense that a slight increase in b_i will not increase the objective value obtainable.[1] If (3-73) holds, the ith primal constraint is *not binding*; it is not preventing the objective function from attaining a greater value. This means that for small perturbations

$$\frac{\delta f}{\delta b_i} \equiv u_{0i} = 0 \tag{3-74}$$

But because $\mathbf{x}_0$ is feasible, for every primal constraint i, either $(\mathbf{A}\mathbf{x}_0)_i = b_i$ or (3-73) holds, in which case $u_{0i} = 0$. It follows that (3-72) is an equality, and

$$\mathbf{u}_0{}^T\mathbf{A}\mathbf{x}_0 = \mathbf{u}_0{}^T\mathbf{b} \tag{3-75}$$

Finally, from (3-70) and (3-75) we may write

$$\mathbf{u}_0{}^T\mathbf{b} = \mathbf{c}^T\mathbf{x}_0$$

as was to be shown.

[1] And obviously a small decrease in b_i would not decrease the optimal objective value because the constraint (3-73) would remain satisfied.

Having begun with the hypothesis that $\mathbf{x}_0$ is a feasible and optimal solution to P, we have constructed $\mathbf{u}_0$, a feasible solution to D, and have shown that the primal and dual objective functions evaluated at these solutions are equal. This proves the duality theorem. We know from Theorem 3.5 that our $\mathbf{u}_0$ must also be an optimal solution to D.

We now turn our attention to a simple example that will illustrate the theoretical development of this section.

EXAMPLE 3.2. Consider the primal problem in canonical form,

$$\begin{aligned} &\text{Max } z = 2x_1 + x_2 \\ &\text{subject to} \quad 2x_1 - x_2 \leq 4 \\ &\qquad\qquad\qquad\quad x_2 \leq 6 \\ &\text{and} \qquad\qquad x_1, x_2 \geq 0 \end{aligned}$$

which is graphed in Figure 3-1(a). Its optimal solution is easily seen to be $x_1 = 5$, $x_2 = 6$, $z = 16$; accordingly, using the scalar function $f(\boldsymbol{\beta})$ defined in the foregoing proof, we can say that $f(4,6) = 16$. The reader may want to verify that $f(1,2) = 5$; that is, if the right-hand-side vector were (1,2) instead of (4,6), the optimal value of z would be 5.

Because the primal problem has an optimal solution with $z_{\text{opt}} = 16$, we know from the duality theorem that the dual problem must have also an optimal

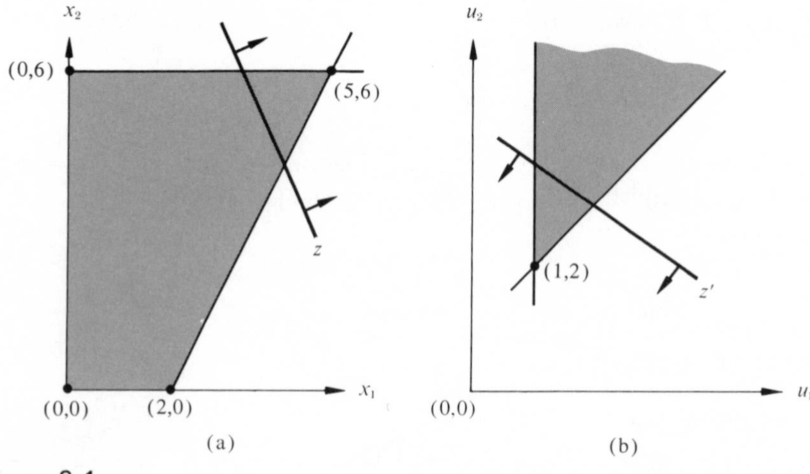

Figure 3-1

solution $\mathbf{u}_0$ whose objective value is 16. In fact, we can deduce the optimal values of the dual variables (provided that they are unique) from the identity

$$u_{0i} \equiv \frac{\delta f}{\delta b_i} \tag{3-66}$$

In the case of u_{01}, a little arithmetic suffices to show that

$$f(3,6) = 15 \qquad f(4,6) = 16 \qquad \text{and} \qquad f(5,6) = 17$$

Therefore, when $b_2 = 6$ and b_1 is in the vicinity of 4, as in the given primal problem,

$$u_{01} \equiv \frac{\delta f}{\delta b_1} = 1$$

Similarly, we can show that

$$f(4,5) = 14 \qquad f(4,6) = 16 \qquad \text{and} \qquad f(4,7) = 18$$

so that $u_{02} = 2$. Thus, we have determined that the optimal solution to the dual problem must be $u_1 = 1$, $u_2 = 2$, $z' = 16$. This can be verified by writing the dual problem

$$\begin{aligned} &\text{Min } z' = 4u_1 + 6u_2 \\ &\text{subject to} \quad 2u_1 \qquad \geq 2 \\ &\qquad\qquad -u_1 + u_2 \geq 1 \\ &\text{and} \qquad\qquad u_1, u_2 \geq 0 \end{aligned}$$

and solving it graphically—see Figure 3-1(b).

Notice that the duality theorem does not apply to unbounded primal problems because its hypothesis specifically requires an optimal solution $\mathbf{x}_0$. In fact, we shall see later on that when the primal problem is unbounded, the dual has no feasible solution.

3.9 ECONOMIC INTERPRETATION OF PRIMAL AND DUAL

In the past few sections, we have taken a linear programming problem in canonical form, used inputs from it to construct another LPP, and established that (unless the given problem has no optimal solution) their objective functions must have the same optimal value. This was an elegant mathematical exercise and furnishes useful theoretical results, but it is somewhat frustrating for the student, who might prefer a logical explanation of why it had to work out that way, what the dual problem means, and how it is related to the primal. Unfortunately, it is impossible—and that word is used cautiously—to communicate in simple terms a complete economic explanation of duality. We offer instead a qualitative discussion of some of the implications.

The example of Section 3.1 is an ideal illustration of a primal linear program in canonical form. Its general representation, repeated once more, is

$$\text{Max } z = c_1x_1 + \cdots + c_nx_n \tag{3-76}$$

$$\text{subject to} \quad a_{i1}x_1 + \cdots + a_{in}x_n \leq b_i \qquad i = 1, \ldots, m \tag{3-77}$$

$$\text{and} \qquad x_j \geq 0 \qquad j = 1, \ldots, n \tag{3-78}$$

and it can be viewed quite easily in economic terms. The b_i are the total supplies of inputs or available raw materials, and the x_j are the unknown amounts of various outputs, or products, that will be manufactured from the available inputs. Each a_{ij} is the amount of input i needed to make one unit of output j, so that the constraints express the production limitations caused by the scarcity of the inputs. The factory manager's objective is simply to maximize overall profit or sales income, where each c_j is the income from one unit of product j. This is a very natural interpretation—indeed, many LPPs arising in practice are of precisely this form.

By virtue of the duality theorem and of (3-66), in particular, the identity of the dual variables associated with the given linear programming problem stand revealed: They are the rates of change of the optimal value of the primal objective function with respect to the right-hand-side elements of the primal constraints. Suppose the value of the ith dual variable in the optimal solution to the dual problem were u_{0i}. This would tell us that, if the right-hand side b_i of the ith primal constraint were increased by a sufficiently small ϵ, then the optimal value of the primal objective would be increased by ϵu_{0i} (except for certain values of $\mathbf{b}$ at which the first partial derivative $\delta f/\delta b_i$ is discontinuous). In other words, if one extra small unit ($\epsilon = 1$) of the ith input were available, then an extra u_{0i} dollars of sales income could be realized by the factory. Therefore, u_{0i} is the maximum price that the manager would be willing to pay for an additional unit of the ith input; this is also known as the *shadow price* or the *marginal value* of the input. Equivalently, u_{0i} can be thought of also as the loss incurred when one unit of the available ith input is lost or destroyed; hence it gives the correct insurance valuation only for the *last* unit of input i (after one unit has been lost, the primal problem would be different, and so might be the optimal value of the ith dual variable).

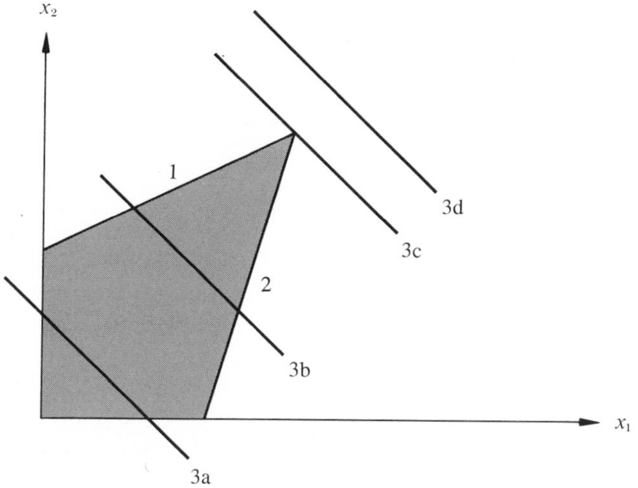

Figure 3-2

Our remarks can be illustrated with a diagram of a two-variable problem. Suppose that the (primal) linear program of Figure 3-2 has three constraints, labeled numerically, as well as the usual nonnegativity restrictions. In particular, suppose constraint 3 is

$$x_1 + x_2 \leq b_3 \tag{3-79}$$

and let the objective be to maximize $z = x_2$. When b_3 is quite small, so that this constraint is in the position 3a shown in Figure 3-2, the optimal solution occurs at ($x_1 = 0, x_2 = b_3$), where 3a intersects the x_2 axis. Here the constraint is satisfied as an equality and, therefore, is said to be *binding*:[1] It binds down the objective function and prevents it from assuming a greater value, as it would be free to do in the absence of 3a. Similarly, we say that (when b_3 is small) constraints 1 and 2 are *not binding*.

Observing Figure 3-2, it is clear that if (3-79) were *relaxed*—that is, if b_3 were increased—the feasible region would be enlarged and the objective function $z = x_2$ could take on a greater value than before. For a while, as b_3 increases, the optimal point will simply climb up the x_2 axis, with the optimal value of z (which equals b_3) gaining one unit for each unit of increase in b_3. Thus, over this range of b_3 values, the marginal value u_{03} will equal 1. For somewhat larger values of b_3, when constraint (3-79) has reached, say, position 3b, the optimum will lie at the intersection of constraints 1 and 3, both of which will then be

[1] By definition, a constraint is binding if and only if the optimal value of its slack or surplus variable is zero. An equality constraint is considered to be automatically binding.

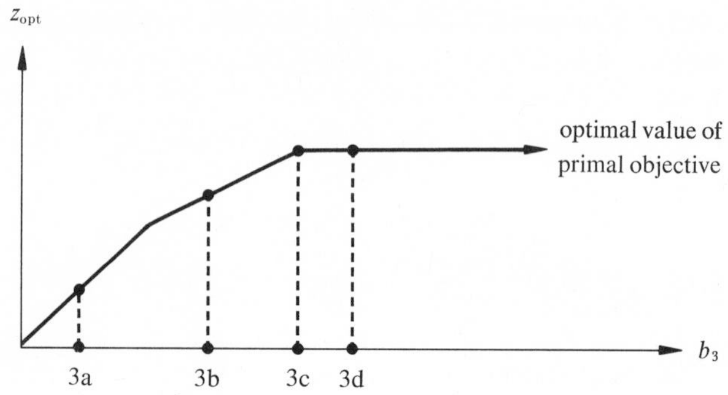

Figure 3-3

binding. As b_3 increases further, the growth of the optimal objective value will continue in linear proportion, at a rate given by u_{03} (which will now be less than 1). When the constraint reaches position 3c, however, further relaxation can no longer raise the optimal value of z; constraints 1 and 2 have met and completely bounded the feasible region, so that additional increases in b_3 cannot enlarge it. Therefore, when constraint (3-79) is, say, in position 3d, the optimal value of the third dual variable u_{03} (that is, the marginal value of the third input) must be zero; in this position the constraint is *superfluous*, in the sense of not forming any part of the boundary of the feasible region.[1] Similarly, when (3-79) was in its original position 3a, the optimal values of the first and second dual variables were both zero.

If we imagine that this entire linear programming problem (i.e., maximize $z = x_2$ subject to nonnegativity and to constraints 1, 2, and 3) is held fixed with the single exception that b_3 is permitted to vary, we can plot z_{opt}, the optimal value of z, as a function of b_3. This is done in Figure 3-3; the labeled points are the values of b_3 for which the constraint (3-79) assumes the various positions shown in Figure 3-2. The slope of this piecewise linear function at any point b_3, *except at the "break points" where line segments meet*, is just dz_{opt}/db_3, which is the optimal value of the third dual variable.

With the aid of Figure 3-3 the student should be able to see that each of the following statements is almost, but not quite, correct:

(a) The optimal value u_{03} of the third dual variable equals the change in z_{opt} per "unit" change in b_3.
(b) u_{03} is the maximum price that our factory manager would be just willing to pay for an extra "unit" of input 3.
(c) u_{03} is the correct insurance valuation for the last available "unit" of input 3.

[1] It would be careless to call such a constraint *redundant* because in Chapter 2 we defined as redundant any constraint that could be expressed as a linear combination of the others. This is not true of constraint 3 in the example we are considering.

In all three cases the inaccuracy lies in the fact that, for any unit of finite size, a one-unit change in b_3 might result in a shift from one segment of the piecewise linear function to another; when this happens, the change in z_{opt} is not equal to u_{03}. The optimal value of the third dual variable

$$u_{03} \equiv \frac{dz_{opt}}{db_3} \equiv \frac{\delta f}{\delta b_3}$$

should instead be regarded as an instantaneous rate of change, with the further understanding that u_{03} is not uniquely defined at the break points between linear segments (one such point is 3c). This is obvious from Figure 3-3: The first derivative of the function is discontinuous at those points. It turns out that when constraint (3-79) of our linear program is in the break-point position 3c, the dual problem has more than one optimal solution; in particular, u_{03} then can be equal to the slope of *either* of the line segments in Figure 3-3 that meet at position 3c. This phenomenon of alternate dual optima will be illustrated in Exercise 3-12.

Figure 3-3 makes good sense from an economic point of view. Returning to our factory manager's problem, suppose we held everything fixed except the available amount b_3 of the third input, and suppose we then graphed the maximum obtainable sales income as a function of b_3. We would obtain a piecewise linear curve similar to that of Figure 3-3. If input 3 were essential to all products and none of it were on hand, then no output whatsoever could be produced by the factory and no income would accrue (of course, some product might well require none of input 3, so that, in general, $z_{opt} \geq 0$ for $b_3 = 0$). If a small amount of input 3 were available in inventory, it would be used along with certain other inputs in the most profitable combination possible, say, to manufacture product Q. The marginal value u_{03} of input 3 would then be relatively high, because each additional unit of it would yield a relatively large increment of sales income. As b_3 increased, however, it would eventually be impossible to continue deriving product Q from it because one of the other necessary inputs would be totally depleted. At this break point, further increments of input 3 would have to be devoted to the production of some less profitable output, and z_{opt} would begin to increase at a slower rate. Thus the marginal value of one of these latter units of input 3 would be less than the marginal value of the first unit. By this argument, we can see that for each input i, $i = 1, \ldots, m$, the piecewise linear function we have been discussing must continue to flatten out as the abscissa b_i increases: The slope can never increase and must eventually reach zero.[1]

[1] This statement is not true for inputs that are themselves salable outputs. If some particular input k can be sold directly, in the absence of all other inputs, at a price of θ per unit, then the slope of its piecewise linear function will eventually fall, not to zero, but to θ. This reflects the fact that, even when all other inputs have been exhausted, the total sales income obtainable still increases by θ for each additional available unit of input k.

Having discussed the dual variables at some length, let us shift our attention to the constraints upon them. What economic realities or truths do the dual constraints reflect? That is, why *should* they hold? Can we find an interpretation according to which they seem "natural"? The answer is yes. Substituting (3-66) into the jth dual constraint yields

$$a_{1j}\frac{\delta f}{\delta b_1} + a_{2j}\frac{\delta f}{\delta b_2} + \cdots + a_{mj}\frac{\delta f}{\delta b_m} \geq c_j \tag{3-80}$$

Remember that, for any given j, the a_{ij} are the various amounts of the inputs i that go into producing one unit of the jth output. If the manufacturer were suddenly given an additional amount a_{ij} of each input, $i = 1, \ldots, m$, he could (among other options) produce one more unit of the jth output. Therefore, (3-80) argues, the total marginal values of all these inputs must add up to *at least* c_j, the sale price of that extra output unit. Of course, by assigning the additional inputs cleverly and producing outputs other than j, it is possible that the manufacturer could further increase his total sales income. If so, the marginal values of the various inputs would be large enough to make the left-hand side of (3-80) greater than c_j; hence the inequality.

The economic interpretation of the dual constraints is quite satisfactory, but the meaning of the dual objective—at least for our resource-allocation-example problem—is rather vague. As before, we substitute (3-66) into the objective in order to get a better look at it:

$$\text{Min } z = b_1\frac{\delta f}{\delta b_1} + \cdots + b_m\frac{\delta f}{\delta b_m} \tag{3-81}$$

If it were true that the marginal value of *all* units of input i were the same as that of the *last* available unit, then z would represent the total amount of marginal value in the available supplies of all inputs. Unfortunately, the hypothesis is not true; different units of input i have different marginal values. There is, however, another way of looking at the objective (3-81). If a primal production scheme—that is, a feasible solution $\mathbf{x}$—is "unbalanced" in some sense, so that some inputs are totally consumed while others are hardly touched, then the marginal values of those exhausted inputs are quite high. (A high marginal value means that an added increment of the input could immediately combine with other unexhausted, freely available inputs to produce an additional valuable unit of output.) Apparently, an input whose initial supply is large should not be carelessly exhausted and driven to a high marginal value. The dual objective (3-81) appears to seek balanced production schemes, where all inputs are more or less exhausted and thus marginal values are low.

Finally, the duality theorem has an economic interpretation that, in our resource-allocation context, is only partially satisfying, at best. It seems proper that the maximum profit obtainable should equal the total gross amount of value in the inputs because "value" should be defined in terms of profit that can be derived. The $\delta f/\delta b_i$ in (3-81), however, are *marginal* values valid only for the last units of the various inputs, and this considerably obscures the picture. We shall not belabor any further these economic interpretations of the dual problem. Unresolved ambiguities will be left to the reader's imagination, along with the consoling reminder that one is perfectly free simply to ignore the whole question of interpretation and to stick to the mathematical results, where the ground is firmer.

3.10 THE EXISTENCE THEOREM

We shall now pick up the theoretical theme again and derive the second major result of this chapter. Mathematics is inevitably concerned with the existence of solutions, and the field of linear programming is no exception. The existence theorem is extremely important and has far-reaching implications. Because it derives so easily from the duality theorem, however, it is more accurately a corollary rather than a theorem of the first magnitude. Nevertheless,

THEOREM 3.7 (Existence Theorem). (a) A linear program has a finite optimal solution if and only if both it and its dual have feasible solutions.

(b) If the primal problem has an unbounded maximum, then the dual problem has no feasible solution.

(c) If the dual problem has no feasible solution but the primal problem has, then the primal problem has an unbounded maximum.

Proof. (a) If the primal has an optimal feasible solution, then the duality theorem guarantees that the dual has a feasible solution.

Going the other way, assume that the primal and dual have feasible solutions $\hat{\mathbf{x}}$ and $\hat{\boldsymbol{u}}$; therefore, $\mathbf{c}^T\hat{\mathbf{x}}$ and $\mathbf{b}^T\hat{\boldsymbol{u}}$ are finite. From Theorem 3.4, $\mathbf{b}^T\hat{\boldsymbol{u}}$ serves as an upper bound on the primal objective $\mathbf{c}^T\mathbf{x}$, although not necessarily a least upper bound. At any rate, the primal must have a finite optimum, Q.E.D.

(b) Let the primal be unbounded, in which case it has a feasible solution. If the dual also had a feasible solution, then, according to Theorem 3.7(a), the primal would have to have a finite optimum, which is a contradiction.

(c) Any feasible solution $\hat{\mathbf{x}}$ to the primal would have objective value $\mathbf{c}^T\hat{\mathbf{x}}$. This cannot be an optimum because, if it were, the duality theorem would dictate that a feasible solution to the dual problem must exist, contrary to our hypothesis. It follows that no feasible solution to the primal can be optimal, so that the primal is unbounded.

These results are summarized as follows:

	Primal problem is feasible	*Primal problem is not feasible*
Dual problem is feasible	Both optima exist	Dual problem unbounded
Dual problem is not feasible	Primal problem unbounded	May occur

As an example of how primal unboundedness is associated with dual infeasibility, consider the linear programming problem

$$\text{Max } z = x_1 + x_2$$
$$\text{subject to} \quad x_1 - x_2 \leq 1$$
$$3x_1 - x_2 \leq 10$$
$$\text{and} \qquad x_1, x_2 \geq 0$$

Holding $x_1 = 0$, let x_2 increase without bound. Because all constraints continue to be satisfied, z can approach positive infinity and the problem is thus unbounded. The dual problem is

$$\text{Min } z' = u_1 + 10u_2$$
$$\text{subject to} \quad u_1 + 3u_2 \geq 1$$
$$-u_1 - u_2 \geq 1$$
$$\text{and} \qquad u_1, u_2 \geq 0$$

The second constraint can never be satisfied by nonnegative values of the dual variables, so no feasible solution exists.

We close this section with an example of primal *and* dual infeasibility. Let the primal be

$$\text{Max } z = x_1 + 2x_2$$
$$\text{subject to} \quad x_1 - x_2 \leq 0$$
$$-x_1 + x_2 \leq -1$$
$$\text{and} \qquad x_1, x_2 \geq 0$$

Then the dual problem must be

$$\text{Min } z' = -u_2$$

$$\text{subject to} \quad u_1 - u_2 \geq 1$$

$$-u_1 + u_2 \geq 2$$

$$\text{and} \quad u_1, u_2 \geq 0$$

Neither problem has a feasible solution.

3.11 COMPLEMENTARY SLACKNESS

In this section we shall derive some more very provocative results about the primal and dual optimal solutions. Recall from Section 3.5 that an inequality constraint may be transformed into an equality by the introduction of a slack or surplus variable, which then is restricted to being nonnegative. This maneuver was illustrated in (3-23) and (3-24). Let us return to the familiar canonical primal problem P and transform the constraints (3-28) into equalities. Each equation is given a slack variable, which appears only in that one equation; hence the primal may be written as follows:

$$\text{Max } z = \mathbf{c}^T\mathbf{x} \tag{3-82}$$

$$\text{subject to} \quad \mathbf{Ax} + \mathbf{x}_s = \mathbf{b} \tag{3-83}$$

$$\text{and} \quad \mathbf{x}, \mathbf{x}_s \geq \mathbf{0} \tag{3-84}$$

where $\mathbf{x}_s$ is a (column) vector consisting of the m nonnegative slack variables. The dual problem D also may be converted to standard form, by means of surplus variables:

$$\text{Min } z' = \mathbf{b}^T\mathbf{u} \tag{3-85}$$

$$\text{subject to} \quad \mathbf{A}^T\mathbf{u} - \mathbf{u}_s = \mathbf{c} \tag{3-86}$$

$$\text{and} \quad \mathbf{u}, \mathbf{u}_s \geq \mathbf{0} \tag{3-87}$$

with the n nonnegative surplus variables collected into $\mathbf{u}_s$.

At this point, we observe that (3-83) implies

$$\mathbf{u}^T\mathbf{Ax} + \mathbf{u}^T\mathbf{x}_s = \mathbf{u}^T\mathbf{b}$$

for any feasible primal solution $(\mathbf{x},\mathbf{x}_s)$ and for any $m \times 1$ vector of values $\mathbf{u}$. This is equivalent to

$$\mathbf{x}^T\mathbf{A}^T\mathbf{u} + \mathbf{u}^T\mathbf{x}_s = \mathbf{b}^T\mathbf{u} \tag{3-88}$$

two scalar numbers having been transposed. Similarly, (3-86) leads to

$$\mathbf{x}^T\mathbf{A}^T\mathbf{u} - \mathbf{x}^T\mathbf{u}_s = \mathbf{c}^T\mathbf{x} \tag{3-89}$$

for any feasible dual solution $(\mathbf{u},\mathbf{u}_s)$ and for any n-component vector $\mathbf{x}$.

Suppose now that optimal solutions $(\mathbf{x}_0,\mathbf{x}_{0s})$ and $(\mathbf{u}_0,\mathbf{u}_{0s})$ exist for the primal and dual problems, where $\mathbf{x}_{0s}$ and $\mathbf{u}_{0s}$ are the optimal values of the primal slack vector $\mathbf{x}_s$ and dual surplus vector $\mathbf{u}_s$. Theorems 3.5 and 3.6 together insure that

$$\mathbf{c}^T\mathbf{x}_0 = \mathbf{b}^T\mathbf{u}_0 \tag{3-90}$$

(cost coefficients of all slack and surplus variables are zero). Optimal solutions are by definition feasible, so (3-88) and (3-89) must hold for them:

$$\mathbf{x}_0{}^T\mathbf{A}^T\mathbf{u}_0 + \mathbf{u}_0{}^T\mathbf{x}_{0s} = \mathbf{b}^T\mathbf{u}_0 \tag{3-91}$$

and

$$\mathbf{x}_0{}^T\mathbf{A}^T\mathbf{u}_0 - \mathbf{x}_0{}^T\mathbf{u}_{0s} = \mathbf{c}^T\mathbf{x}_0 \tag{3-92}$$

Then, because of (3-90), we can equate (3-91) and (3-92), arriving at

$$\mathbf{u}_0{}^T\mathbf{x}_{0s} + \mathbf{x}_0{}^T\mathbf{u}_{0s} = 0 \tag{3-93}$$

All components of all four vectors are nonnegative, so we can permit none of the $m + n$ terms in the sum (3-93) to be nonzero. The implications are stated in the theorem of *complementary slackness*:

THEOREM 3.8. Given any pair of optimal solutions to a linear programming problem and its dual, the following hold:

(a) For each i, $i = 1, \ldots, m$, the product of the ith primal slack variable and the ith dual variable is zero; that is, if either is positive, the other must be zero.

(b) For each j, $j = 1, \ldots, n$, the product of the jth primal variable and the jth dual surplus variable is zero.

(In this theorem and in the discussion below, any allusion to a variable being positive or zero is understood to mean that the variable's *optimal value* is positive or zero.)

If Theorem 3.8(a) and (b) hold, we say that complementary slackness prevails. It is important to observe that if, say, a primal slack variable is zero it does *not* follow that the corresponding dual variable is positive. Both may be zero because the requirement is simply that their product be zero.

The economic implications of complementary slackness should be fairly obvious after Section 3.9. Consider Theorem 3.8(a), for example: If the ith primal slack variable is positive, then the optimal production scheme leaves the ith input unexhausted. In these circumstances a little bit more of it clearly would be useless; its marginal value and corresponding dual variable must be zero. Conversely, if the ith dual variable is positive, then the ith ingredient has some marginal value. If any of it were still available, it would be immediately used, in order to extract that value. Hence, in the optimal production scheme it must have been completely exhausted, and the ith primal slack is zero. Notice that it is possible for the ith dual variable to be zero while the ith primal constraint is binding. This could occur if there were several other binding constraints; in that case a slight increment of input i might not increase profits because the other binding constraints would continue to "freeze" the optimal solution.

One important result remains to be derived:

THEOREM 3.9. If $(\mathbf{x}_0,\mathbf{x}_{0s})$ and $(\mathbf{u}_0,\mathbf{u}_{0s})$ are feasible solutions to the primal and dual problems and complementary slackness prevails, then $(\mathbf{x}_0,\mathbf{x}_{0s})$ and $(\mathbf{u}_0,\mathbf{u}_{0s})$ are both optimal solutions.

The proof reverses the derivation of the complementary slackness conditions, 3.8(a) and 3.8(b). We are given

$$\mathbf{u}_0{}^T\mathbf{x}_{0s} + \mathbf{x}_0{}^T\mathbf{u}_{0s} = 0 \tag{3-94}$$

which yields

$$\mathbf{u}_0{}^T\mathbf{x}_{0s} = -\mathbf{u}_{0s}{}^T\mathbf{x}_0 \tag{3-95}$$

after transposing the scalar right-hand side. Adding $\mathbf{u}_0{}^T\mathbf{A}\mathbf{x}_0$ to both sides of (3-95) produces

$$\mathbf{u}_0{}^T(\mathbf{A}\mathbf{x}_0 + \mathbf{x}_{0s}) = (\mathbf{u}_0{}^T\mathbf{A} - \mathbf{u}_{0s}{}^T)\mathbf{x}_0 \tag{3-96}$$

Now, $(\mathbf{x}_0,\mathbf{x}_{0s})$ is a feasible solution to the primal, so we may substitute the right-hand side of the primal constraints (3-83) into the above, and similarly for the right side of the dual constraints; thus

$$\mathbf{u}_0{}^T\mathbf{b} = \mathbf{c}^T\mathbf{x}_0 \tag{3-97}$$

By Theorem 3.5 both solutions are optimal.

Although this result has no bearing on the solution of linear programming problems via the simplex method, it does form the theoretical foundation of certain other solution algorithms (including the so-called "primal-dual" [5] and "out-of-kilter" [10] algorithms). These methods, which will not be covered in our text, proceed by setting up a linear program and its dual and then operating on both of them simultaneously, perturbing first one variable, then another, and aiming for a pair of feasible solutions that satisfy complementary slackness. Theorem 3.9 guarantees that such a pair of solutions optimizes both the primal and the dual problems.

EXAMPLE 3.3. Consider the primal problem

$$\begin{aligned} \text{Max } z &= 3x_1 - 3x_2 \\ \text{subject to} \quad x_1 + 2x_2 &\leq 8 \\ 3x_1 - x_2 &\leq 3 \\ \text{and} \quad x_1, x_2 &\geq 0 \end{aligned}$$

The feasible region is shown in Figure 3-4(a), and the optimum evidently lies at $x_1 = 1$, $x_2 = 0$, which yields an objective value of $z = 3$. We now form the dual:

$$\begin{aligned} \text{Min } z' &= 8u_1 + 3u_2 \\ \text{subject to} \quad u_1 + 3u_2 &\geq 3 \\ 2u_1 - u_2 &\geq -3 \\ \text{and} \quad u_1, u_2 &\geq 0 \end{aligned}$$

It can be seen in Figure 3-4(b) that the optimal solution to the dual is $u_1 = 0$, $u_2 = 1$, and $z' = 3$; note that the two optimal objective values are equal, as

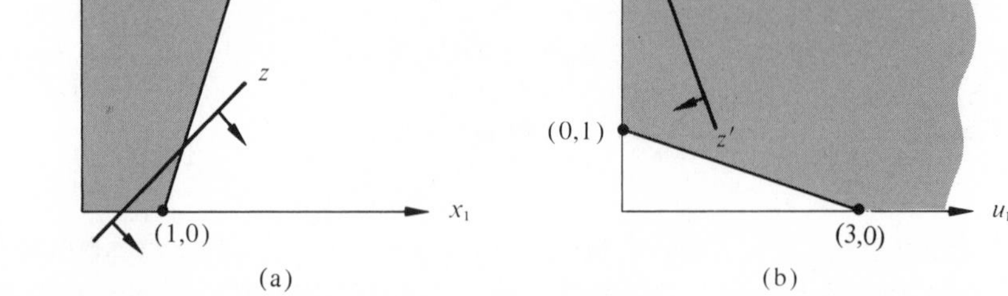

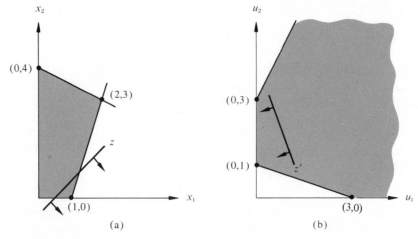

Figure 3-4

required by the duality theorem. The second primal constraint is binding, but the first is not. Complementary slackness dictates, therefore, that $u_1 = 0$ at the optimum. Similarly, because $u_2 \neq 0$, the second primal constraint must have zero slack. Complementary slackness between primal variables and dual constraints also holds in our example.

Suppose we now alter the objective function of the primal problem, making it parallel to the second constraint, as follows:

$$\begin{aligned} &\text{Max } z = 3x_1 - x_2 \\ &\text{subject to} \quad x_1 + 2x_2 \leq 8 \\ &\qquad\qquad\quad 3x_1 - x_2 \leq 3 \\ &\text{and} \qquad\qquad x_1, x_2 \geq 0 \end{aligned}$$

As can be verified from Figure 3-4(a), any point on the line segment between (and including) the extreme points (1, 0) and (2, 3) is now an optimal solution with an objective value of 3; for example, the nonextreme point ($\frac{3}{2}, \frac{3}{2}$) is optimal. The new dual problem is

$$\begin{aligned} &\text{Min } z' = 8u_1 + 3u_2 \\ &\text{subject to} \quad u_1 + 3u_2 \geq 3 \\ &\qquad\qquad\quad 2u_1 - u_2 \geq -1 \\ &\text{and} \qquad\qquad u_1, u_2 \geq 0 \end{aligned}$$

and it is easy to show graphically that the *unique* optimal solution is $u_1 = 0$, $u_2 = 1$, and $z' = 3$, with *both dual constraints binding*. Consider now the pair of optimal solutions

$$\mathbf{x} = (1,0) \qquad \text{and} \qquad \mathbf{u} = (0,1)$$

Complementary slackness prevails throughout, but notice, in particular, that both the second primal variable and the second dual surplus variable equal zero; this illustrates our earlier remark that when one of a pair of corresponding (in the complementary slackness sense) variables is zero, it does not necessarily follow that the other is positive. Similarly, if the optimal solutions are taken to be

$$\mathbf{x} = (2,3) \qquad \text{and} \qquad \mathbf{u} = (0,1)$$

both the first primal slack and the first dual variable are zero. The occurrence of this "double-zero" condition is due to the fact that the dual optimum lies at the intersection of three hyperplanes—the two dual constraints and the $\mathbf{u}_2$ axis—rather than two; thus an extra zero-valued variable is associated with the dual optimum. The extra zero can be neutralized by choosing a primal optimum that is not an extreme point:

$$\mathbf{x} = (3/2, 3/2) \qquad \text{and} \qquad \mathbf{u} = (0, 1)$$

At this primal optimum neither variable is zero and only one constraint is binding, and in checking the complementary slackness conditions, we find no cases of double zero occurring.

EXERCISES

Section 3.4

3-1. Transform each of the following problems into canonical form:

(a) Min $z = 2x_1 - x_2 + x_3$

$$\begin{aligned} \text{subject to} \quad 2x_1 + x_2 - x_3 &\leq 8 \\ -x_1 \quad\quad + x_3 &\geq 1 \\ x_1 + 2x_2 + 3x_3 &= 9 \\ \text{and} \quad x_1, x_2 &\geq 0 \end{aligned}$$

(b) Min $z = x_1$

$$\begin{aligned} \text{subject to} \quad x_1 + x_2 + x_3 &= 13 \\ x_2 &\geq x_3 \\ 5 - x_3 &\leq x_1 \\ x_1 &\leq 0 \\ \text{and} \quad x_3 &\leq 10 \end{aligned}$$

3-2. Given the linear program

$$\begin{aligned} \text{Max } z &= \mathbf{c}^T\mathbf{x} \\ \text{subject to} \quad \mathbf{A}\mathbf{x} &\leq \mathbf{b} \\ \text{and} \quad \mathbf{x} &\geq \mathbf{0} \end{aligned}$$

state simple conditions involving $\mathbf{b}$ and/or $\mathbf{c}$ that will make each of the following statements true.

(a) $\mathbf{x} = \mathbf{0}$ is a feasible solution.

(b) $\mathbf{x} = \mathbf{0}$ is an optimal solution.

(c) $\mathbf{x} = \mathbf{0}$ is feasible, but not optimal.

Section 3.6

3-3. Using the definition of the dual problem, derive the duals of the following linear programs. Simplify the dual constraints and variables where appropriate and then check your answers with the mnemonic table at the end of Section 3.6.

(a) Max $z = x_1$

$$\begin{aligned} \text{subject to} \quad & x_1 - x_2 - x_3 \leq 10 \\ & x_2 \leq 3 \\ & x_3 \leq 4 \\ \text{and} \quad & x_1, x_2, x_3 \geq 0 \end{aligned}$$

(b) Min $z = 2x_1 + x_2 - x_3$

$$\begin{aligned} \text{subject to} \quad & x_1 + x_2 + x_3 \leq 12 \\ & 3x_1 - 2x_3 \leq 4 \\ & -x_1 + 4x_2 + x_3 \leq 10 \\ \text{and} \quad & x_1, x_2, x_3 \geq 0 \end{aligned}$$

(c) Min $z = 6x_1 - 3x_2$

$$\begin{aligned} \text{subject to} \quad & -x_1 + 4x_2 \geq 12 \\ & 2x_1 + 6x_2 = 20 \\ \text{and} \quad & x_1, x_2 \geq 0 \end{aligned}$$

(d) Max $z = -2x_1 - x_2$

$$\begin{aligned} \text{subject to} \quad & x_1 + x_2 \leq 10 \\ & x_1 - 2x_2 = -8 \\ & x_1 + 3x_2 \geq 9 \\ \text{and} \quad & x_1 \geq 0 \end{aligned}$$

3-4. Write the duals of the following problems:

(a) Max $z = x_1$

$$\begin{aligned} \text{subject to} \quad & x_1 \leq 9 \\ \text{and} \quad & x_1 \geq 4 \end{aligned}$$

(b) Min $z = 3x_1 - 2x_2$

$$\begin{aligned} \text{subject to} \quad & -4x_1 + 3x_2 \leq 10 \\ \text{and} \quad & x_1 \geq 0 \end{aligned}$$

(c) Exercise 3-1(a)

(d) Exercise 3-1(b)

3-5. Using the primal problem

$$\begin{aligned} \text{Max } z = {} & x_1 + 2x_2 + 3x_3 \\ \text{subject to} \quad & 3x_1 - x_2 + 2x_3 = 12 \\ & x_1 + 3x_2 + x_3 \leq 15 \\ \text{and} \quad & x_1, x_2 \geq 0 \end{aligned}$$

show that the dual of the dual is the primal.

3-6. Show that the linear program

$$\begin{aligned} \text{Max } z = {} & \mathbf{c}^T\mathbf{x} \\ \text{subject to} \quad & \mathbf{A}\mathbf{x} \leq \mathbf{b} \\ \text{and} \quad & \mathbf{x} \geq \mathbf{0} \end{aligned}$$

has the same dual problem regardless of whether $\mathbf{x} \geq \mathbf{0}$ is treated as a nonnegativity restriction or as an additional group of constraints.

3-7. Use our definition rigorously to show how the dual of the linear program

$$\begin{aligned} &\text{Max } z = \mathbf{c}^T\mathbf{x} \\ &\text{subject to} \quad \mathbf{Ax} \geq \mathbf{b} \\ &\text{and} \qquad\qquad \mathbf{x} \geq \mathbf{0} \end{aligned}$$

can be written in terms of nonpositive variables.

3-8. Show that if the same constraint is inadvertently included *twice* in the set of primal constraints, the optimal solution to the dual problem is not materially affected.

3-9. What is the effect on the optimal solution $\mathbf{u}^*$ of the dual problem when one of the primal constraints is multiplied through by a constant k? Assume the primal is in standard form.

3-10. Find graphically the optimal value of z in the following problem:

$$\begin{aligned} &\text{Min } z = 2x_1 + 9x_2 + 6x_3 + 8x_4 \\ &\text{subject to} \quad x_1 + x_2 + x_3 - x_4 \geq 1 \\ &\qquad\qquad\quad x_1 + 2x_2 - x_3 + 2x_4 \geq 0 \\ &\text{and} \qquad\qquad x_1, x_2, x_3, x_4 \geq 0 \end{aligned}$$

3-11. Graph the following primal linear program and locate its optimal solution:

$$\begin{aligned} &\text{Max } z = x_1 + x_2 \\ &\text{subject to} \quad 2x_1 + x_2 \leq 12 \\ &\qquad\qquad\quad -x_1 + 2x_2 \leq 4 \\ &\text{and} \qquad\qquad x_1, x_2 \geq 0 \end{aligned}$$

Deduce the optimal solution to the dual problem by evaluating the functions $\delta f/\delta b_i$ in the vicinity of the primal optimum. Check by solving the dual problem graphically.

Section 3.9

3-12. Diagram and solve graphically the primal linear program

$$\begin{aligned} &\text{Max } z = x_2 \\ &\text{subject to} \quad x_1 + x_2 \leq 3 \\ &\qquad\qquad\qquad x_2 \leq 2 \\ &\qquad\qquad\quad -x_1 + x_2 \leq 1 \\ &\text{and} \qquad\qquad x_1, x_2 \geq 0 \end{aligned}$$

Calculate the rate of change of the optimal value of z with respect to (a) increases and (b) decreases in the right-hand side of the first constraint ($b_1 = 3$). Now write the dual problem and obtain *all* optimal solutions. Discuss the relationship between dz_{opt}/db_1 and the optimal value(s) of the first dual variable.

Section 3.10

3-13. Use the duality theorem to prove Farkas' lemma, as stated in Section 3.8.

3-14. For any $m \times n$ matrix $\mathbf{A}$ and any n-component vector $\mathbf{c}$, prove that there exists some $\mathbf{x}$ satisfying $\{\mathbf{Ax} \leq \mathbf{0}, \mathbf{c}^T\mathbf{x} = 1\}$ if and only if there exists *no* $\mathbf{u}$ satisfying $\{\mathbf{A}^T\mathbf{u} = \mathbf{c}, \mathbf{u} \geq \mathbf{0}\}$.

Section 3.11

3-15. Prove that $x_1 = 8/5$, $x_2 = 12/5$, $x_3 = 43/5$ is an optimal solution to the linear program

$$\begin{aligned}
&\text{Min } z = 10x_1 + 26x_2 + x_3\\
&\text{subject to} \quad x_1 + x_2 \geq 4\\
&\qquad\qquad\quad x_1 + 6x_2 \geq 16\\
&\text{and} \qquad\quad x_1 + 2x_2 + x_3 \geq 15\\
&\text{with} \quad x_1, x_2, x_3 \text{ unrestricted}
\end{aligned}$$

3-16. Suppose that $x_1 = x_3 = 0$, $x_2 = 10.4$, $x_4 = 0.4$ is given as the optimal solution to the linear program

$$\begin{aligned}
&\text{Max } z = 2x_1 + 4x_2 + 3x_3 + x_4\\
&\text{subject to} \quad 3x_1 + x_2 + x_3 + 4x_4 \leq 12\\
&\qquad\qquad\quad x_1 - 3x_2 + 2x_3 + 3x_4 \leq 7\\
&\qquad\qquad\quad 2x_1 + x_2 + 3x_3 - x_4 \leq 10\\
&\text{and} \qquad\qquad x_1, x_2, x_3, x_4 \geq 0
\end{aligned}$$

Use this information to find the optimal solution to the dual problem.

3-17. Consider the linear program

$$\text{Max } x = \mathbf{c}^T\mathbf{x} \qquad \text{subject to } \mathbf{Ax} \leq \mathbf{b} \text{ and } \mathbf{x} \geq \mathbf{0}$$

where $\mathbf{A}$ is an $m \times n$ matrix. Representing the slack variables by the vector $\mathbf{x}_s$, suppose the optimal solution $(\mathbf{x}^*,\mathbf{x}_s^*)$ is a *nondegenerate* basic solution—that is, exactly m of the variables and slack variables are positive (with the rest being zero). Show that if $(\mathbf{x}^*,\mathbf{x}_s^*)$ is known, the optimal solution to the dual problem can always be obtained by solving a square set of simultaneous linear equations. Now suppose $(\mathbf{x}^*,\mathbf{x}_s^*)$ is *degenerate*; how does this affect the above procedure for solving the dual problem? What can be said about the optimal solution to the dual?

3-18. Sketch out an argument within the familiar context of allocating resources for manufacturing to justify Theorem 3.8(b).

General

3-19. Consider a centrally planned national economy satisfying the Leontief model. The economy is to produce an unknown amount x_i of each of n products in the upcoming year. Production of a unit of the jth good, or output, requires inputs of a_{ij} units of each ith good, $i = 1, \ldots, n$. In addition, there must be enough of each ith good left over to satisfy an exogenous demand b_i. If it costs c_i dollars to produce one unit of the ith output from its various inputs, show that the central planners can find the cheapest production scheme by solving a linear program.

Now form and discuss the dual problem, interpreting the dual variables as *prices* to be charged for each of the goods. In particular, verify that the dual constraints dictate that prices must be set so that no profits may be made in the production of any good.

3-20. Suppose that k units of some commodity, where $k > 0$, are stored at *node* A (which may be a warehouse, a town, etc.) in the following *network* diagram:

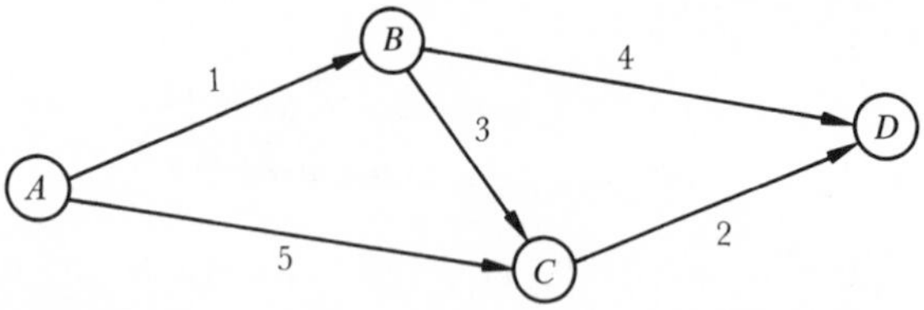

They all must be shipped to node D (a retail store, perhaps) along one-way roads or paths represented by the *directed arcs* in the network. No charge is assessed for shipping through the intermediate nodes B and C, which merely serve to define points where arcs branch apart or come together. Associated with each arc in the network is a number giving the per-unit cost of shipping commodity along the arc. If the number of units to be shipped from A to B is represented by x_{AB}, and so on, verify that the shipping plan that gets the k units to D at minimum cost is obtained by solving the following linear program:

$$\begin{aligned}
\text{Min } z = x_{AB} + 5x_{AC} + 3x_{BC} + 4x_{BD} + 2x_{CD} \\
\text{subject to} \quad -x_{AB} - x_{AC} \qquad\qquad\qquad\qquad &= -k \\
x_{AB} \qquad - x_{BC} - x_{BD} \qquad\quad &= 0 \\
x_{AC} + x_{BC} \qquad\quad - x_{CD} &= 0 \\
x_{BD} + x_{CD} &= k \\
\text{and all} \qquad\qquad\qquad\qquad x_{IJ} &\geq 0
\end{aligned}$$

Will all k units necessarily be shipped over the same route from A to D?

Now write the dual of this LPP and suggest an interpretation of the dual variables, which are called *node numbers*. What are the implications of the dual constraints and objective function and of complementary slackness? How does your interpretation account for the equality of the optimal values of the primal and dual objectives? Show that the primal constraints stated above are redundant, and discuss the impact of this fact upon the dual problem and its optimal solution.

(*Warning:* The following problem is difficult and should not be attempted until after the student has pondered Exercise 3-20.)

3-21. Suppose the network of Exercise 3-20 is now *capacitated*; that is, the number of units shipped over any arc (I to J) is required to be no greater than some upper bound c_{IJ}, which is called the *capacity*. The constraints

$$x_{IJ} \leq c_{IJ} \qquad \text{for all arcs } (I \text{ to } J)$$

now must be added to the linear program. Write the dual of the new LPP and discuss the dual variables, complementary slackness, etc.

3-22. Consider a simple zero-sum, two-person game in which each player has two possible strategies or moves. A single play of the game consists of each player choosing simultaneously one of his two strategies. If Row chooses his ith strategy and Column chooses his jth, Row collects from Column a payment a_{ij} as given in the following payoff matrix:

		Column's strategies 1	Column's strategies 2
Row's strategies	1	$a_{11} = 1$	$a_{12} = 4$
	2	$a_{21} = 3$	$a_{22} = 2$

The game is repeated many times. If Row always chooses 2, Column will soon deduce this and choose 2, holding Row to winnings of 2 units per play. If Row always chooses 1 he will do even worse. John von Neumann showed that it is optimal for Row to play a mixed strategy: At every play he uses a random device that chooses strategy 1 or 2 with probabilities x_1 and x_2, respectively. Define V, the value of the game, to be the maximum expected winnings per play that Row can achieve, given that Column is playing his own best strategy also. Show that Row can obtain his optimal probabilities, or optimal mixed strategy, by solving the linear program

$$\text{Max } z = V$$

$$\begin{aligned} \text{subject to} \quad V - x_1 - 3x_2 &\leq 0 \\ V - 4x_1 - 2x_2 &\leq 0 \\ x_1 + x_2 &= 1 \\ \text{and} \quad x_1, x_2 &\geq 0 \end{aligned}$$

Now derive and interpret the dual problem.

3-23. Repeat the preceding problem when the payoff matrix is

	1	2
1	1	2
2	4	3

Discuss as completely as possible. In this example Row's strategy 2 is called a *dominant strategy*. How would the game be played if both players had dominant strategies?

4

THE SIMPLEX METHOD

4.1 INTRODUCTION TO THE METHOD

In Chapter 3 we saw that any given linear programming problem, after suitable algebraic manipulation, can be represented in *standard form*:

$$\begin{aligned}
&\text{Max } z = \mathbf{c}^T\mathbf{x} \\
&\text{subject to} \quad \mathbf{A}\mathbf{x} = \mathbf{b} \\
&\text{and} \qquad\qquad \mathbf{x} \geq \mathbf{0}
\end{aligned}$$

To attain this formulation, all inequality constraints appearing in the original problem are converted to equations through the addition of nonnegative slack or surplus variables, while simple transformations are used to substitute nonnegative variables for unrestricted ones. These modifications do not materially alter the problem because they change neither the set of feasible solutions nor the optimum.

We are now ready to introduce the *simplex method*, which, in conjunction with certain auxiliary procedures, is capable of solving any linear program in standard form, and therefore, ultimately, any linear program. Leaving aside the auxiliary procedures, however, as we shall for the duration of this chapter, the term "simplex method" refers specifically to an iterative computational algorithm that can obtain the optimal solution of any standard-form problem for which the following two conditions hold:

(1) None of the constraints are redundant.
(2) An initial *basic feasible solution* (BFS) has been found.

The first of these is necessary because the simplex method, as we shall see, generates a sequence of invertible submatrices using the columns of $\mathbf{A}$; it could not do so if one of the rows was a linear combination of the others. The second condition introduces the notion of a basic feasible solution: This is a basic solution (as defined in Section 2.3) to the constraint set $\mathbf{Ax} = \mathbf{b}$ that also satisfies $\mathbf{x} \geq \mathbf{0}$. Inasmuch as the simplex method can begin to operate only upon an initial basic feasible solution to the standard-form LPP, we shall prove in Section 4.4 that any linear program with a feasible solution actually *has* a BFS. This will allow us to conclude that any feasible LPP can potentially be solved via the simplex method (although it will not tell us how to find the initial BFS).

A third condition might be added to the two listed above, namely, that the number of variables in the standard-form LPP must exceed the number of constraints in order to make the problem "interesting." If the number of variables is less than the number of constraints, a basic feasible solution—as required by condition 2—cannot be found. And if the number of variables exactly equals the number of constraints—none of which, according to condition 1, may be redundant—then a *unique* feasible solution must exist: $\mathbf{x} = \mathbf{A}^{-1}\mathbf{b}$. In such a case there is no need to choose an optimum, hence no need to apply the simplex method.

At this point, it is not at all obvious how to deal with a linear programming problem in standard form when conditions 1 and 2 do *not* hold. As the reader might suspect, both the elimination of redundant constraints and the determination of an initial BFS involve significant amounts of computational labor. It turns out that a rather clever trick is available whereby the simplex method can itself be used to perform these two tasks—and then used a second time to solve the original problem! These matters will be discussed in detail in Chapter 5. What we shall show in this chapter is how the simplex method, starting with a basic feasible solution to any linear program in standard form, proceeds to the optimum.

4.2 EXTREME POINTS AND BASIC FEASIBLE SOLUTIONS

The geometry of linear programming was introduced in Chapters 1 and 2 to provide the student with a mental picture that would illuminate and clarify the algebraic development. In this section, however, we shall make direct use of the geometric concept of an extreme point in order to obtain an important theoretical result, namely, that *some basic feasible solution to a bounded LPP is optimal.* Proving this proposition with linear algebra alone would be rather difficult and not particularly edifying; instead, we can state it as a corollary to the two theorems of this section.

Recall from Section 2.4 that an *extreme point* $\mathbf{x}^*$ of a convex set X is a point in X that *cannot* be expressed as

$$\mathbf{x}^* = \lambda \mathbf{x}_1 + (1 - \lambda)\mathbf{x}_2 \tag{4-1}$$

where $\mathbf{x}_1$ and $\mathbf{x}_2$ are two different points in X and $0 < \lambda < 1$. An extreme point on the feasible region of a linear program corresponds to our ordinary idea of a corner.

THEOREM 4.1. At least one optimal solution to any feasible and bounded linear programming problem occurs at an extreme point of the convex set of feasible solutions.

Assume that the maximum attainable value (proof for a minimization LPP is analogous) of the objective function $z = \mathbf{c}^T\mathbf{x}$ is z_0 and that one of the points $\mathbf{x}_0$ on the optimal objective hyperplane $z_0 = \mathbf{c}^T\mathbf{x}$ is an interior point of the feasible set. There must then be some positive ϵ such that any point within a distance ϵ of $\mathbf{x}_0$ also is in the feasible region. We now move away from $\mathbf{x}_0$ a distance ϵ in the direction of $\mathbf{c}$, bringing us to the point

$$\mathbf{x}^* = \mathbf{x}_0 + \frac{\epsilon}{|\mathbf{c}|}\mathbf{c}$$

where $|\mathbf{c}|$ is the length of the cost vector. Because $\mathbf{x}^*$ is only ϵ away from $\mathbf{x}_0$, it is feasible, and

$$\mathbf{c}^T\mathbf{x}^* = \mathbf{c}^T\mathbf{x}_0 + \frac{\epsilon}{|\mathbf{c}|}\mathbf{c}^T\mathbf{c}$$

Because $\mathbf{c}^T\mathbf{c}$ is the sum of the squares of the components of $\mathbf{c}$ and because ϵ and $|\mathbf{c}|$ are positive

$$\mathbf{c}^T\mathbf{x}^* > \mathbf{c}^T\mathbf{x}_0$$

We have arrived at a contradiction: The interior point $\mathbf{x}_0$ cannot be optimal, and it follows that every optimal point must lie on the boundary of the feasible region.

Let $\hat{\mathbf{x}}$ be one such optimal point. If it is an extreme point, the proof is complete. If not, $\hat{\mathbf{x}}$ may lie on some edge E, in which case the optimal hyperplane containing $\hat{\mathbf{x}}$ must be coincident with E in order to avoid entering the interior. Thus every point on E is optimal, including its two end points, which are extreme points. Finally, if $\hat{\mathbf{x}}$ does not lie on an edge, then it necessarily lies in some hyperplane H that forms part of the boundary of the feasible region. In that case, the optimal hyperplane must be coplanar with H, again to avoid entering the interior, and, therefore, must pass through the end points of whatever edges lie in H. This completes the proof.

Theorem 4.1 holds for bounded problems, which include LPPs with unbounded feasible regions whose optima are nevertheless finite. In an unbounded problem, however, no specifiable extreme point can possibly be optimal, because such a point would have a finite objective value.

THEOREM 4.2. The geometric coordinates of any extreme point of the feasible set of solutions to an LPP form a basic feasible solution.

For this proof we assume that the linear program is in standard form, as required by the simplex method:

$$\text{Max } z = \mathbf{c}^T\mathbf{x} \tag{4-2}$$

$$\text{subject to} \quad \mathbf{A}\mathbf{x} = \mathbf{b} \tag{4-3}$$

$$\text{and} \qquad \mathbf{x} \geq \mathbf{0} \tag{4-4}$$

where $\mathbf{A}$ is an $m \times n$ matrix. Our proof follows Hadley [15]. Given some extreme point $\mathbf{x}^* = (x_1^*, \ldots, x_n^*)$ of the feasible region, let the variables and their associated columns of the $\mathbf{A}$ matrix be reordered so that $x_1^*, x_2^*, \ldots,$ and x_k^* are positive, but $x_{k+1}^*, \ldots,$ and x_n^* are zero. Assume now that the first k columns of $\mathbf{A}$ are linearly *dependent*; thus there exist scalar numbers y_i not all zero, satisfying

$$\sum_{i=1}^{k} y_i \mathbf{a}_i = \mathbf{0} \tag{4-5}$$

where $\mathbf{a}_i$ is the ith column of $\mathbf{A}$.

Define a positive number

$$\alpha = \min_i \frac{x_i^*}{|y_i|} \qquad \text{where } 1 \leq i \leq k \textit{ and } y_i \neq 0 \tag{4-6}$$

Then we may choose a positive ϵ smaller than α, resulting in

$$x_i^* + \epsilon y_i > 0 \tag{4-7}$$

and

$$x_i^* - \epsilon y_i > 0 \tag{4-8}$$

for $i = 1, 2, \ldots, k$. Define an n-component column vector $\mathbf{y} \neq \mathbf{0}$, which has the y_i in the first k positions and zero in the rest. Let

$$\mathbf{x}_1 = \mathbf{x}^* + \epsilon\mathbf{y} \tag{4-9}$$

and

$$\mathbf{x}_2 = \mathbf{x}^* - \epsilon\mathbf{y} \tag{4-10}$$

Because of (4-7) and (4-8), $\mathbf{x}_1$ and $\mathbf{x}_2$ must be nonnegative vectors. Furthermore, from our linear dependence assumption (4-5) and the fact that $y_i = 0$, $i = k + 1, \ldots, n$, we can write

$$\mathbf{A}\mathbf{y} \equiv \sum_{i=1}^{n} y_i\mathbf{a}_i = \mathbf{0}$$

This implies

$$\mathbf{A}\mathbf{x}_1 = \mathbf{A}\mathbf{x}^* = \mathbf{b} \tag{4-11}$$

and

$$\mathbf{A}\mathbf{x}_2 = \mathbf{A}\mathbf{x}^* = \mathbf{b} \tag{4-12}$$

because $\mathbf{x}^*$ is a feasible point.

Therefore $\mathbf{x}_1$ and $\mathbf{x}_2$ are feasible solutions different from each other and from $\mathbf{x}^*$; moreover,

$$\mathbf{x}^* = \tfrac{1}{2}\mathbf{x}_1 + \tfrac{1}{2}\mathbf{x}_2 \tag{4-13}$$

as can be checked from (4-9) and (4-10). This contradicts the fact that $\mathbf{x}^*$ is an extreme point, and it follows that the columns of $\mathbf{A}$ associated with the nonzero components of any extreme point must be linearly independent. There can be, at most, m linearly independent columns and nonzero variables, so the components of an extreme point constitute a basic solution by definition. Because the extreme point belongs to the feasible set, it is a BFS, as was to be shown.

Theorems 4.1 and 4.2 together imply that *an exhaustive search over all basic feasible solutions must yield an optimal solution to any bounded linear programming problem!* This is precisely how the simplex method operates. Beginning at some extreme point, it identifies another extreme point adjacent to the first, but having a better objective value—here "adjacent" is used in its technical sense,

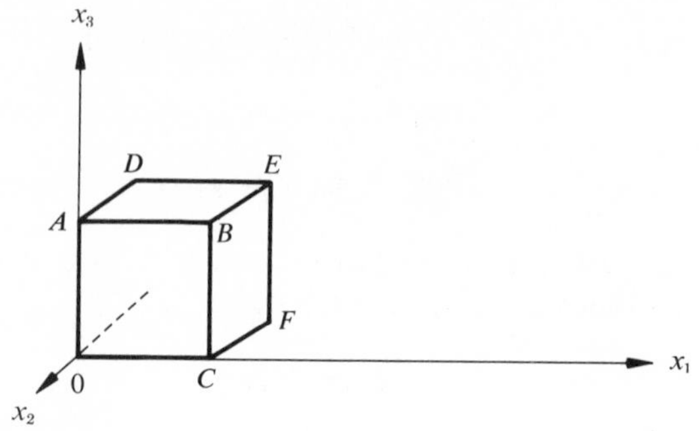

Figure 4-1

as defined in Section 2.5. Thus if the feasible region is the unit cube in E^3, shown in Figure 4-1, the simplex method might step from the initial BFS at the origin 0 to C, then to F, and finally to the optimum at E. These steps to adjacent extreme points are called *pivots.*

The simplex method is unable to pivot directly from 0 to E or from 0 to B to E. In a large problem it might seem to be faster and more efficient to leap ahead to nonadjacent extreme points, thereby reaching the optimum in fewer steps. Modifications of the simplex method, in fact, have been devised to permit these longer jumps, but they have proved to save little or no time, the reason being that the simple adjacent pivots made by the simplex method require far less computation.

4.3 DESIRABLE PROPERTIES OF A SOLUTION METHOD

In algebraic terms, the simplex method pivots from one basic feasible solution (corresponding to an extreme point) to another until it reaches the optimum, which is also a BFS (from Theorems 4.1 and 4.2). Therefore, we would like to have a computationally efficient procedure for performing these pivots. Time would be wasted in pivoting to a less favorable extreme point, so the algorithm should include some test or rule for choosing favorable pivots and proving that they, in fact, do improve the objective function. Finally, some criterion of optimality must be stated so that the algorithm will "know" when to stop, either because the optimum has been reached or because it has been discovered that no optimum exists, that is, that the problem is unbounded. As we shall see, the simplex method satisfies each of these requirements.

The next section establishes the existence of a basic feasible solution whenever a feasible solution exists, provided that the set of constraints is not redundant. This guarantees that any feasible LPP can be solved, in theory, by the simplex method, although it does not specify how to remove redundant constraints or how to find the necessary starting BFS. And what about solving (or, rather, identifying) an *infeasible* LPP? It turns out that the same auxiliary procedures (to be discussed in Chapter 5) that are used to identify redundant constraints and obtain the initial basic feasible solution also can determine when a given linear program is infeasible, namely, by *failing* to find an initial BFS for it. The simplex method, with all these desirable properties and capabilities, was developed principally by George Dantzig in the two or three years immediately following World War II; its conception was a monumental event in the history of operations research.

4.4 EXISTENCE OF A BASIC FEASIBLE SOLUTION

The following result is known traditionally as the "fundamental theorem" of linear programming, although that phrase is really more descriptive of the duality theorem.

THEOREM 4.3. If a linear programming problem with no redundant constraints has a feasible solution, then it has a basic feasible solution.

Notice that this result is *more* than a trivial corollary of Theorems 4.1 and 4.2, which together imply that at least one optimal (therefore, feasible) solution to any feasible *and bounded* LPP occurs at a BFS. Theorem 4.3 establishes the existence of a basic feasible solution for any unbounded linear program as well, thereby guaranteeing that it is possible to get the simplex method started on *any* feasible LPP. Moreover, in the course of proving this result we shall also develop a computational method for actually obtaining a BFS from any given feasible solution. The reader is urged, therefore, to study the following proof carefully and to take heart from the fact that it will be the only one of any great length in this chapter.

We begin by assuming that the problem has been converted into standard form. The constraints are

$$\mathbf{Ax} = \mathbf{b} \tag{4-14}$$

and

$$\mathbf{x} \geq \mathbf{0}$$

where $\mathbf{A}$ is $m \times n$. By hypothesis the constraints are neither inconsistent nor redundant, so the rows of $\mathbf{A}$ must be linearly independent with $m \leq n$. Let there be a feasible solution with exactly p positive variables, renumbering if necessary so that the first p variables are the ones that are positive. Denote the values of these variables by x_1 through x_p. The constraints then reduce to

$$\sum_{j=1}^{p} x_j \mathbf{a}_j = \mathbf{b} \tag{4-15}$$

where the $\mathbf{a}_j$ in the sum are the columns of $\mathbf{A}$ associated with the positive variables. This collection of columns will be called *set K*.

Suppose first that these p column vectors are linearly independent, in which case $p \leq m$. Choose a basis of m linearly independent columns of $\mathbf{A}$ (Theorem 2.2 guarantees that it is possible to do so) and let L be the *union* of this basis and the set K—that is, if column $\mathbf{a}_j$ is in the basis or in K, then it is in L. Thus L spans E^m and contains anywhere from m to $m + p$ vectors, m of which must be linearly independent. Now, if L has exactly m members, they immediately form a basis that includes all the columns of K. In view of (4-15), the values of the associated basic variables must be $x_1, x_2, \ldots, x_p, 0, \ldots, 0$. These constitute a basic feasible solution, as was to be obtained (note that the BFS is degenerate unless $p = m$).

In general, however, L will contain more than m column vectors, in which case they will be linearly dependent (recall property 5 of Section 2.2); there must then exist λ_i not all zero such that

$$\sum_{i \in L} \lambda_i \mathbf{a}_i = \mathbf{0} \tag{4-16}$$

We now claim that some λ_j in (4-16) corresponding to a column $\mathbf{a}_j$ *not in the original set K* must be nonzero. If all these λ_i were 0, then (4-16) would reduce to

$$\sum_{i \in K} \lambda_i \mathbf{a}_i = \mathbf{0}$$

contradicting the linear independence of the members of K. Therefore, we can divide through (4-16) by some λ_j corresponding to a column not in K, yielding

$$\mathbf{a}_j = -\sum_{\substack{i \in L \\ i \neq j}} \frac{\lambda_i}{\lambda_j} \mathbf{a}_i$$

Because $\mathbf{a}_j$ is a linear combination of vectors in L, we can remove $\mathbf{a}_j$ from L and it will still span E^m. Though L now contains one less vector, it still contains all the vectors of K. If it now has m members, we have again found a basis whose variables have the values $x_1, x_2, \ldots, x_p, 0, \ldots, 0$; as before, these constitute a basic feasible solution. If L still has more than m members, we return to equation (4-16) and eliminate another vector from it, continuing in this way until we reduce its membership to m. We have shown that if the columns associated with the p positive variables in our feasible solution are linearly independent, then they participate in a *basic* feasible solution, which may be degenerate.

Suppose instead that the p columns $\mathbf{a}_j$ *in (4-15) are linearly dependent*. We then seek another feasible solution to (4-14) that will have fewer positive variables. By the linear dependence of the $\mathbf{a}_j$, there exist y_j not all zero such that

$$\sum_{j=1}^{p} y_j \mathbf{a}_j = \mathbf{0} \tag{4-17}$$

Multiplying by -1 if necessary, we can assume that at least one of the coefficients in (4-17) is positive. Let some $y_r > 0$ be chosen. We then may express $\mathbf{a}_r$ as a linear combination of the remaining $p - 1$ columns,

$$\mathbf{a}_r = -\sum_{\substack{j=1 \\ j \neq r}}^{p} \frac{y_j}{y_r} \mathbf{a}_j \tag{4-18}$$

and substitution of (4-18) into (4-15) yields

$$\sum_{\substack{j=1 \\ j \neq r}}^{p} \left(x_j - x_r \frac{y_j}{y_r}\right) \mathbf{a}_j = \mathbf{b} \tag{4-19}$$

Equation (4-19) represents a new solution to $\mathbf{Ax} = \mathbf{b}$ with no more than $p - 1$ nonzero variables. Had we selected $\mathbf{a}_r$ at random from the group of columns having positive y_j in (4-17), several of the new variables

$$\hat{x}_j = x_j - x_r \frac{y_j}{y_r} \tag{4-20}$$

might have turned out to be negative. Suppose, however, we specifically chose $\mathbf{a}_r$ such that[1]

$$\frac{x_r}{y_r} = \min_j \left(\frac{x_j}{y_j}, \quad y_j > 0\right) \tag{4-21}$$

It is evident from (4-20) that where $y_j = 0$ the new variable $\hat{x}_j$ equals the old x_j, which was positive by assumption. Further, where $y_j < 0$ the new variable is greater than the old, because x_r and y_r are positive; therefore, again $\hat{x}_j$ is positive. Finally, if y_j is positive, the criterion for choosing $\mathbf{a}_r$ guarantees that the new variable $\hat{x}_j$ is nonnegative, because

$$\frac{x_r}{y_r} \leq \frac{x_j}{y_j} \qquad \text{whenever } y_j > 0 \tag{4-22}$$

This follows from (4-21).

Thus, equation (4-19) demonstrates a *feasible* solution to (4-14) with no more than $p - 1$ nonzero variables. By repeating this procedure, we can continue to produce feasible solutions with fewer and fewer positive variables, until all of the associated columns $\mathbf{a}_j$ are linearly independent (at worst, the number of positive variables will have to be reduced to 1, since any one column $\mathbf{a}_j \neq \mathbf{0}$

[1] Equation (4-21) is read, "Choose from among all columns j having a positive y_j that one for which the quotient x_j/y_j is smallest, calling it column r." Those columns for which $y_j \leq 0$ are not considered "candidates" for this choice.

is linearly independent by definition). At that point, we can return to the first step of the proof and, ultimately, arrive at a basic feasible solution, Q.E.D.

The implications of Theorem 4.3 should be underscored. Because any linear program can be converted to standard form, a basic feasible solution must exist for any LPP whose set of feasible solutions is not empty. Because the simplex method operates upon BFSs, it is capable of solving any linear programming problem having a feasible solution. As we remarked earlier, the actual finding of the initial BFS will be discussed in Chapter 5; the present chapter is concerned with how to proceed from it to the optimum.

The following example illustrates the proof of Theorem 4.3. The student is advised to work it through, as it reviews several concepts with which he should become familiar.

EXAMPLE 4.1. Consider the equations

$$2x_1 + 2x_2 + x_3 = 13 \tag{4-23}$$

and

$$4x_1 + x_2 - 4x_3 = 5$$

In vector notation we may write

$$x_1\begin{bmatrix}2\\4\end{bmatrix} + x_2\begin{bmatrix}2\\1\end{bmatrix} + x_3\begin{bmatrix}1\\-4\end{bmatrix} = x_1\mathbf{a}_1 + x_2\mathbf{a}_2 + x_3\mathbf{a}_3 = \mathbf{b} = \begin{bmatrix}13\\5\end{bmatrix}$$

One possible solution to these equations is $x_1 = 4$, $x_2 = 1$, $x_3 = 3$. Because the **a** vectors have only two components, any three of them are linearly dependent. In fact,

$$3\mathbf{a}_1 - 4\mathbf{a}_2 + 2\mathbf{a}_3 = \mathbf{0}$$

That is, $y_1 = 3$, $y_2 = -4$, $y_3 = 2$. We now choose $\mathbf{a}_r$, the column whose associated variable x_r is to be driven to zero, according to the criterion (4-21):

$$\min_j \left(\frac{x_j}{y_j}, y_j > 0\right) = \min\left(\frac{x_1}{y_1}, \frac{x_3}{y_3}\right) = \min\left(\frac{4}{3}, \frac{3}{2}\right) = \frac{4}{3}$$

Thus, x_1 will be driven to zero; from (4-20), using $r = 1$, the new values of the variables become

$$\hat{x}_1 = 4 - 4(3/3) = 0$$

$$\hat{x}_2 = 1 - 4(-4/3) = 19/3$$

and

$$\hat{x}_3 = 3 - 4(2/3) = 1/3$$

Thus, an alternate solution to the given equations is $x_1 = 0$, $x_2 = 19/3$, $x_3 = 1/3$. There are only two nonzero variables, both positive. Their associated columns $\mathbf{a}_2$ and $\mathbf{a}_3$ are linearly independent, so we have found a nondegenerate basic feasible solution.

4.5 DEFINITIONS

For the rest of this chapter we shall be concerned with the problem P:

$$\text{Max } z = \mathbf{c}^T\mathbf{x} \tag{4-24}$$

$$\text{subject to} \quad \mathbf{Ax} = \mathbf{b} \tag{4-25}$$

$$\text{and} \quad \mathbf{x} \geq \mathbf{0} \tag{4-26}$$

where $\mathbf{A}$ is $m \times n$. The label P stands for "primal." Although it is problem P that is to be solved directly via the simplex method, several references will be made also to its dual, which, as a "side effect," will be solved indirectly. We shall assume throughout that the conditions of Section 4.1 hold: In particular, $\mathbf{Ax} = \mathbf{b}$ *contains no redundant constraints* and $m < n$.

Any basic feasible solution (BFS) to problem P includes m basic variables, which can be collected into an $m \times 1$ vector $\mathbf{x_B}$. The remaining $(n - m)$ nonbasic variables, whose value is required to be zero (by definition of a basis), are contained in the vector $\mathbf{x_R}$. The ordering of the variables in $\mathbf{x_B}$ is not necessarily related to their original order in $\mathbf{x}$; thus, for $m = 3$ the vector $\mathbf{x_B}$ might be (x_7, x_2, x_4). It should be emphasized that $\mathbf{x_B}$ is a vector of variables and does *not* represent the values assumed by those variables. Moreover, $\mathbf{x_B}$ does not always include the same set of variables; in the example just cited the third basic variable $x_{\mathbf{B}3}$ is x_4, but in some other basis (perhaps arrived at after a sequence of pivot steps) $x_{\mathbf{B}3}$ might be x_5 or x_1.

The columns of $\mathbf{A}$ associated with the variables of a basic feasible solution $\mathbf{x_B}$ are assembled into the $m \times m$ *basis matrix* $\mathbf{B}$. The columns, like their associated variables, are said then to be *in the basis*. These *basic columns* must be arranged in the same order, from left to right, as are their associated variables in the vector $\mathbf{x_B}$. Recall that $\mathbf{Ax} = \mathbf{b}$ also can be written as

$$x_1\mathbf{a}_1 + x_2\mathbf{a}_2 + \cdots + x_n\mathbf{a}_n = \mathbf{b} \tag{4-27}$$

where the $\mathbf{a}_j$ are the columns of $\mathbf{A}$. If we cross out the terms in (4-27) corresponding to the nonbasic variables, whose value is zero, it is evident that the sum of the remaining basic terms can be written

$$\mathbf{Bx_B} = \mathbf{b} \tag{4-28}$$

By the definition of a BFS, the basis matrix $\mathbf{B}$ is invertible. Premultiplication of (4-28) by $\mathbf{B}^{-1}$ yields

$$\mathbf{x_B} = \mathbf{B}^{-1}\mathbf{b} \tag{4-29}$$

Evidently we can evaluate directly the components of any basic feasible solution if we can find the inverse of its associated basis matrix.

For any basis $\mathbf{B}$ and for any column $\mathbf{a}_j$ of $\mathbf{A}$, *whether or not it is a basic column*, we define

$$\mathbf{y}_j \equiv \mathbf{B}^{-1}\mathbf{a}_j \tag{4-30}$$

Because $\mathbf{B}^{-1}$ is $m \times m$ and $\mathbf{a}_j$ is $m \times 1$, $\mathbf{y}_j$ also must be an m-component column vector. The definition (4-30) implies

$$\mathbf{a}_j = \mathbf{B}\mathbf{y}_j \tag{4-31}$$

and we may expand (4-31) to obtain

$$\mathbf{a}_j = y_{1j}\mathbf{b}_1 + y_{2j}\mathbf{b}_2 + \cdots + y_{mj}\mathbf{b}_m \tag{4-32}$$

where y_{ij} is the ith component of $\mathbf{y}_j$ and $\mathbf{b}_i$ is the ith column of $\mathbf{B}$. Thus, the y_{ij}, $i = 1, \ldots, m$, *are the scalar coefficients in the expression of* $\mathbf{a}_j$ *as a linear combination of the basic columns of* $\mathbf{B}$.

For any basic feasible solution $\mathbf{x_B}$ the components of the cost vector $\mathbf{c}$ that correspond to the basic variables may be collected into the m-component column vector $\mathbf{c_B}$. The ordering of the components of $\mathbf{c_B}$ must correspond to the order within $\mathbf{x_B}$. Similarly, $\mathbf{c_R}$ contains the cost coefficients of the nonbasic variables. Now, for any BFS and for any column $\mathbf{a}_j$, basic or not, we define

$$z_j \equiv \mathbf{c_B}^T\mathbf{y}_j = \mathbf{c_B}^T\mathbf{B}^{-1}\mathbf{a}_j \tag{4-33}$$

substituting (4-30) into the definition. Although z_j by itself has no particular name, the expression $(z_j - c_j)$ is known, for reasons that will appear later, as the *reduced cost* of the jth column or of the jth variable.

Notice that if the variable x_j is in the basis, say, in the pth position, then $\mathbf{y}_j$ must be a unit vector with a 1 in the pth position (all other components being zero). This is evident from (4-32); $\mathbf{a}_j$ simply equals the pth basic vector, and this expression of $\mathbf{a}_j$ as a linear combination of the basis vectors must be unique. Furthermore, if x_j is the pth basic variable, then from (4-33) we have

$$z_j = c_j \tag{4-34}$$

because the unit vector $\mathbf{y}_j$ will pick out the pth component of $\mathbf{c_B}$, which is the jth component of the original cost vector.

As a matter of terminology, we remark that $\mathbf{y}_j$ and z_j are associated with the variable x_j as well as with the column $\mathbf{a}_j$. In fact, we shall refer to x_j and $\mathbf{a}_j$ interchangeably as being nonbasic, pivoting into the basis, having a negative $(z_j - c_j)$, and so on.

4.6 INCREASING THE OBJECTIVE VALUE: THE OPTIMALITY THEOREM

In order to develop logically the rules by which the simplex method will work, suppose we have found some particular set of m variables whose columns $\mathbf{a}_j$ form a basis for the linear programming problem P. By reordering the original problem variables, along with their associated columns $\mathbf{a}_j$ and cost coefficients c_j, we can partition the original vector of variables into basic and nonbasic pieces:

$$\mathbf{x} = \begin{bmatrix} \mathbf{x_B} \\ \mathbf{x_R} \end{bmatrix} \tag{4-35}$$

In accordance with (4-35), the constraint matrix $\mathbf{A}$ then can be partitioned into $[\mathbf{B} \quad \mathbf{R}]$, where $\mathbf{B}$ is the basis matrix associated with $\mathbf{x_B}$ and $\mathbf{R}$ is the $m \times (n - m)$ matrix of nonbasic columns. The values of the variables for this particular BFS are given by

$$\mathbf{x_B} = \mathbf{B}^{-1}\mathbf{b} \qquad \text{and} \qquad \mathbf{x_R} = \mathbf{0} \tag{4-36}$$

Noting that the associated partition of the cost vector is

$$\mathbf{c}^T = [\mathbf{c_B}^T \quad \mathbf{c_R}^T] \tag{4-37}$$

we now can rewrite the original linear programming problem P as

$$\text{Max } z = \mathbf{c_B}^T\mathbf{x_B} + \mathbf{c_R}^T\mathbf{x_R} \tag{4-38}$$

$$\text{subject to} \quad \mathbf{B}\mathbf{x_B} + \mathbf{R}\mathbf{x_R} = \mathbf{b} \tag{4-39}$$

$$\text{and} \qquad \mathbf{x_B}, \mathbf{x_R} \geq \mathbf{0} \tag{4-40}$$

The left-hand side of (4-39) was obtained via the "block-wise" multiplication of the partitioned matrix $\mathbf{A}$ and vector $\mathbf{x}$.

Although we know of one feasible solution to this problem, namely, (4-36), suppose we seek other values of $\mathbf{x_B}$ and $\mathbf{x_R}$ that satisfy (4-39) and (4-40)—we do not restrict ourselves to basic solutions. For convenience, we rewrite the constraints (4-39) as

$$\mathbf{B}\mathbf{x_B} + \sum_{j \in J} x_j \mathbf{a}_j = \mathbf{b} \tag{4-41}$$

where J is the set of subscripts of the nonbasic variables (that is, j is in J if $\mathbf{x_R}$ includes x_j). Because $\mathbf{B}$ is invertible, (4-41) yields

$$\mathbf{x_B} = \mathbf{B}^{-1}\mathbf{b} - \sum_{j \in J} x_j \mathbf{y}_j \tag{4-42}$$

where we have used the definition (4-30). Substituting this into the objective function (4-38) produces

$$z = \mathbf{c_B}^T\mathbf{B}^{-1}\mathbf{b} - \sum_{j \in J} x_j(\mathbf{c_B}^T\mathbf{y}_j) + \sum_{j \in J} c_j x_j \tag{4-43}$$

Now, using (4-33),

$$z = \mathbf{c_B}^T\mathbf{B}^{-1}\mathbf{b} - \sum_{j \in J} (z_j - c_j)x_j \tag{4-44}$$

What we have done is to eliminate the m basic variables by solving the m constraint equations (4-39) for $\mathbf{x_B}$ and substituting into (4-38). Thus, the original linear program has been transformed into an equivalent LPP in the $(n - m)$ nonbasic variables x_j:

$$\begin{aligned} &\text{Max } z = \mathbf{c_B}^T\mathbf{B}^{-1}\mathbf{b} - \sum_{j \in J} (z_j - c_j)x_j \\ &\text{subject to } \sum_{j \in J} x_j\mathbf{y}_j \le \mathbf{B}^{-1}\mathbf{b} \\ &\text{and} \qquad x_j \ge 0 \quad j \in J \end{aligned}$$

Bear in mind that $\mathbf{c_B}$, $\mathbf{B}$, and the $\mathbf{y}_j$ are constants, determined by our partition (4-35); the only variables in this new problem are the x_j. The objective function is just (4-44), the nonnegativity restrictions are taken from the original problem, and the constraints reflect the requirement that $\mathbf{x_B}$, given by (4-42), must continue to be nonnegative.

At this point, let us consider the current solution, the one that corresponds to the partition we have been using. The values of the variables, $x_j = 0$ for all j in J, are given by (4-36) and at the moment

$$z = \mathbf{c_B}^T\mathbf{B}^{-1}\mathbf{b} \tag{4-45}$$

We would like to improve this solution (maximize further) *by changing the value of one of the nonbasic variables* x_j, $j \in J$. Nonnegativity prevents us from decreasing the value of any of the x_j, so we must evidently select one with a negative reduced cost $(z_j - c_j)$ and increase it above its present zero level. Equation (4-44) guarantees that this maneuver will increase the value of z. We want to increase z as fast as possible, and one obvious and reasonable strategy is to increase the variable x_j having the most negative (algebraically smallest) reduced cost; this is exactly what the simplex method does.

SIMPLEX BASIS ENTRY CRITERION. The nonbasic variable x_k is chosen to increase and enter the basis if and only if $(z_k - c_k)$ is negative and

$$(z_k - c_k) = \min_{j \in J} (z_j - c_j) \tag{4-46}$$

In case of a tie among two or more nonbasic variables, the actual basis entrant is chosen arbitrarily; a computer code might typically select the variable whose $(z_j - c_j)$ was scanned first. We do not claim that our "obvious and reasonable" strategy will always work out best; some discussion of this point can be found in Section 4.13.

The variable x_k chosen via (4-46) will be increased (except in the case of degeneracy, to be covered in Section 4.9) and, therefore, can no longer be nonbasic. From (4-42) we can see that this change in x_k will cause the values of the current basic variables to change as well; some will increase and others will decrease. In general, we shall decrease one of the basic variables precisely to zero, ejecting it from the basis and making room for x_k. Consequently, the number of nonzero variables will remain at (or below) m and the new solution again will be basic.

The question immediately suggests itself: If none of the $(z_j - c_j)$ is negative, is the current basic feasible solution optimal? The answer is yes:

THEOREM 4.4 (Optimality Theorem). The basic feasible solution associated with the basis $\mathbf{B}$ is optimal if $(z_j - c_j)$ is nonnegative for every nonbasic variable x_j.

The proof begins with the formation of the dual of the original primal LPP (4-24) through (4-26). The dual is

$$\begin{aligned} &\text{Min } z' = \mathbf{b}^T\mathbf{u} \\ &\text{subject to} \quad \mathbf{A}^T\mathbf{u} \geq \mathbf{c} \\ &\text{with} \quad \mathbf{u} \text{ unrestricted} \end{aligned} \tag{4-47}$$

Note that the jth dual constraint is

$$\mathbf{a}_j^T\mathbf{u} \geq c_j$$

where $\mathbf{a}_j$ is the jth column of $\mathbf{A}$; this is equivalent to

$$\mathbf{u}^T\mathbf{a}_j - c_j \geq 0 \qquad j = 1, \ldots, n \tag{4-48}$$

Suppose now that for some given basis $\mathbf{B}$, the reduced cost $(z_j - c_j)$ is nonnegative for every nonbasic variable; we know from (4-34) that $z_j - c_j = 0$ for every *basic* variable, so we may write

$$z_j - c_j = \mathbf{c_B}^T\mathbf{B}^{-1}\mathbf{a}_j - c_j \geq 0 \qquad j = 1, \ldots, n \tag{4-49}$$

By comparing (4-48) and (4-49), we can see that the set of values

$$\mathbf{u}_0{}^T \equiv \mathbf{c_B}^T\mathbf{B}^{-1} \tag{4-50}$$

must be a feasible solution to the dual problem (4-47). But the value of the dual objective for the solution (4-50), after transposing, is $\mathbf{c_B}^T\mathbf{B}^{-1}\mathbf{b}$, which is the same as the value of the primal objective $z = \mathbf{c_B}^T\mathbf{x_B} + \mathbf{c_R}^T\mathbf{x_R}$ at the current primal solution

$$\mathbf{x_B} = \mathbf{B}^{-1}\mathbf{b} \qquad \text{and} \qquad \mathbf{x_R} = \mathbf{0} \tag{4-36}$$

Therefore, by Theorem 3.5, both solutions are optimal, Q.E.D.

We now have the important result that optimality has been achieved when all $(z_j - c_j)$ are nonnegative. This is the simplex criterion that will tell us when we have finished solving a problem. As an example, consider the primal linear program

$$\text{Max } z = x_1 \tag{4-51}$$

$$\begin{aligned} \text{subject to} \quad 2x_1 + x_2 &\leq 4 \\ -x_1 + 2x_2 &\leq 2 \end{aligned} \tag{4-52}$$

$$\text{and} \qquad x_1, x_2 \geq 0$$

The feasible region is shown in Figure 4-2. The constraints are converted to standard form by the addition of slack variables:

$$\begin{aligned} 2x_1 + x_2 + x_3 \qquad &= 4 \\ -x_1 + 2x_2 \qquad + x_4 &= 2 \end{aligned} \tag{4-53}$$

$$\text{and} \qquad x_j \geq 0 \qquad j = 1, \ldots, 4$$

Suppose we are told (or have discovered) that the point (2, 0) is an extreme point in (x_1,x_2) space; the values of all the variables are then $x_1 = 2$, $x_2 = 0$, $x_3 = 0$, $x_4 = 4$. If we take as our basis $\mathbf{x_B} = (x_1,x_4)$, so that $\mathbf{c_B} = (1, 0)$, we then have

$$\mathbf{B} = \begin{bmatrix} 2 & 0 \\ -1 & 1 \end{bmatrix} \qquad \text{and} \qquad \mathbf{B}^{-1} = \begin{bmatrix} 0.5 & 0 \\ 0.5 & 1 \end{bmatrix}$$

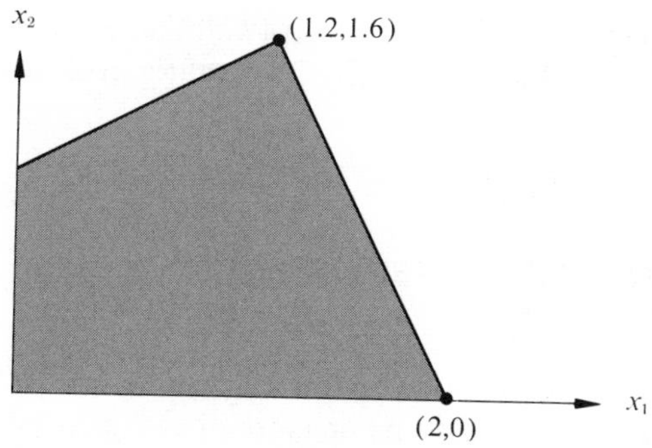

Figure 4-2

For the nonbasic variable x_2,

$$\mathbf{y}_2 = \mathbf{B}^{-1}\mathbf{a}_2 = \begin{bmatrix} 0.5 & 0 \\ 0.5 & 1 \end{bmatrix} \cdot \begin{bmatrix} 1 \\ 2 \end{bmatrix} = \begin{bmatrix} 0.5 \\ 2.5 \end{bmatrix}$$

and $$z_2 - c_2 = \mathbf{c_B}^T\mathbf{y}_2 - c_2 = 0.5 - 0 = 0.5$$

while for the other nonbasic variable x_3,

$$\mathbf{y}_3 = \mathbf{B}^{-1}\mathbf{a}_3 = \begin{bmatrix} 0.5 & 0 \\ 0.5 & 1 \end{bmatrix} \cdot \begin{bmatrix} 1 \\ 0 \end{bmatrix} = \begin{bmatrix} 0.5 \\ 0.5 \end{bmatrix}$$

and $$z_3 - c_3 = \mathbf{c_B}^T\mathbf{y}_3 - c_3 = 0.5 - 0 = 0.5$$

Both reduced costs are nonnegative; we therefore conclude (as is obvious from Figure 4-2) that the present solution is optimal. To check this, form the dual of the primal problem (4-51) and (4-53):

$$\text{Min } z = 4u_1 + 2u_2$$

$$\begin{aligned} \text{subject to} \quad & 2u_1 - u_2 \geq 1 \\ & u_1 + 2u_2 \geq 0 \\ & u_1 \geq 0 \\ \text{and} \quad & u_2 \geq 0 \end{aligned}$$

The dual solution (4-50) is given by

$$\mathbf{u}_0^T = [u_{01}, u_{02}] = \mathbf{c_B}^T\mathbf{B}^{-1} = [1 \quad 0]\begin{bmatrix} 0.5 & 0 \\ 0.5 & 1 \end{bmatrix} = [0.5 \quad 0]$$

and the reader can check that this does satisfy all the dual constraints. Both the primal and the dual solutions have objective values of 2, so by Theorem 3.5, both must be optimal.

4.7 PIVOTING TO A NEW BASIC FEASIBLE SOLUTION

Having established an optimality condition, let us return to the simplex basis entry criterion and suppose that one or more of the $(z_j - c_j)$ are negative. We therefore apply (4-46) and choose some variable x_k, with associated column $\mathbf{a}_k$, to *pivot into* (enter) the basis. If we want to arrive at a new basic feasible solution, one of the current basic variables must be ejected. Which one?

In order to answer this question, we begin by recalling

$$\mathbf{x_B} = \mathbf{B}^{-1}\mathbf{b} - \sum_{j \in J} x_j \mathbf{y}_j \tag{4-42}$$

which was obtained as an alternate expression of the original constraints (4-39). During the pivot, all nonbasic variables except x_k will be held at zero, so (4-42) may be simplified to

$$\mathbf{x_B} = \mathbf{B}^{-1}\mathbf{b} - x_k \mathbf{y}_k \tag{4-54}$$

The student is again reminded that $\mathbf{x_B}$ denotes the current set of basic variables, *not* the values taken on by those variables. The same is true of x_k. However, $\mathbf{B}^{-1}$ and $\mathbf{y}_k$ are constants determined by the specific partition that we have been considering. Thus, the values of the m variables that constitute the current basis are represented by (4-54) as functions of the single variable x_k.

Initially x_k is zero and the values of the basic variables are given by $\mathbf{x_B} = \mathbf{B}^{-1}\mathbf{b}$. When x_k is brought into the basis, causing the values of these variables to change, the relationship (4-54) must continue to hold; this guarantees that all the original constraints of the LPP will remain satisfied. In particular, after the pivot has been completed, (4-54) will still have to be satisfied by the new values of x_k and $\mathbf{x_B}$, which we shall denote by using hats:

$$\hat{\mathbf{x}}_\mathbf{B} = \mathbf{B}^{-1}\mathbf{b} - \hat{x}_k \mathbf{y}_k$$

Letting $x_{\mathbf{B}i}$ represent the initial value (before the pivot) of the ith basic variable, we may write the ith component of this equation as follows:

$$\hat{x}_{\mathbf{B}i} = x_{\mathbf{B}i} - \hat{x}_k y_{ik} \tag{4-55}$$

Now, because we want the new basic solution at which we are aiming to be *nonnegative*, we are not free to increase x_k as much as we like; that is, we require

$$\hat{x}_{\mathbf{B}i} = x_{\mathbf{B}i} - \hat{x}_k y_{ik} \geq 0 \qquad i = 1, \ldots, m \tag{4-56}$$

Consider, therefore, how the values of the original set of basic variables will change as x_k increases. For those variables whose y_{ik} is negative or zero there is no problem: Because x_k is to be increased from zero into positive territory, $\hat{x}_{\mathbf{B}i}$ cannot be less than $x_{\mathbf{B}i}$ and the nonnegativity requirement can never be violated. Suppose all $y_{ik} \leq 0$, $i = 1, \ldots, m$. Then it is evident from (4-56) that x_k could be increased by any arbitrary positive amount without forcing any variable to become negative. From (4-44), because $(z_k - c_k)$ is negative, an infinite increase in x_k would carry z to infinity. We have proved the following theorem.

THEOREM 4.5 (Unboundedness Theorem). If for any nonbasic variable x_k the reduced cost $(z_k - c_k)$ is negative and all y_{ik} are nonpositive, then x_k can be increased without bound while all other variables remain nonnegative, and the given linear program has an unbounded maximum.

Returning to (4-56), if one or more y_{ik} is positive, then we cannot allow x_k to be increased arbitrarily. In order to prevent some $x_{\mathbf{B}i}$ from becoming negative we must dictate that, whenever $y_{ik} > 0$,

$$\hat{x}_k \leq \frac{x_{\mathbf{B}i}}{y_{ik}} \tag{4-57}$$

Because this inequality must hold for *all i* for which $y_{ik} > 0$, we may permit x_k to increase only up to the ceiling established by the smallest of these ratios. Suppose the smallest ratio is associated with the rth basis variable $x_{\mathbf{B}r}$; then we can satisfy nonnegativity by setting

$$\hat{x}_k = \frac{x_{\mathbf{B}r}}{y_{rk}} \tag{4-58}$$

in which case it follows from (4-55) that $\hat{x}_{\mathbf{B}r} = 0$. This is very convenient because it permits us to remove $x_{\mathbf{B}r}$ from the basis.

Now we have collected a new set of variables that differs from the old basic set only in that the former $x_{\mathbf{B}r}$ has been replaced by x_k. Because $y_{rk} > 0$, we know from Theorem 2.1 that the columns associated with this new set of variables form another basis. The new basic variables are the same as the old (and in the same order) except that the rth is now x_k. The substitution of only a single basic variable from one BFS to the next is what was interpreted geometrically as pivoting to an *adjacent* extreme point; the student will be asked to prove this assertion in Exercise 4-8.

To summarize, the m variables of the new basis will all have nonnegative values, provided that the pivot is made according to the following rule.

SIMPLEX BASIS EXIT CRITERION. Given that the variable x_k is to enter the basis, the column $\mathbf{b}_r$ and variable $x_{\mathbf{B}r}$ must leave, where

$$\frac{x_{\mathbf{B}r}}{y_{rk}} = \min_i \left(\frac{x_{\mathbf{B}i}}{y_{ik}}, y_{ik} > 0\right) \tag{4-59}$$

Referring to (4-58), we note that the minimum ratio in (4-59) is exactly the value assumed by x_k in the new basis.

The new nonnegative values of the variables satisfy the constraints of the linear program and yield an increased value of the objective function z—*unless* $x_{\mathbf{B}r} = 0$. In this case, $\hat{x}_k = 0$ and it is clear from (4-44) and (4-56) that neither the objective z nor the values of the variables change. This "zero-for-zero" pivot has accomplished something, however, for the basis now has a new column $\mathbf{a}_k$. Therefore, $\mathbf{B}^{-1}$ is different and, from (4-30), so are all the $\mathbf{y}_j$; these changes may well permit subsequent profitable pivots to take place.

Recall from Section 2.3 that, when a basic variable is zero, the BFS is said to be *degenerate*. The preceding paragraph described a continuation of degeneracy: $x_{\mathbf{B}r} = 0$ before the pivot and $\hat{x}_{\mathbf{B}r} = \hat{x}_k = 0$ afterward. How does degeneracy arise in the first place? It must happen whenever the minimum ratio (4-59) is not unique. In that case, two (or more) basic variables will unavoidably be driven to zero when x_k enters. One of them will exit from the basis (again, in most computer codes this choice is arbitrary; typically, the column with the smallest subscript might be ejected), but the other, or others, must remain in the basis "at a zero level." This is, in fact, the only way in which a degenerate basic feasible solution can arise.

EXAMPLE 4.2. As an example of basis entry and exit, let us return to the problem discussed at the end of Section 4.7. Leaving the feasible region as before (see Figure 4-2), we modify only the objective function:

$$\text{Max } z = x_1 + 2x_2$$

$$\begin{aligned} \text{subject to} \quad & 2x_1 + x_2 + x_3 = 4 \\ & -x_1 + 2x_2 + x_4 = 2 \\ \text{and} \quad & x_j \geq 0 \quad j = 1, \ldots, 4 \end{aligned}$$

Suppose that the current basic feasible solution is again $x_1 = 2$, $x_2 = 0$, $x_3 = 0$, $x_4 = 4$, with $\mathbf{x_B} = (x_1, x_4) = (2, 4)$. We saw that

$$\mathbf{y}_2 = \mathbf{B}^{-1}\mathbf{a}_2 = \begin{bmatrix} 0.5 \\ 2.5 \end{bmatrix} \quad \text{and} \quad \mathbf{y}_3 = \begin{bmatrix} 0.5 \\ 0.5 \end{bmatrix}$$

Changing the objective function does not affect these values. The reduced costs, however, are different:

$$z_2 - c_2 = \mathbf{c_B}^T\mathbf{y}_2 - c_2 = [1 \quad 0]\begin{bmatrix} 0.5 \\ 2.5 \end{bmatrix} - 2 = -1.5$$

and
$$z_3 - c_3 = \mathbf{c_B}^T\mathbf{y}_3 - c_3 = 0.5 - 0 = 0.5$$

The variable x_2 has the only negative reduced cost and therefore enters the basis, so that $k = 2$. In order to determine which variable exists, we apply (4-59):

$$\min_i \left(\frac{x_{\mathbf{B}i}}{y_{i2}}, y_{i2} > 0\right) = \min\left(\frac{x_{\mathbf{B}1}}{y_{12}}, \frac{x_{\mathbf{B}2}}{y_{22}}\right) = \min\left(\frac{2}{0.5}, \frac{4}{2.5}\right) = \frac{4}{2.5}$$

and the second basic variable x_4 departs. The new value of the entering variable x_2 is given by (4-58),

$$\hat{x}_2 = \frac{x_{\mathbf{B}r}}{y_{r2}} = \frac{x_{\mathbf{B}2}}{y_{22}} = \frac{4}{2.5} = 1.6$$

while the new values of the original basic variables are, using (4-56),

$$\hat{x}_{\mathbf{B}1} = \hat{x}_1 = x_{\mathbf{B}1} - \hat{x}_2 y_{12} = 2 - 1.6(0.5) = 1.2$$

and
$$\hat{x}_{\mathbf{B}2} = \hat{x}_4 = x_{\mathbf{B}2} - \hat{x}_2 y_{22} = 4 - 1.6(2.5) = 0$$

We knew that the new value of x_4 would be zero because it was chosen to exit from the basis. The new value $\hat{z}$ of the objective function can be calculated from (4-44):

$$\hat{z} = \mathbf{c_B}^T\mathbf{B}^{-1}\mathbf{b} - \sum_{j \in J} (z_j - c_j)\hat{x}_j = [1 \quad 0] \cdot \begin{bmatrix} 2 \\ 4 \end{bmatrix} - (z_2 - c_2)\hat{x}_2$$

$$= 2 - (-1.5)(1.6) = 4.4$$

This is readily checked by direct substitution into the objective function:

$$\hat{z} = \hat{x}_1 + 2\hat{x}_2 = 1.2 + 2(1.6) = 4.4$$

4.8 COMPUTATIONAL PROCEDURES

It is evident from the previous section that, in order to decide which variables enter and leave the basis, we need to keep track of the current values of all $x_{\mathbf{B}i}$, all y_{ij}, and all $(z_j - c_j)$. It would be convenient also, though not necessary, to update the value of the objective function as we go along.

Assume, then, that the current basic feasible solution is not optimal and that the current values of all the variable quantities mentioned before are known. *Given that the next pivot consists of the variable x_k entering and $x_{\mathbf{B}r}$ leaving the basis*, we want to determine all the new values in terms of the old. We remark first that the new values of the basic variables $x_{\mathbf{B}i}$ essentially have been derived. We have seen that

$$\hat{x}_{\mathbf{B}r} = \hat{x}_k = \frac{x_{\mathbf{B}r}}{y_{rk}} \tag{4-60}$$

where $\hat{x}_{\mathbf{B}r}$ here represents the new value of the variable x_k that occupies the rth position in the basis after the pivot, and $x_{\mathbf{B}r}$ represents the old value of the variable that occupied that position before the pivot. Substitution of (4-60) into (4-55) produces

$$\hat{x}_{\mathbf{B}i} = x_{\mathbf{B}i} - \frac{x_{\mathbf{B}r}}{y_{rk}} y_{ik} \qquad \text{for } i = 1, \ldots, r-1, r+1, \ldots, m \tag{4-61}$$

Thus the new $\hat{x}_{\mathbf{B}i}$ can be calculated simply and directly from various old, already known values.[1] We continue labeling all new values with hats.

In order to determine the $\hat{y}_{ij}$—that is, the y elements implied by the new basis—we begin by writing the expression for any column of $\mathbf{A}$ in terms of the old basic columns:

$$\mathbf{a}_j = y_{1j}\mathbf{b}_1 + y_{2j}\mathbf{b}_2 + \cdots + y_{mj}\mathbf{b}_m = \sum_{i=1}^{m} y_{ij}\mathbf{b}_i \tag{4-32}$$

In particular, for the entering column $\mathbf{a}_k$,

$$\mathbf{a}_k = \sum_{i=1}^{m} y_{ik}\mathbf{b}_i$$

which yields

$$\mathbf{b}_r = \frac{1}{y_{rk}}\mathbf{a}_k - \sum_{i \neq r} \frac{y_{ik}}{y_{rk}}\mathbf{b}_i \tag{4-62}$$

[1] For completeness we should add that the new value of the variable that previously occupied the rth position in the basis is now zero because that variable has become nonbasic.

The simplex exit criterion insures that $y_{rk} \neq 0$. Substituting this into (4-32) yields, after some manipulation,

$$\mathbf{a}_j = \frac{y_{rj}}{y_{rk}} \mathbf{a}_k + \sum_{i \neq r} \left(y_{ij} - y_{rj} \frac{y_{ik}}{y_{rk}} \right) \mathbf{b}_i \qquad (4\text{-}63)$$

where the summation index in the last two equations moves from 1 to m, excluding r.

The column vectors on the right-hand side of (4-63) are precisely those contained in the new basis. From property 4 of Section 2.2 only one linear combination of them can add up to any given $\mathbf{a}_j$, so we have shown that the scalar coefficients in (4-63) constitute the new $\hat{y}_{ij}$:

$$\hat{y}_{rj} = \frac{y_{rj}}{y_{rk}} \qquad (4\text{-}64)$$

and $$\hat{y}_{ij} = y_{ij} - y_{rj} \frac{y_{ik}}{y_{rk}} \qquad \text{for } i = 1, \ldots, r-1, r+1, \ldots, m \qquad (4\text{-}65)$$

Remember that the vector $\mathbf{y}_j$ is defined for all variables x_j, basic or not. We maintain and update these $\mathbf{y}_j$ because any nonbasic variable is a potential candidate for entry into the basis, in which event its y_{ij} are needed for the exit criterion.

To get at the new $(\hat{z}_j - c_j)$ we use the definition of z_j to write

$$\hat{z}_j - c_j = \hat{\mathbf{c}}_{\mathbf{B}}{}^T \hat{\mathbf{y}}_j - c_j = \sum_{i=1}^{m} \hat{c}_{\mathbf{B}i} \hat{y}_{ij} - c_j \qquad (4\text{-}66)$$

Now all $\hat{c}_{\mathbf{B}i} = c_{\mathbf{B}i}$ because the order of the basic variables is maintained during a pivot, except that $\hat{c}_{\mathbf{B}r} = c_k$. Using (4-64) and (4-65), this yields

$$\hat{z}_j - c_j = \sum_{i \neq r} c_{\mathbf{B}i} \left(y_{ij} - y_{rj} \frac{y_{ik}}{y_{rk}} \right) + c_k \frac{y_{rj}}{y_{rk}} - c_j \qquad (4\text{-}67)$$

But in the summation the missing term for $i = r$ would be

$$c_{\mathbf{B}r} \left(y_{rj} - y_{rj} \frac{y_{rk}}{y_{rk}} \right) = 0 \qquad (4\text{-}68)$$

so that we may sum from $i = 1$ to $i = m$ without omitting $i = r$. Expanding (4-67), we obtain

$$\hat{z}_j - c_j = \left(\sum_{i=1}^{m} c_{\mathbf{B}i} y_{ij} \right) - c_j - \frac{y_{rj}}{y_{rk}} \left(\sum_{i=1}^{m} c_{\mathbf{B}i} y_{ik} - c_k \right) \qquad (4\text{-}69)$$

This is just

$$\hat{z}_j - c_j = (z_j - c_j) - \frac{y_{rj}}{y_{rk}}(z_k - c_k) \tag{4-70}$$

and we have obtained an expression for the new $(\hat{z}_j - c_j)$ in terms of the old values.

Finally, to determine $\hat{z}$, the new value of the objective function, we begin with

$$\hat{z} = \hat{\mathbf{c}}_{\mathbf{B}}{}^T\hat{\mathbf{x}}_{\mathbf{B}} \equiv \sum_{i=1}^{m} \hat{c}_{\mathbf{B}i}\hat{x}_{\mathbf{B}i} \tag{4-71}$$

which follows from the fact that nonbasic variables contribute zero terms to the objective $z = \mathbf{c}^T\mathbf{x}$. Substituting (4-60) and (4-61) into this produces

$$\hat{z} = \sum_{i \neq r} c_{\mathbf{B}i}\left(x_{\mathbf{B}i} - \frac{x_{\mathbf{B}r}}{y_{rk}}y_{ik}\right) + c_k\frac{x_{\mathbf{B}r}}{y_{rk}} \tag{4-72}$$

For the same reason as before, we omit the restriction $i \neq r$. Thus (4-72) becomes

$$\hat{z} = \sum_{i=1}^{m} c_{\mathbf{B}i}x_{\mathbf{B}i} - \left(\sum_{i=1}^{m} c_{\mathbf{B}i}y_{ik}\frac{x_{\mathbf{B}r}}{y_{rk}}\right) + c_k\frac{x_{\mathbf{B}r}}{y_{rk}} \tag{4-73}$$

which reduces to

$$\hat{z} = z - \frac{x_{\mathbf{B}r}}{y_{rk}}(z_k - c_k) \tag{4-74}$$

This completes our derivation of the new post-pivot values of some important variables and quantities associated with the linear program in standard form. All of them are updated at each pivot of the simplex method. For convenience, these *transformation equations*, which express the new values in terms of the old, are collected in Table 4-1.

There is a rather straightforward way of handling all these calculations, whether they are done by hand or by computer. The current values of the relevant variables and quantities are stored in a skeleton diagram called a *simplex tableau*, as shown in Figure 4-3. All variables x_j, basic or not, are listed across the top; the current basic columns $\mathbf{b}_j$ are listed "by name" at the left (if the current $x_{\mathbf{B}1}$ were x_7, then the label $\mathbf{a}_7$ would be entered in the position occupied in the diagram by $\mathbf{b}_1$). The first column of the tableau contains the values of the basic variables and the objective function, which is here being maximized. Each of the remaining columns belongs to (or is associated with) some variable x_j and contains its $\mathbf{y}_j$ vector and its $(z_j - c_j)$.

Table 4-1

If the variable x_k replaces $x_{\mathbf{B}r}$ in the basis, then

$$\hat{x}_{\mathbf{B}r} = \hat{x}_k = \frac{x_{\mathbf{B}r}}{y_{rk}} \qquad (4\text{-}60)$$

$$\hat{x}_{\mathbf{B}i} = x_{\mathbf{B}i} - \frac{x_{\mathbf{B}r}}{y_{rk}} y_{ik} \qquad \text{for } i = 1, \ldots, r-1, r+1, \ldots, m \qquad (4\text{-}61)$$

$$\hat{y}_{rj} = \frac{y_{rj}}{y_{rk}} \qquad (4\text{-}64)$$

$$\hat{y}_{ij} = y_{ij} - \frac{y_{rj}}{y_{rk}} y_{ik} \qquad \text{for } i = 1 \ldots, r-1, r+1, \ldots, m \qquad (4\text{-}65)$$

$$\hat{z}_j - c_j = (z_j - c_j) - \frac{y_{rj}}{y_{rk}} (z_k - c_k) \qquad (4\text{-}70)$$

$$\hat{z} = z - \frac{x_{\mathbf{B}r}}{y_{rk}} (z_k - c_k) \qquad (4\text{-}74)$$

At each pivot the old values are read from the current tableau and the transformation equations are applied. The newly calculated values are then entered in a new simplex tableau. Although a fair amount of algebraic manipulation is necessary in any case, we can minimize it and reduce our chance of error by arranging the computations in an orderly and symmetric manner. Suppose we have selected the variable x_k to enter and $x_{\mathbf{B}r}$ to leave the basis. Let us form a new vector $\mathbf{\Phi}$ from the tableau column belonging to x_k:

$$\mathbf{\Phi} = \left(-\frac{y_{1k}}{y_{rk}}, \ldots, -\frac{y_{r-1,k}}{y_{rk}}, \frac{1}{y_{rk}} - 1, -\frac{y_{r+1,k}}{y_{rk}}, \ldots, -\frac{y_{mk}}{y_{rk}}, -\frac{(z_k - c_k)}{y_{rk}}\right) \qquad (4\text{-}75)$$

Note that the last element, the $(z_k - c_k)$, is treated in the same way as all the y_{ik}, with the exception of y_{rk}. In the old tableau this element y_{rk} lies in the column

		x_1	x_2	$\cdots$	x_n
$\mathbf{b}_1$	$x_{\mathbf{B}1}$	y_{11}	y_{12}	$\cdots$	y_{1n}
$\mathbf{b}_2$	$x_{\mathbf{B}2}$	y_{21}	y_{22}	$\cdots$	y_{2n}
$\vdots$	$\vdots$	$\vdots$	$\vdots$		$\vdots$
$\mathbf{b}_m$	$x_{\mathbf{B}m}$	y_{m1}	y_{m2}	$\cdots$	y_{mn}
	Max z	$z_1 - c_1$	$z_2 - c_2$	$\cdots$	$z_n - c_n$

Figure 4-3

of the newly entering variable x_k and in the row of the departing basic variable $x_{\mathbf{B}r}$; it is known as the *pivot element*.

In the new tableau the various labels x_j and $\mathbf{b}_i$ will be the same as in the old, except that the rth basic column $\mathbf{b}_r$ now will be $\mathbf{a}_k$. We now assert that *each column of entries for the new tableau is derived from its corresponding column in the old tableau by means of the following transformation*:

$$(\text{new column } j) = (\text{old column } j) + y_{rj}\mathbf{\Phi} \tag{4-76}$$

When (4-76) is used to transform the left-most column, which contains the values of the basic variables and of the objective function, it is understood that $x_{\mathbf{B}r}$ replaces y_{rj} on the right-hand side. The reader will find it easy to verify that all the simplex transformations are legitimately performed by (4-76).

EXAMPLE 4.3. Suppose that the tableau of Figure 4-4 represents some stage in the application of the simplex method to a maximization linear programming problem. The LPP evidently has six variables and three constraints (as well as the standard nonnegativity requirements). We first observe that the current basic feasible solution is $x_{\mathbf{B}1} = x_2 = 0$, $x_{\mathbf{B}2} = x_5 = 2$, $x_{\mathbf{B}3} = x_3 = 3$, and it is degenerate. As was remarked in Section 4.5, the $\mathbf{y}_j$ of the basic variables x_j are unit vectors and the $(z_j - c_j)$ are zero. Next we see that optimality has not been achieved yet because some of the $(z_j - c_j)$ are negative. The simplex basis entry rule tells us that the new basic column will be $\mathbf{a}_4$ because $-3 < -1$. Applying the exit criterion,

$$\min_i \left(\frac{x_{\mathbf{B}i}}{y_{ik}}, y_{ik} > 0\right) = \min\left(\frac{x_{\mathbf{B}2}}{y_{24}}, \frac{x_{\mathbf{B}3}}{y_{34}}\right) = \min\left(\frac{2}{3}, \frac{3}{5}\right) = \frac{3}{5}$$

so the variable x_3 leaves the basis. (If none of the elements of $\mathbf{y}_4$ had been positive, we would have concluded from Theorem 4.5 that the problem was unbounded.) We now evaluate the $\mathbf{\Phi}$ vector directly from the tableau column of x_4: Substituting $k = 4$ and $r = 3$ into (4-75) yields

$$\mathbf{\Phi} = (+0.2, -0.6, -0.8, +0.6)$$

		x_1	x_2	x_3	x_4	x_5	x_6
$\mathbf{a}_2$	$x_{\mathbf{B}1} = x_2 = 0$	1.2	1	0	−1	0	4.8
$\mathbf{a}_5$	$x_{\mathbf{B}2} = x_5 = 2$	−2.4	0	0	3	1	1.6
$\mathbf{a}_3$	$x_{\mathbf{B}3} = x_3 = 3$	2	0	1	5	0	−1
	Max $z = 6$	2	0	0	−3	0	−1

Figure 4-4

		x_1	x_2	x_3	x_4	x_5	x_6
$\mathbf{a}_2$	$x_{\mathbf{B}1} = x_2 = 0.6$	1.6	1	0.2	0	0	4.6
$\mathbf{a}_5$	$x_{\mathbf{B}2} = x_5 = 0.2$	-3.6	0	-0.6	0	1	2.2
$\mathbf{a}_4$	$x_{\mathbf{B}3} = x_4 = 0.6$	0.4	0	0.2	1	0	-0.2
	Max $z = 7.8$	3.2	0	0.6	0	0	-1.6

Figure 4-5

Now we can use the vector transform equation (4-76) to generate the new tableau, which is shown in Figure 4-5. The objective value has improved from 6 to 7.8; because $z_6 - c_6 = -1.6$, optimality has still not been attained. Notice that degeneracy has disappeared; this occurred because x_2, the zero-valued basic variable, had a negative y entry in the tableau column of x_4 and, therefore, was not a candidate for removal from the basis.

Having gotten our feet wet with a single simplex pivot, we shall now plunge in to solve a linear programming problem completely, from start to finish. The student is advised to work through the example carefully and in particular to take note of the manner in which the initial basic feasible solution so conveniently arises.

EXAMPLE 4.4. Find values of all variables x_i, $i = 1, \ldots, 4$, in order to

$$\begin{aligned} \text{Maximize } z &= 2x_1 + 4x_2 + 3x_3 + x_4 \\ \text{subject to }\quad 3x_1 + x_2 + x_3 + 4x_4 &\leq 12 \\ x_1 - 3x_2 + 2x_3 + 3x_4 &\leq 7 \\ 2x_1 + x_2 + 3x_3 - x_4 &\leq 10 \\ \text{and}\qquad x_1, x_2, x_3, x_4 &\geq 0 \end{aligned}$$

We begin by converting the problem from canonical to standard form:

$$\begin{aligned} &\text{Max } z = 2x_1 + 4x_2 + 3x_3 + x_4 \\ &\text{subject to }\quad 3x_1 + x_2 + x_3 + 4x_4 + x_5 = 12 \\ &\qquad x_1 - 3x_2 + 2x_3 + 3x_4 + x_6 = 7 \\ &\qquad 2x_1 + x_2 + 3x_3 - x_4 + x_7 = 10 \\ &\text{and}\qquad x_i \geq 0 \qquad i = 1, \ldots, 7 \end{aligned}$$

The problem still has three constraints, but now it has seven variables instead of four. The last three are slacks and contribute nothing to the objective function, but they are full-fledged variables nonetheless; for example, any three of the seven whose columns are linearly independent form a legitimate basic solution.

Observe that the columns associated with the slack variables x_5, x_6, and x_7 are unit vectors that form an identity matrix I_3, which equals its own inverse. This gives us a basis. Since the right-hand sides of all the constraints are nonnegative, the solution associated with this basis, which is

$$\mathbf{x_B} \equiv \begin{bmatrix} x_5 \\ x_6 \\ x_7 \end{bmatrix} = \mathbf{B}^{-1}\mathbf{b} = \mathbf{b} \equiv \begin{bmatrix} 12 \\ 7 \\ 10 \end{bmatrix}$$

is also nonnegative. Thus we have an initial basic feasible solution.

Like our example, a great many real-world linear programming problems arise in canonical form. This is extremely convenient because an identity matrix of slack variables is immediately available; the slacks constitute an initial basic feasible solution whenever the right-hand-side vector **b** is nonnegative. When the LPP is not given in canonical form it is usually necessary to use a special modification of the simplex method (described in Chapter 5) in order to find an initial BFS; for small problems, however, a little algebraic manipulation may suffice (see Exercises 4-15 and 4-16).

We are ready to write the initial simplex tableau, which takes a very simple form when the initial basis is an identity matrix. Because $\mathbf{B}^{-1} = I_3$, we have

$$\begin{aligned} \mathbf{x_B} &= \mathbf{b} && \\ \mathbf{y}_j &= \mathbf{a}_j && \text{for all } j \\ \mathbf{c_B} &= \mathbf{0} && \\ z &= \mathbf{c_B}^T\mathbf{x_B} = 0 && \\ z_j &= \mathbf{c_B}^T\mathbf{y}_j = 0 && \text{for all } j \end{aligned}$$

and $$z_j - c_j = -c_j \qquad \text{for all } j \qquad j = 1, \ldots, 7$$

The initial tableau becomes

		x_1	x_2	x_3	x_4	x_5	x_6	x_7
$\mathbf{a}_5$	$x_{\mathbf{B}1} = x_5 = 12$	3	1	1	4	1	0	0
$\mathbf{a}_6$	$x_{\mathbf{B}2} = x_6 = 7$	1	−3	2	3	0	1	0
$\mathbf{a}_7$	$x_{\mathbf{B}3} = x_7 = 10$	2	1	3	−1	0	0	1
	Max $z = 0$	−2	−4	−3	−1	0	0	0

Several reduced costs are negative, so our initial BFS is not optimal (as it would be if we were *minimizing* z). Because $z_2 - c_2 = -4 < 0$ and

$$(z_2 - c_2) = \min_j (z_j - c_j, \quad j \text{ nonbasic})$$

the variable x_2 enters the basis. Now

$$\min_i \left(\frac{x_{Bi}}{y_{i2}}, y_{i2} > 0\right) = \min \left(\frac{x_{B1}}{y_{12}}, \frac{x_{B3}}{y_{32}}\right) = \frac{x_{B3}}{y_{32}} = \frac{10}{1}$$

so the variable $x_{B3} = x_7$ exits. We form the $\mathbf{\Phi}$ vector around the pivot element $y_{32} = 1$:

$$\mathbf{\Phi} = (-1, 3, 0, 4)$$

Using the fact that each (new column j) = (old column j) + $y_{3j}\mathbf{\Phi}$, we can write the new tableau:

		x_1	x_2	x_3	x_4	x_5	x_6	x_7
$\mathbf{a}_5$	$x_{B1} = x_5 = 2$	1	0	−2	5	1	0	−1
$\mathbf{a}_6$	$x_{B2} = x_6 = 37$	7	0	11	0	0	1	3
$\mathbf{a}_2$	$x_{B3} = x_2 = 10$	2	1	3	−1	0	0	1
	Max $z = 40$	6	0	9	−5	0	0	4

The solution is not yet optimal; the variable x_4 enters the basis and x_5 exits. The pivot element is $y_{14} = 5$ and $\mathbf{\Phi} = (-0.8, 0.0, 0.2, 1.0)$ and the new tableau is

		x_1	x_2	x_3	x_4	x_5	x_6	x_7
$\mathbf{a}_4$	$x_{B1} = x_4 = 0 4$	0.2	0.0	−0.4	1.0	0.2	0.0	−0.2
$\mathbf{a}_6$	$x_{B2} = x_6 = 37.0$	7.0	0.0	11.0	0.0	0.0	1.0	3.0
$\mathbf{a}_2$	$x_{B3} = x_2 = 10.4$	2.2	1.0	2.6	0.0	0.2	0.0	0.8
	Max $z = 42.0$	7.0	0.0	7.0	0.0	1.0	0.0	3.0

Because all reduced costs are now nonnegative, optimality has been attained. The optimal BFS has $x_2 = 10.4$, $x_4 = 0.4$, $x_6 = 37.0$, with all other variables zero, and $z = 42.0$.

4.9 DEGENERACY AND CONVERGENCE OF THE SIMPLEX METHOD

Toward the end of Section 4.7 we mentioned that degeneracy arises initially when the minimum of the ratios $x_{\mathbf{B}i}/y_{ik}$ in the simplex exit criterion (4-59) is not unique. In that case two or more basic variables are driven to zero, but only one can exit. Inspection of the transformation formulas shows that the entire system changes as usual. If, at the next iteration, some other column is chosen by the exit criterion, then the basic variable that was zero can be restored to a positive level.

We also noted in Section 4.7 that whenever the departing basic variable equals zero, degeneracy is continued in a zero-for-zero pivot. Suppose that x_k has been chosen to enter the basis and that some basic variable $x_{\mathbf{B}p}$ equals zero. If its y element y_{pk} happens to be positive, then the ratio $x_{\mathbf{B}p}/y_{pk}$ is zero and $x_{\mathbf{B}p}$ will necessarily be chosen to leave the basis (unless *another* zero-valued $x_{\mathbf{B}i}$ is chosen). This follows from the fact that all of the ratios $x_{\mathbf{B}i}/y_{ik}$ considered by the simplex exit criterion (4-59) are nonnegative, so that the minimum possible value is zero. When the exiting variable $x_{\mathbf{B}r} = 0$, the transform formulas then show that $\hat{z} = z$ and all $\hat{x}_{\mathbf{B}i} = x_{\mathbf{B}i}$—that is, the values of the variables and the objective remain the same—while the $\hat{y}_{ij}$ and $(\hat{z}_j - c_j)$ are, in general, different than before.

The failure of the objective function to change when the departing variable $x_{\mathbf{B}r}$ has a zero value suggests the important and disturbing possibility that the simplex basis-pivot procedure could fail to arrive at the optimal solution within a finite number of pivot steps. This possibility must concern both the real-world problem-solver, with his computer churning away at \$400 per hour, and the mathematician, who demands the theoretical elegance of a proof of *convergence* in order to validate a solution method. Were it not for degeneracy, it would be easy to prove that the simplex method must *converge*—that is, must either demonstrate unboundedness or discover an optimal solution within a finite number of pivots.[1] Consider the transformation formula for the value of the objective:

$$\hat{z} = z - \frac{x_{\mathbf{B}r}}{y_{rk}}(z_k - c_k) \tag{4-74}$$

If we assume (unrealistically) *that degeneracy never occurs*, then every simplex pivot must increase the value of z because $(z_k - c_k)$ is negative and $x_{\mathbf{B}}$ and y_{rk} are

[1] More generally, convergence means getting arbitrarily close to the desired solution, say, within any given small distance ϵ, in a finite period of time or after a finite number of steps. This more mathematical definition applies to solution methods for the nonlinear programming problem, whose optimum can lie anywhere inside some continuous region of space. The LPP optimum, on the other hand, occurs at some discrete extreme point that is reached directly by a long-distance pivot, so that the idea of continuous convergence by smaller and smaller steps is not really applicable.

positive. This means that as soon as a basic feasible solution—which determines a value of z—has been disrupted by a pivot, that basis can never reappear in any subsequent iteration. Its reappearance would require a return to a lower value of z, which is prohibited by the simplex entry criterion (for maximization problems). If no basis can ever be repeated, the number of simplex pivots required to solve any given linear programming problem must be less than or equal to the number of different basic feasible solutions it has [recall from Section 2.3 that an upper bound on this latter number is $n!/m!\,(n - m)!$]. Thus the simplex method must eventually converge, arriving at one of two possible conclusions: After some pivot step either all the $(z_j - c_j)$ are nonnegative, in which case (optimality) Theorem 4.4 guarantees that an optimal solution has been found, or some variable x_j has its $(z_j - c_j)$ negative but all its y_{ij} nonpositive, in which case the problem has an unbounded solution (recall Theorem 4.5).

If we drop the nondegeneracy provision, however, the possibility arises that a sequence of zero-for-zero pivots, each of which causes degeneracy to continue and leaves z unchanged, conceivably could lead to a repetition of some basis. If this occurred, an endless cycle would ensue because a current basis, by determining the next entering and exiting variables, uniquely determines its successor (assuming that the same tie-breaking rules are used throughout). But notice that, insofar as convergence is concerned, degeneracy can be tolerated *so long as no basic feasible solution is ever repeated*, because then the search must be finite. It is possible to guarantee that no basis is ever repeated by performing certain auxiliary calculations at every pivot (the Charnes perturbation method, introduced in [2] and discussed in [15]) or by adopting a complicated lexicographic procedure (see [6] or [24] for details). The simplex method with either of these modifications attached can be proved to converge in a finite number of pivots to the optimal solution of any linear programming problem.

It is an interesting and fortunate fact, however, that *cycling has never occurred in a practical problem!* Degeneracy has persisted for several successive bases, but has never led to a repeated basis (though two or three artificial examples have been constructed to prove that cycling is theoretically possible). Because of this experience, the anticycling precautions mentioned above, which would involve extra computer time and storage, are simply not used. From a statistical point of view, these time-consuming tests are too high a price to pay for insurance against the remote risk of an infinite cycling disaster.

In practice, the simplex method has proved to be quite efficient—so efficient that it has maintained its supremacy through 25 years that have seen startling advances in most other areas of applied mathematics. As an extremely crude rule of thumb, a linear program with m constraints can be expected to require something like $2m$ simplex pivots for solution, the number of variables n being relatively unimportant. This insensitivity to n holds over a surprisingly wide range, which includes most problems arising in the real world; solving a standard-form linear program with m constraints and $3m$ variables is likely to require almost as many pivots as solving one with m constraints and $10m$ variables.

One geometrical note should be added. If a zero-for-zero pivot occurs, with none of the values of the variables being changed, then the old and new bases both correspond to the same extreme point on the convex set of feasible solutions. Thus, in the case of degeneracy one extreme point corresponds to more than one basic feasible solution.

4.10 UNIQUENESS AND MULTIPLE OPTIMA

In most real-world decision problems that are modeled as linear programs, the optimal solution, unless it is unbounded, is *unique*; that is, there exists only a single optimal set of values for the variables. When multiple optima do exist, however, it is generally in the interest of the decision-maker to discover them all. Even though the solutions $\mathbf{x}_1$, $\mathbf{x}_2$, and $\mathbf{x}_3$ all may yield the same optimal value of the objective, there might be marginal, but perfectly sound, reasons for preferring one over the others—reasons that may not have lent themselves to mathematical expression when the problem was being modeled. (Of course, if one of the co-optimal solutions is *strongly* preferred over the others, either the proper values were not assigned to the cost coefficients c_j or a linear programming model should not have been selected in the first place.)

Unfortunately, the simplex method, as we have outlined it in this chapter—and as it is usually coded in computer programs—cannot determine by itself all optimal solutions to a given LPP because it stops as soon as a single optimal extreme point has been discovered. Therefore, in order to obtain multiple optima (or, for that matter, to establish that a single known optimal solution is unique), it is necessary to graft certain supplementary algebraic procedures onto the end of the simplex method. These procedures include both the computations involved in pivoting to alternate optima and the logical testing required to search through all possible pivots and to determine finally that no more optimal solutions remain to be found. The logical details are not particularly edifying, and we shall content ourselves with a brief discussion of the computational aspects; one well-ordered and exhaustive logical scheme for identifying all optima may be found in Chapter 5 of [15].

As far as a decision-maker is concerned, it is really the alternate optimal extreme points, as opposed to alternate basic feasible solutions, that are of interest. These are signaled by the presence of zero-valued reduced costs in the optimal simplex tableau. From the statement of Theorem 4.4 it is clear that the simplex method can lead to an optimal BFS for which some nonbasic variable x_k has a reduced cost $(z_k - c_k)$ equal to zero. In such a case, assuming that $\mathbf{y}_k$ had at least one positive component, it would be possible to perform an ordinary simplex pivot to bring x_k into the basis. Unless the basic variable replaced by x_k had a zero value, the transform formulas show that, while z would remain the same, the values of the basic variables would all change. This would result in an alternate optimal solution, located at a different extreme point.

Moreover, it is easy to see that if two (or more) extreme points are optimal, so is any convex combination of them. Let the extreme points $\mathbf{x}_1$ and $\mathbf{x}_2$ be optimal, so that $\mathbf{c}^T\mathbf{x}_1 = \mathbf{c}^T\mathbf{x}_2 = z^*$, where z^* is the optimal value of the objective function. Since $\mathbf{x}_1$ and $\mathbf{x}_2$ are feasible, so is any point $\mathbf{x}_3 = \lambda\mathbf{x}_1 + (1 - \lambda)\mathbf{x}_2$, with $0 \leq \lambda \leq 1$; this follows from the fact that the feasible region is a convex set. But

$$\mathbf{c}^T\mathbf{x}_3 = \lambda\mathbf{c}^T\mathbf{x}_1 + (1 - \lambda)\mathbf{c}^T\mathbf{x}_2 = \lambda z^* + (1 - \lambda)z^* = z^*$$

Hence $\mathbf{x}_3$ also is an optimum. We have shown that when alternate optimal extreme points exist, there must be an infinite number of different optimal solutions; usually, however, the extreme points are of prime importance.

Having seen how a zero-valued reduced cost can introduce alternate optima, can we conclude that a current solution must be a unique maximum if $(z_j - c_j) > 0$ for every nonbasic variable x_j? The answer depends on what the uniqueness in question refers to: Under the given condition no other set of values for the variables (i.e., no other extreme point) can be optimal, but some other basic feasible solution might be. If the optimal BFS is degenerate, then any nonbasic variable x_j, whatever its $(z_j - c_j)$, can be substituted into the basis for any zero-valued $x_{\mathbf{B}i}$, provided only that y_{ij} is nonzero. The resulting zero-for-zero pivot, as we noted in the previous section, yields another degenerate basic feasible solution corresponding to the same extreme point. To be sure, the new BFS is optimal; however, because all variables have the same values as before, an alternate optimum, at least for decision-making purposes, has not really been introduced.

4.11 MINIMIZATION PROBLEMS

When the LPP is to be *minimized* rather than maximized, it is possible to make explicit use of all the results of this chapter by changing the given problem to a maximization. This is achieved, of course, simply by maximizing $z' = -z = -\mathbf{c}^T\mathbf{x}$ whenever the given objective is to minimize $z = \mathbf{c}^T\mathbf{x}$. The optimal values of the variables are the same under either objective, while the optimal z' obtained by maximizing must be multiplied by -1 to give the correct optimal z. Computer programs convert all linear programming problems to one type or the other.

If it is desired to solve a minimization problem directly, we must return to equation (4-44). We now want decreases in z, so only those columns with positive $(z_j - c_j)$ are candidates for basis entry; the greatest of these is selected. Because nonnegativity of the variables must be preserved, *the simplex exit criterion is precisely the same* as for the maximization problem, and so are all the transformation formulas. Conditions that identify optimal and unbounded solutions are altered in obvious ways.

4.12 ECONOMIC INTERPRETATION OF THE SIMPLEX METHOD

Having treated in some detail the computational aspects of the simplex method, we now offer a rather straightforward explanation of it in economic terms. As before, our purpose in providing this sort of interpretation is to enrich the student's understanding rather than to derive specific results.

Recalling the manufacturing context of Sections 3.1 and 3.9, we can regard any current, but not necessarily optimal, basic feasible solution as a *production policy*, in which the values of the variables x_j give the amounts of each of the outputs to be produced. The fact that a BFS or production policy is *feasible* implies that it actually can be carried out with the inputs available; it will yield a certain amount of sales income, as given by the objective function. Consider now what would happen if we were to alter the current production policy to permit the manufacture of some amount of the kth output, which at present is not scheduled to be produced at all (that is, x_k is not in the basis). Initially, the constraints of the problem are being satisfied via

$$x_{\mathbf{B}1}\mathbf{b}_1 + x_{\mathbf{B}2}\mathbf{b}_2 + \cdots + x_{\mathbf{B}m}\mathbf{b}_m + x_k\mathbf{a}_k = \mathbf{b} \qquad (4\text{-}77)$$

where x_k now equals zero. The unique representation of $\mathbf{a}_k$ as a linear combination of basis vectors is given by

$$\mathbf{a}_k = y_{1k}\mathbf{b}_1 + y_{2k}\mathbf{b}_2 + \cdots + y_{mk}\mathbf{b}_m \qquad (4\text{-}78)$$

If we increase x_k from zero into positive territory, we must make room for it by adjusting the basic variables $x_{\mathbf{B}i}$. The amount of adjustment necessary is discovered by substituting (4-78) into (4-77):

$$(x_{\mathbf{B}1} + x_k y_{1k})\mathbf{b}_1 + \cdots + (x_{\mathbf{B}m} + x_k y_{mk})\mathbf{b}_m = \mathbf{b} \qquad (4\text{-}79)$$

The representation of a given vector $\mathbf{b}$ in terms of a given set of basis vectors $\mathbf{b}_i$ is unique, so these coefficients $(x_{\mathbf{B}i} + x_k y_{ik})$ must continue to have the same values they had under the former production policy; that is,

$$\hat{x}_{\mathbf{B}i} + \hat{x}_k y_{ik} = x_{\mathbf{B}i} + 0 \qquad i = 1, \ldots, m \qquad (4\text{-}80)$$

where hats denote new values, as before. This relationship should look familiar: In Section 4.7 we derived

$$\hat{x}_{\mathbf{B}i} = x_{\mathbf{B}i} - \hat{x}_k y_{ik} \tag{4-55}$$

It is evident that, for every unit of the kth output that the manufacturer proposes to produce, he must produce y_{ik} *fewer* units, algebraically speaking, of the ith basic output (for every i, $i = 1, \ldots, m$). Thus, when $y_{ik} > 0$, these two outputs must be competing for some of the same scarce inputs. We may think of x_k and $x_{\mathbf{B}i}$ as competing for elbow room in the basis: The greater x_k becomes, the more $x_{\mathbf{B}i}$ must decrease in compensation. Of course, it is impossible to produce fewer than zero units of the ith basic output, so $\hat{x}_{\mathbf{B}i}$ must be nonnegative; it follows from (4-55) that

$$\hat{x}_k \leq \frac{x_{\mathbf{B}i}}{y_{ik}} \qquad \text{whenever } y_{ik} > 0 \tag{4-81}$$

A negative y_{ik}, on the other hand, reverses this natural relationship. It implies that x_k and $x_{\mathbf{B}i}$ are somehow complementary—the greater x_k becomes, the more $x_{\mathbf{B}i}$ is also permitted to increase. Such increases do not, of course, threaten to result in negative amounts of the ith basic output in the production policy. Therefore, when deciding by how much x_k can be increased, it is only necessary to examine the basic variables having positive and competitive y_{ik} and to be sure that (4-81) holds for all of them. This will be recognized as the simplex exit criterion.

Recall now the definition

$$z_j = \mathbf{c_B}^T \mathbf{y}_j \tag{4-33}$$

We said that for every unit of the kth output (currently nonbasic) the manufacturer proposes to produce, he must produce y_{ik} fewer units of the ith basic output, implying that his sales income from that output will decrease by $c_{\mathbf{B}i} y_{ik}$ dollars. Summing over the entire basis, the total decrease in income derived from all basic outputs then must be

$$\sum_{i=1}^{m} c_{\mathbf{B}i} y_{ik} = \mathbf{c_B}^T \mathbf{y}_k = z_k \tag{4-82}$$

To compensate for this, each new unit of the kth output itself will yield c_k dollars of income. Therefore, *the overall net loss resulting from the production of one unit of the kth output is* $(z_k - c_k)$, which is commonly known as the *reduced cost* associated with the variable x_k. In a maximization problem, x_k should be considered for entry into the basis only if this net loss is negative, that is, only if the net gain is positive. If none of the nonbasic variables x_j has a negative $(z_j - c_j)$, then none of the nonbasic variables can be introduced profitably into the basis, and we conclude that the current production policy must be optimal. These are precisely the ideas incorporated in the simplex basis entry criterion and in the optimality theorem.

4.13 TWO REMARKS

We recall that, once the variable x_k has been chosen to enter the basis, the nonnegativity requirement automatically determines which variable must leave [except when the minimum ratio found by the exit criterion (4-59) is not unique]. Therefore, when a linear program is being solved via simplex pivoting from basis to basis, freedom of choice is available only in the selection of the entering variable at each step. One way to resolve this choice is to use the simplex entry criterion (4-46): Bring into the basis the variable with the most negative reduced cost $(z_j - c_j)$. In Section 4.6 we remarked that this simple strategy is obvious and reasonable; it requires no auxiliary computations and, as can be seen from (4-44), it has the merit of providing the fastest *rate of increase* of z with respect to increase in the entering variable x_k (in other words, the simplex basis entry criterion selects that x_j that maximizes $\Delta z/\Delta x_j$).

We noted also in Section 4.6 that the simplex strategy would not work out best for every linear program; of course, neither would any other generalized strategy—any reasonable procedure would work well for some problems and poorly for others. Competing methods can be evaluated only from a *statistical* point of view, by measuring average computation times for large numbers of randomly chosen problems. Over the years, after much testing of this sort, the ordinary simplex method with the obvious and reasonable entry criterion (4-46) has established and maintained its supremacy over a wide variety of challengers.

This is not the place for a catalog of the various rules that have been proposed for choosing a sequence of bases.[1] The reader may be interested, however, in one particular variation, which differs from the simplex method of this chapter in only a single respect: Instead of choosing for basis entry the variable that maximizes the *rate* of increase of z, it chooses the one that maximizes the actual *amount* of increase. Recalling the simplex transformation formula

$$\hat{z} = z - \frac{x_{Br}}{y_{rk}}(z_k - c_k) \tag{4-74}$$

or

$$\hat{z} = z - \hat{x}_k(z_k - c_k)$$

the quantity to be maximized at each pivot step is evidently

$$-\hat{x}_k(z_k - c_k)$$

[1] There have been methods, for example, that do not maintain $\mathbf{x_B} \geq \mathbf{0}$, others that change two or more basic variables per pivot, still others that begin by calculating upper bounds for all the variables, and so on.

This requires several applications of the simplex exit criterion; for every non-basic variable x_k having a negative reduced cost it is necessary to determine which basic variable would be replaced by x_k in order to calculate the hypothetical post-pivot value $\hat{x}_k$. Although this variation causes z to increase in larger steps and requires (on the average) slightly fewer pivots, it turns out that all the extra calculation per pivot far outweighs that small advantage and easily tips the scales in favor of the simpler entry criterion (4-46).

We conclude this chapter with a brief geometric picture of unboundedness. From Theorem 4.5, we know that unboundedness occurs when some nonbasic x_k has its $(z_k - c_k) < 0$ and all its $y_{ik} \leq 0$. The negative reduced cost insures that it is profitable overall to increase x_k, but because all $y_{ik} \leq 0$, it is clear from

$$\hat{x}_{\mathbf{B}i} = x_{\mathbf{B}i} - \hat{x}_k y_{ik} \tag{4-55}$$

that none of the $x_{\mathbf{B}i}$ will be driven to zero as x_k increases toward infinity. From the geometric viewpoint, the current basic feasible solution is some extreme point on the convex set of feasible solutions. For every unit of increase in x_k, the value of the ith basic variable is increased by a nonnegative amount $-y_{ik}$. The same is true of all the basic variables, so that each increase of one unit in x_k causes $\mathbf{x_B}$ to increase by $-\mathbf{y}_k$, a nonnegative vector; that is,

$$\mathbf{x_B} = \mathbf{B}^{-1}\mathbf{b} - x_k \mathbf{y}_k \tag{4-54}$$

a result we have seen before. As x_k increases to 1, then 2, and 3, the values of the basic variables increase to $\mathbf{x_B} - \mathbf{y}_k$, $\mathbf{x_B} - 2\mathbf{y}_k$, $\mathbf{x_B} - 3\mathbf{y}_k$, and so on. Therefore the values $-y_{ik}$, in conjunction with a value of 1 associated with x_k and a value of 0 associated with every other nonbasic variable, *define the slope of the extreme ray* in E^n that is anchored at $\mathbf{x_B}$ and extends toward infinity in a direction of increasing z.

EXERCISES

Section 4.2

4-1. Find all extreme points and all basic solutions of the following pair of equations:

$$\begin{aligned} 2x_1 + x_2 - 2x_3 - x_4 &= 4 \\ x_1 + 5x_2 - x_3 + 3x_4 &= 20 \end{aligned}$$

4-2. Prove the converse of Theorem 4.2—that any basic feasible solution of the linear program

$$\begin{aligned} &\text{Max } z = \mathbf{c}^T\mathbf{x} \\ &\text{subject to} \quad \mathbf{Ax} = \mathbf{b} \\ &\text{and} \qquad\qquad \mathbf{x} \geq \mathbf{0} \end{aligned}$$

corresponds to an extreme point of the feasible region. (*Hint:* Assume that a given BFS $\mathbf{x}^*$ lies between two other feasible points $\mathbf{x}_1$ and $\mathbf{x}_2$ and then show that $\mathbf{x}^* = \mathbf{x}_1 = \mathbf{x}_2$.) Why was Theorem 4.2 proved in the text rather than this converse?

4-3. Prove or disprove the following: Given any extreme point $\mathbf{x}^*$ on a convex set X, it is possible to find an objective function whose optimal value within X occurs at $\mathbf{x}^*$.

Section 4.4

4-4. Given the system of linear equations

$$\mathbf{Ax} \equiv \begin{bmatrix} 3 & 2 & 1 & 1 & -2 \\ 1 & 0 & 1 & 0 & 1 \\ -1 & 1 & 0 & -1 & 1 \end{bmatrix} \begin{bmatrix} x_1 \\ x_2 \\ x_3 \\ x_4 \\ x_5 \end{bmatrix} = \begin{bmatrix} 4 \\ 9 \\ 2 \end{bmatrix}$$

verify that $\mathbf{x} = (1, 2, 3, 4, 5)$ is a feasible solution and that

$$3\mathbf{a}_1 + \mathbf{a}_2 - 6\mathbf{a}_3 + \mathbf{a}_4 + 3\mathbf{a}_5 = \mathbf{0}$$

where $\mathbf{a}_j$ is the jth column of $\mathbf{A}$. Use the methods of this section to obtain a non-negative basic solution to these equations. In determining the y values required for the second iteration use $y_5 = 1$ (do not bother renumbering the variables).

Section 4.6

4-5. (a) Suppose we have a basic feasible solution $\mathbf{x_B}$, with associated basis matrix $\mathbf{B}$, to some linear program P:

$$\begin{aligned} &\text{Max } z = \mathbf{c}^T\mathbf{x} \\ &\text{subject to} \quad \mathbf{Ax} = \mathbf{b} \\ &\text{and} \qquad\qquad \mathbf{x} \geq \mathbf{0} \end{aligned}$$

If $\hat{\mathbf{x}}$ is any other feasible solution (not necessarily basic), prove that

$$x_{\mathbf{B}i} = \sum_{j=1}^{n} \hat{x}_j y_{ij} \qquad i = 1, \ldots, m$$

where y_{ij} is the ith component of the vector $\mathbf{y}_j = \mathbf{B}^{-1}\mathbf{a}_j$.

(b) Use part (a) to prove that a basic feasible solution to problem P is optimal if all nonbasic reduced costs $(z_j - c_j)$ are nonnegative.

4-6. Exercise 4-5 constitutes a valid proof of the optimality theorem that does *not* rely on the concept of duality. Use this result, along with Theorem 3.5, to prove the duality theorem.

Section 4.7

4-7. Suppose that a linear program with four constraints is being solved by the simplex method and that the current values of the basic variables are $x_{B1} = 2$, $x_{B2} = 4$, $x_{B3} = 0$, and $x_{B4} = 7$. Given that the variable x_k has just been chosen to enter the basis, identify the exiting variable (if any) and compute the new values of all x_{Bi} when $\mathbf{y}_k$ has the following values:

$$\text{(a) } \mathbf{y}_k = \begin{bmatrix} 2 \\ 3 \\ -3 \\ 4 \end{bmatrix} \qquad \text{(b) } \mathbf{y}_k = \begin{bmatrix} 1 \\ 0 \\ 0 \\ 1 \end{bmatrix} \qquad \text{(c) } \mathbf{y}_k = \begin{bmatrix} 1 \\ 2 \\ 0 \\ 3 \end{bmatrix}$$

$$\text{(d) } \mathbf{y}_k = \begin{bmatrix} 1 \\ 2 \\ 13 \\ -1 \end{bmatrix} \qquad \text{(e) } \mathbf{y}_k = \begin{bmatrix} -1 \\ 0 \\ 0 \\ -2 \end{bmatrix}$$

4-8. Prove that, if it is possible to get from one extreme point to another by a single simplex pivot, then those two extreme points are *adjacent*, that is, connected by an *edge* of the convex set (recall Section 2.5). This can be done by characterizing the two extreme points as $G = (g_1, g_2, \ldots, g_m, 0, 0, \ldots, 0)$ and $H = (0, h_2, \ldots, h_m, h_{m+1}, 0, \ldots, 0)$ and then proving the following sequence of statements:

(a) The hyperplane in E_n

$$x_{m+2} + x_{m+3} + \cdots + x_n = 0$$

passes through G and H and is a bounding hyperplane of the feasible region.

(b) Any other feasible point $\hat{\mathbf{x}} = (x_1, x_2, \ldots, x_m, x_{m+1}, 0, \ldots, 0)$ on this hyperplane must be collinear with G and H.

(c) Furthermore, $\hat{\mathbf{x}}$ must be on the line segment between G and H, which therefore satisfies the definition of an edge.

Section 4.8

4-9. Can a variable that has been removed from the basis reenter on the next pivot? Can a variable that has just entered the basis be removed on the next pivot?

4-10. Suppose that the simplex method is to be used to solve a standard-form linear program having m constraints and n variables.

(a) Given that x_k has been chosen to replace x_{Br} in the current basis and Φ has been calculated, what further computations are involved in determining the new values for the tableau column of the exiting variable?

(b) Multiplications and divisions by $+1$, 0, and -1 are said to be "trivial". Given that x_k has just been chosen to enter the current basis, what is the maximum possible number of nontrivial multiplications and divisions that may be required to fill in the new simplex tableau?

4-11. (a) In solving a linear program by the simplex method, under what circumstances would it be possible for the **y** column of some nonbasic variable x_p to be a unit vector? What would this condition imply about the formulation of the problem? Under what circumstances, if any, could such an x_p enter the basis in some later pivot?

(b) Suppose that the sth row of some current simplex tableau has a 1 in the column of x_q and 0's everywhere else, where x_q is the sth basic variable. Under what circumstances, if any, might x_q leave the basis? Can any of the values in the sth row of the tableau ever change?

Exercises 4-12 through 4-16. (a) Find an initial BFS and use the simplex method to solve each of the following linear programming problems, adding slack or surplus variables and altering the format as desired. (b) Write the dual problem and find its optimal solution (or determine that no feasible solution exists).

4-12. Max $z = 4x_1 + 5x_2$

subject to

$$\begin{aligned} x_1 + 2x_2 &\le 10 \\ -x_1 + 2x_2 &\le 4 \\ 3x_1 - x_2 &\le 9 \end{aligned}$$

and $x_1, x_2 \ge 0$

4-13. Min $z = 3x_1 - x_2 + 2x_3$

subject to

$$\begin{aligned} 4x_1 - 3x_2 + 2x_3 &\le 4 \\ 3x_1 + 2x_2 - 5x_3 &\le 1 \\ -x_1 + x_2 - 3x_3 &\le 0 \end{aligned}$$

and $x_1, x_2, x_3 \ge 0$

4-14. Max $z = x_1$

subject to

$$\begin{aligned} x_1 &\le 3x_2 \\ x_1 &\le 4x_3 \\ 2x_2 + 3x_3 &\le 12 \end{aligned}$$

and $x_1, x_2, x_3 \ge 0$

4-15. Max $z = x_1 + x_3$

subject to

$$\begin{aligned} -2x_1 + x_2 - x_3 &\ge -6 \\ x_1 - x_3 &\ge 2 \\ 4x_1 + x_2 + 3x_3 &\le 20 \end{aligned}$$

and $x_1, x_2, x_3 \ge 0$

4-16. Max $z = 5x_1 + 2x_2 + 3x_3 + 4x_4$

subject to

$$\begin{aligned} 4x_1 + x_2 + x_3 + 2x_4 &= 20 \\ x_1 + x_2 + 2x_3 - x_4 &= 4 \end{aligned}$$

and $x_1, x_2, x_3, x_4 \ge 0$

4-17. Obtain the solution to the following linear program by solving its *dual* via the simplex method:

$$\begin{aligned} \text{Min } z &= 7x_1 + 6x_2 + 5x_3 \\ \text{subject to} \quad x_1 + 2x_2 - x_3 &= 1 \\ x_1 - x_2 + 3x_3 &= 2 \\ \text{and} \quad x_1, x_2, x_3 &\geq 0 \end{aligned}$$

It is necessary to use the transformation (3-19) to produce a set of nonnegative variables.

4-18. Suppose that a linear program is being solved by the simplex method and that the initial basis contained slack variables only (i.e., the initial basis matrix was an identity matrix).

(a) Show that the basis inverse associated with any current BFS can be read directly from the simplex tableau.

(b) Show that the reduced costs of the slack variables *in the optimal tableau* are the optimal values of the dual variables.

Section 4.10

Exercises 4-19 and 4-20. Use the simplex method to solve each problem, obtaining all optimal basic solutions and extreme points.

4-19. $\text{Max } z = 2x_1 + x_2 + 3x_3$

$$\begin{aligned} \text{subject to} \quad 3x_1 - x_2 - 2x_3 &\leq 1 \\ 4x_1 + 2x_2 + 6x_3 &\leq 15 \\ \text{and} \quad x_1, x_2, x_3 &\geq 0 \end{aligned}$$

4-20. $\text{Max } z = x_1 + 2x_2$

$$\begin{aligned} \text{subject to} \quad -x_1 + 2x_2 + x_3 \qquad\qquad &= 2 \\ x_2 \qquad + x_4 \qquad &= 2 \\ x_1 + 2x_2 \qquad\qquad + x_5 &= 6 \\ \text{and} \quad x_i \geq 0 \quad i = 1, \ldots, 5 & \end{aligned}$$

4-21. Tell whether each of the following statements is true or false.

(a) The reduced costs associated with an optimal basic feasible solution of a maximization problem all must be nonnegative.

(b) If an optimal basic feasible solution is nondegenerate and the reduced cost of some nonbasic variable equals zero, then there must be two or more optimal extreme points.

(c) If $\mathbf{x_B}$ is an optimal basic solution and one or more other extreme points are optimal, then at least one of them can be reached by a single pivot from $\mathbf{x_B}$.

Section 4.13

4-22. A pivot in which two or more new variables simultaneously enter the basis is known as a *block pivot*. Discuss the computational aspects of solving a linear program by block-pivoting, say, two variables at a time. Would the two exiting variables automatically be determined by the necessity of maintaining $\mathbf{x_B} \geq \mathbf{0}$? What strategies might be used to choose the entering variables?

4-23. Show how to obtain the slopes of all extreme rays of the convex set

$$\{\mathbf{x} \mid \mathbf{Ax} \leq \mathbf{b}, \quad \mathbf{x} \geq \mathbf{0}\}$$

though not necessarily the extreme points at which they are anchored, by finding the slopes of all extreme rays of the set

$$\{\mathbf{x} \mid \mathbf{Ax} \leq \mathbf{0}, \quad \mathbf{x} \geq \mathbf{0}\}$$

General

4-24. So far we have not learned how to solve linear programs with constraints of the form $\mathbf{Ax} = \mathbf{b}$ when no convenient identity matrix is available to use as an initial basis. Discuss the implications of solving the LPP

$$\begin{aligned} &\text{Min } z' = \mathbf{1}\cdot\mathbf{w} \\ &\text{subject to} \quad \mathbf{Ax} + \mathbf{w} = \mathbf{b} \\ &\text{and} \qquad\qquad \mathbf{x}, \mathbf{w} \geq \mathbf{0} \end{aligned}$$

where $\mathbf{w} = (w_1, \ldots, w_m)$ is a vector of auxiliary variables and $\mathbf{1}$ is a row vector of 1's. What would an optimal value of $z' = 0$ indicate? What if the optimal value of z' were positive?

5

THE TWO-PHASE ALGORITHM AND THE REVISED SIMPLEX METHOD

5.1 DESIRABLE ADJUNCTS TO THE SIMPLEX METHOD

By means of the simplex method developed in Chapter 4 we can now solve any linear programming problem in standard form, provided that redundant constraints have been eliminated and that a basic feasible solution has been identified. Unfortunately, an LPP arising in the real world may not meet these requirements. In particular, unless a basis of slack variables is available, an initial BFS is usually difficult to find. The given problem may even have no feasible solution at all.

In order to be able to solve any standard-form LPP, therefore, we must graft onto the simplex method a preliminary procedure that will identify redundant constraints and find an initial basic feasible solution whenever one exists (in this connection Theorem 4.3 implies that, in the absence of redundancy, a problem having no BFS has no feasible solution at all). These functions are performed by phase 1 of the so-called "two-phase" simplex algorithm [6]. In the first phase, to be discussed shortly, the simplex method is applied to a specially constructed linear program, leading to a final tableau that contains a basic feasible solution to the original problem. Phase 2 then consists of a straightforward application of the simplex method to this initial BFS.

Before going on to describe the two-phase algorithm for solving linear programs, we pause to clear up what may be an ambiguous point. The redundant constraints or constraint sets to be identified in phase 1 are those having the property defined in Section 2.3: A set of k simultaneous linear equations of the form

$$a_{i1}x_1 + a_{i2}x_2 + \cdots + a_{in}x_n = b_i \qquad i = 1, \ldots, k$$

is *redundant* if the row vectors $\mathbf{s}_i \equiv [a_{i1}, a_{i2}, \ldots, a_{in}, b_i]$, $i = 1, \ldots, k$, are linearly dependent, that is, if there exist $\lambda_1, \lambda_2, \ldots, \lambda_k$ not all zero such that

$$\sum_{i=1}^{k} \lambda_i \mathbf{s}_i = \mathbf{0}$$

Any individual equation in the set whose λ_i is nonzero is also said to be redundant. It follows that inequality constraints in an LPP can never be redundant in this sense because each is converted to an equation by the addition of its own slack or surplus variable. Thus, even if an inequality constraint were inadvertently repeated, as in

$$x_1 + x_2 + x_3 \leq 3$$

and

$$x_1 + x_2 + x_3 \leq 3$$

routine addition of slack variables would produce

$$x_1 + x_2 + x_3 + x_4 \qquad = 3$$

and

$$x_1 + x_2 + x_3 \qquad + x_5 = 3$$

which are not redundant. It is worth mentioning that the word "redundant" should not be used to describe the second constraint of the pair

$$x_1 + x_2 + x_3 \leq 3$$

and

$$x_1 + x_2 + x_3 \leq 5$$

To be sure, the second constraint, which is satisfied by any solution satisfying the first, forms no part of the boundary of the feasible region, no matter what the other constraints are. For this reason we may say that it is "superfluous" or "inoperative," but not that it is "redundant."

5.2 PHASE 1: THE INITIAL BASIC FEASIBLE SOLUTION

We introduce this important section with a brief overview of the material to be covered here. The first phase in solving a linear programming problem in standard form begins with the addition of an auxiliary nonnegative variable to the left-hand side of each constraint. The columns associated with these *artificial variables* are unit vectors, and they can be combined to form an initial identity basis for what is now a new set of constraints. Phase 1 then consists of using the simplex method to minimize the sum of the artificial variables, subject to the new constraints. Pivoting continues either until that sum is driven to zero, in which case a basic feasible solution using only the original variables has been obtained, or until the optimal value of the phase-1 objective is shown to be positive, in which case no feasible solution to the original LPP exists. If artificial variables of value zero remain in the basis after the sum has been driven to zero and cannot be removed by subsequent simplex pivots, then the constraints of the original LPP are redundant, as we shall see.

The mathematical development of these ideas begins with a linear programming problem stated in standard form, which we shall call problem P:

$$\text{Max } z = \mathbf{c}^T\mathbf{x} \tag{5-1}$$

$$\text{subject to} \quad \mathbf{A}\mathbf{x} = \mathbf{b} \tag{5-2}$$

$$\text{and} \quad \mathbf{x} \geq \mathbf{0} \tag{5-3}$$

$$\text{with} \quad \mathbf{b} \geq \mathbf{0} \tag{5-4}$$

The matrix $\mathbf{A}$ is taken to be $m \times n$. We do not lose generality by requiring that the right-hand side of the constraint equation be nonnegative because an equality constraint whose b_i is negative can simply be multiplied by -1. Therefore, any LPP in standard form can easily be fitted to the format of problem P.

Given P, we form the *auxiliary problem* Q:

$$\text{Max } z' = -\mathbf{1}\cdot\mathbf{w} = -\sum_{i=1}^{m} w_i \tag{5-5}$$

$$\text{subject to} \quad \mathbf{A}\mathbf{x} + \mathbf{I}_m\mathbf{w} = \mathbf{b} \tag{5-6}$$

$$\text{and} \quad \mathbf{x}, \mathbf{w} \geq \mathbf{0} \tag{5-7}$$

where $\mathbf{I}_m$ is an $m \times m$ identity matrix, $\mathbf{w}$ is an m-component column vector of *artificial variables*—in this context the n members of $\mathbf{x}$ are called the *real variables*—and $\mathbf{1}$ is an m-component row vector consisting entirely of 1's. The columns of $\mathbf{A}$ and of $\mathbf{I}_m$ are called *real* and *artificial columns*, respectively.

Problem Q is a maximization linear program with m constraints and $n + m$ variables (of course, the equivalent objective

$$\text{Min } z' = \mathbf{1} \cdot \mathbf{w}$$

also could have been used). We now observe that the following set of values constitutes a basic feasible solution to problem Q:

$$\mathbf{w} = \mathbf{b} \qquad \mathbf{x} = \mathbf{0} \tag{5-8}$$

This solution is clearly feasible, because it satisfies the constraints (5-6) and because $\mathbf{b}$ is nonnegative. Moreover, it has at most m nonzero variables, namely, the w_i, whose associated constraint columns form an identity matrix. Therefore, the solution (5-8) is also basic and must be a BFS. Using the notation of Chapter 4, the values of the basic variables are given by

$$\mathbf{x_B} \equiv \mathbf{w} = \mathbf{B}^{-1}\mathbf{b} = \mathbf{I}_m^{-1}\mathbf{b} = \mathbf{b} \tag{5-9}$$

Phase 1 of the two-phase algorithm consists of using the simplex method to solve problem Q, beginning with the BFS (5-9). The initial tableau is easily prepared. The basis inverse is an identity matrix, so the values

$$\mathbf{y}_j = \mathbf{a}_j \qquad j = 1, \ldots, n + m \tag{5-10}$$

may be entered in the tableau columns directly; however,

$$\mathbf{c_B}^T = -\mathbf{1} \tag{5-11}$$

so a bit of arithmetic is required for

$$z' = \mathbf{c_B}^T\mathbf{x_B} = \mathbf{c_B}^T\mathbf{b} = -\sum_{i=1}^{m} b_i \tag{5-12}$$

and

$$z_j = \mathbf{c_B}^T\mathbf{y}_j = \mathbf{c_B}^T\mathbf{a}_j = -\sum_{i=1}^{m} a_{ij} \qquad j = 1, \ldots, n + m \tag{5-13}$$

The reduced costs for nonbasic columns are just z_j because for them $c_j = 0$.

Pivoting now can proceed in the usual manner. Because of the nonnegativity of $\mathbf{w}$, it is clear that z', which begins with a negative value, can never become positive. It can attain a value of zero, however, if all the artificial variables can be driven to zero. Thus the application of the simplex method to problem Q will lead to one of three mutually exclusive, collectively exhaustive results:

(1) *The optimal z′ is negative.* This must mean that problem Q has no feasible solution with $\mathbf{w} = \mathbf{0}$; that is, no set of nonnegative values $\mathbf{x}_0$ can be found such that

$$\mathbf{A}\mathbf{x}_0 + \mathbf{I}_m \cdot \mathbf{0} \equiv \mathbf{A}\mathbf{x}_0 = \mathbf{b} \tag{5-14}$$

It follows that *the original linear program P has no feasible solution.* This may be because the given constraints (5-2) are inconsistent or because the only solutions they permit fail to satisfy the nonnegativity restriction (5-3).

(2) *The optimal z′ is zero and no artificial variables are in the basis.* In this case the optimal basis $\mathbf{B}$ for problem Q consists of m linearly independent columns from the $\mathbf{A}$ matrix, so that $\mathbf{B}$ must also be a basis for problem P. Because the values of the basic variables associated with $\mathbf{B}$, which are given by $\mathbf{x_B} = \mathbf{B}^{-1}\mathbf{b}$, constitute an optimal solution to problem Q, they must be nonnegative as well. Therefore, $\mathbf{B}$ *and* $\mathbf{x_B}$ *comprise a legitimate basis matrix and BFS for problem P*; degeneracy may or may not be present. Moreover, we recall from property 6 of Section 2.2 that because $\mathbf{B}$ has an inverse, its rows, considered as vectors, must be linearly independent. Therefore, so must the rows of $\mathbf{A}$, and it follows from the definition stated in Section 2.3 that *none of the constraints of problem P are redundant.*

Phase 1 has now been completed and we enter phase 2, in which the simplex method is used to solve the given problem P, beginning with the initial basic feasible solution discovered in phase 1. Two sets of alterations are required to transform the optimal simplex tableau of problem Q into an initial tableau for P. First, because the artificial variables w_i do not appear in P and thus will not be candidates for basis entry, we simply delete the tableau columns associated with them. This has no effect on future pivots or future basic solutions, nor does it invalidate the condition for optimality. Second, because the objective function of P is entirely different from that of Q, the basic cost vector $\mathbf{c_B}$ changes, and the values of the reduced costs $(z_j - c_j)$ and the objective z must be recalculated. Note, however, that no alterations at all are required for the vectors $\mathbf{y}_j = \mathbf{B}^{-1}\mathbf{a}_j$ because our conversion from problem Q to problem P changes neither the columns $\mathbf{a}_j$ nor the basis inverse $\mathbf{B}^{-1}$. The same is true of the values of the basic variables $\mathbf{x_B} = \mathbf{B}^{-1}\mathbf{b}$.

Having prepared the initial tableau for phase 2, we now examine the newly calculated values of the reduced costs. If all of them are nonnegative, the current BFS is already the optimal solution to problem P; if not, we apply the simplex method until we find it.

(3) *The optimal z′ is zero and at least one artificial variable is in the basis,* necessarily having the value zero. In this case we have found a nonnegative solution to problem Q

$$\mathbf{x} = \mathbf{x}_0 \quad \text{and} \quad \mathbf{w} = \mathbf{0} \tag{5-15}$$

such that (5-14) holds; therefore, $\mathbf{x}_0$ is a feasible solution to problem P, although not yet a basic feasible solution. We are still left with the question of redundant constraints and with the problem of finding a BFS.

Without shifting to phase 2, we continue performing simplex pivots, starting with the phase-1 optimal tableau, in an effort to expel all artificial variables from the basis. To remove some artificial variable $w_s = 0$ from the rth position in the basis (that is, $x_{\mathbf{B}r} = w_s = 0$) and replace it with a real variable it is only necessary to find some nonbasic column $\mathbf{a}_k$, associated with the real variable x_k, such that $y_{rk} \neq 0$. Then a simplex pivot can be performed in which x_k enters the basis and $x_{\mathbf{B}r} = w_s$ exits. As can be seen from the tableau transformation formulas of Chapter 4, this zero-for-zero pivot changes none of the values of the basic variables, although the $\mathbf{y}_j$ and the $(z_j - c_j)$ must be updated. If the new basis still contains an artificial variable (which would still be of zero value), we try to repeat the process. If we succeed in pivoting out all the artificial variables, then the problem belongs in the domain of case (2), discussed earlier. The final basic feasible solution consists entirely of real variables, so it must be a legitimate, though degenerate, BFS for problem P.

But suppose there are exactly q zero-valued artificial variables that we cannot remove from the basis (except by pivoting in other artificial variables to replace them). This implies that $y_{ij} = 0$ for each of these artificial $x_{\mathbf{B}i}$ and for every nonbasic real column $\mathbf{a}_j$. But, recalling from (4-32) that

$$\mathbf{a}_j = \sum_{i=1}^{m} y_{ij}\mathbf{b}_i$$

every nonbasic $\mathbf{a}_j$ then must be expressible as a linear combination of the $m - q$ basic columns that are real—and so must every basic $\mathbf{a}_j$ as well (in this case the linear combination is trivial: $\mathbf{a}_j = 1 \cdot \mathbf{a}_j$). Therefore, any set of m columns of the $m \times n$ matrix $\mathbf{A}$ must be linearly dependent,[1] and it follows from Theorem 2.2 that the rows of $\mathbf{A}$ are linearly dependent. Because we have found a feasible solution to problem P, inconsistency is ruled out, and *the constraints* (5-2) *of problem P must be redundant.*

[1] Perhaps we should prove this statement, on behalf of those readers whose linear algebra is a bit rusty. Choose any m columns of the $\mathbf{A}$ matrix, numbering them $\mathbf{a}_1, \ldots, \mathbf{a}_m$ for convenience, and assume that they are linearly independent, which means that they span E^m. We are given that each of them can be expressed as some linear combination of the real basis vectors (again numbering for convenience) $\mathbf{b}_1, \ldots, \mathbf{b}_{m-q}$, where $0 < (m - q) < m$. We remark that some of these basis vectors may also be members of the set $\mathbf{a}_1, \ldots, \mathbf{a}_m$. Consider the collection of vectors $\{\mathbf{a}_1, \ldots, \mathbf{a}_m, \mathbf{b}_1, \ldots, \mathbf{b}_{m-q}\}$, which we call set T; by our assumption, T spans E^m. But $\mathbf{a}_1$ is a linear combination of $\mathbf{b}_1, \ldots, \mathbf{b}_{m-q}$; hence we can remove $\mathbf{a}_1$ from T and it will span E^m. By the same argument, every one of the $\mathbf{a}_j$ can be removed, reducing T to $\{\mathbf{b}_1, \ldots, \mathbf{b}_{m-q}\}$, and T must still span E^m. This is impossible because no fewer than m vectors can span E^m (property 5 of Section 2.2); therefore our assumption was false, and any set of m columns of $\mathbf{A}$ must be linearly dependent.

5.3 THE REDUNDANT CONSTRAINTS

The following theorem identifies the specific constraints of P that are redundant. It is not essential to the development of this chapter, and the student may prefer merely to note the result and skip the lengthy, rather difficult proof.

THEOREM 5.1. If the optimal value of the phase-1 objective is zero, but the artificial column $\mathbf{e}_s$ remains in the basis and cannot be replaced by any real column in a single simplex pivot, then the sth constraint of problem P is redundant.

Recall from (5-6) that the sth artificial column is a unit vector $\mathbf{e}_s$, consisting of all zeros except for a single 1 in the sth position; repeating a familiar warning, the artificial variable w_s associated with this column is not necessarily $x_{\mathbf{B}s}$. The proof begins with the hypothesis of the theorem that some artificial variable cannot be removed from the basis. As concluded before, the constraints $\mathbf{Ax} = \mathbf{b}$ must be redundant; that is, if $\mathbf{r}_i$ represents the ith row of $\mathbf{A}$, there must exist λ_i not all zero such that

$$\sum_{i=1}^{m} \lambda_i[\mathbf{r}_i, b_i] = \mathbf{0} \qquad (5\text{-}16)$$

or, using only the first n components,

$$\sum_{i=1}^{m} \lambda_i a_{ij} = 0 \qquad j = 1, \ldots, n \qquad (5\text{-}17)$$

Suppose that it has been impossible to remove one or more artificial columns $\mathbf{e}_s$, $s \in S$, from the optimal phase-1 basis $\mathbf{B}$; here S is the set of subscripts of the artificial variables (or columns) in the basis. Assume now that $\lambda_s = 0$ in (5-16) for all $s \in S$. Because only the sth element of the unit vector $\mathbf{e}_s$ is different from zero, this assumption directly implies that

$$\sum_{i=1}^{m} \lambda_i e_{is} = 0 \qquad \text{for every } s \in S \qquad (5\text{-}18)$$

where e_{is} is the ith component of $\mathbf{e}_s$. Every column of the optimal basis $\mathbf{B}$ is either one of the $\mathbf{a}_j$, $j = 1, \ldots, n$, or one of the $\mathbf{e}_s$, s in S. Therefore it follows from (5-17) and (5-18) that there exists a linear combination (with scalar coefficients λ_i) of the rows of $\mathbf{B}$ that adds up to $\mathbf{0}$. This contradicts the definition of a basis, so our assumption must have been false; at least one of the λ_s, say, $\lambda_{\hat{s}}$, is nonzero in (5-16). The corresponding $\mathbf{r}_{\hat{s}}$ is then expressible as a linear combination of the other rows. Because we have found a feasible solution to problem P, it follows that the *$\hat{s}$th constraint is redundant in the set of equations* $\mathbf{Ax} = \mathbf{b}$, where $\hat{s} \in S$.

If:		then:	
	$\mu_1(b_{11}, b_{12}, \quad b_{14}, b_{15})$		$\mu_1(b_{11}, b_{12}, 0, b_{14}, b_{15})$
	$+\mu_2(b_{21}, b_{22}, \quad b_{24}, b_{25})$		$+\mu_2(b_{21}, b_{22}, 0, b_{24}, b_{25})$
	$+\mu_3(b_{31}, b_{32}, \quad b_{34}, b_{35})$		$+\mu_3(b_{31}, b_{32}, 0, b_{34}, b_{35})$
			$+0\ (b_{41}, b_{42}, 1, b_{44}, b_{45})$
	$+\mu_5(b_{51}, b_{52}, \quad b_{54}, b_{55})$		$+\mu_5(b_{51}, b_{52}, 0, b_{54}, b_{55})$
	$= (0, \ 0, \quad 0, \ 0\ )$		$= (0, \ 0, \ 0, 0, \ 0\ )$

Figure 5-1

If S contains only one member, the proof of Theorem 5.1 is finished. If not, let us remove from $\mathbf{B}$ both the redundant $\hat{s}$th row and the artificial column $\mathbf{e}_{\hat{s}}$. We now claim that the rows of the resulting $(m - 1) \times (m - 1)$ matrix $\mathbf{B}'$ are linearly independent. If this were false, some linear combination of them with coefficients μ_i not all zero would add up to $\mathbf{0}_{m-1}$, the $(m - 1)$-component null vector. Then, however, a linear combination of the rows of the original basis $\mathbf{B}$, with scalar coefficients consisting of these μ_i plus $\mu_{\hat{s}} = 0$, would add to $\mathbf{0}_m$, as the student can verify from the 5×5 example diagram of Figure 5-1. Such linear dependence is impossible, because $\mathbf{B}$ is a basis. Therefore, $\mathbf{B}'$, whose rows are linearly independent, is an invertible matrix. It is composed of $m - 1$ columns of the linear programming problem Q', which is formed by deleting the $\hat{s}$th constraint from problem Q; therefore $\mathbf{B}'$ *is a basis for problem* Q'.

Recall now that the optimal phase-1 solution $\mathbf{B}\mathbf{x}_\mathbf{B} = \mathbf{b}$ may be represented as

$$x_{\mathbf{B}1}\mathbf{b}_1 + x_{\mathbf{B}2}\mathbf{b}_2 + \cdots + x_{\mathbf{B},m-1}\mathbf{b}_{m-1} + 0\cdot\mathbf{e}_{\hat{s}} = \mathbf{b} \tag{5-19}$$

where we have reordered the columns $\mathbf{b}_i$ of $\mathbf{B}$ to place the artificial column $\mathbf{e}_{\hat{s}}$ last. The artificial variable $x_{\mathbf{B}m} = w_{\hat{s}}$ has a value of zero. Because vector equations imply equality for all components, we may simply remove the $\hat{s}$th row of (5-19). If we also drop the last term $(0\cdot\mathbf{e}_{\hat{s}})$, the remaining $(m - 1)$ equalities are not disturbed and we can write

$$x_{\mathbf{B}1}\mathbf{b}_1' + x_{\mathbf{B}2}\mathbf{b}_2' + \cdots + x_{\mathbf{B},m-1}\mathbf{b}_{m-1}' = \mathbf{b}' \tag{5-20}$$

where $\mathbf{b}'$ is the right-hand side of the constraint equation of problem Q' and $\mathbf{b}_i$ is the ith column of the basis $\mathbf{B}'$. Therefore, the basic variables of problem Q' associated with the basis $\mathbf{B}'$ have the same values as they had in the optimal solution to problem Q, except that one of them, $w_{\hat{s}} = 0$, has been omitted.

A similar argument shows that the new $\mathbf{y}'$ vectors for the real columns $\mathbf{a}_j'$ are the same as before, with the omission of a single zero component. The $(z_j - c_j)$ and z are unchanged. Therefore, by merely deleting the row of the artificial basic variable $w_{\hat{s}}$ from the optimal phase-1 tableau, we produce a new tableau that corresponds to an optimal solution to the problem Q'. Since the set S contained more than one member, other artificial variables $x'_{\mathbf{B}i} = 0$ are still in the basis $\mathbf{B}'$. The elements y'_{ij} still equal zero for each of these artificial $x'_{\mathbf{B}i}$ and for every nonbasic real column $\mathbf{a}'_j$, so the various $\mathbf{a}'_j$ all must be linear combinations of the fewer than $(m - 1)$ real columns in B'. Therefore, the constraints are redundant in the new problem P', which is formed by deleting the $\hat{s}$th constraint from P. We repeat the above steps, continuing until every constraint corresponding to some $\mathbf{e}_s$, $s \in S$, has been shown to be redundant. This completes the proof.

Theorem 5.1 does *not* imply that if the sth constraint of problem P is redundant, then the artificial column $\mathbf{e}_s$ must appear in the optimal phase-1 basis. Remember that a group G of constraints

$$a_{i1}x_1 + a_{i2}x_2 + \cdots + a_{in}x_n = b_i \qquad i \in G$$

is said to be redundant if there exist scalar coefficients λ_i not all zero such that

$$\sum_{i \in G} \lambda_i \mathbf{s}_i = \mathbf{0}$$

where $\mathbf{s}_i \equiv [a_{i1}, a_{i2}, \ldots, a_{in}, b_i]$. We have also defined as redundant any individual constraint whose associated λ_i in this linear combination is nonzero. In general, however, a group G of redundant constraints may lead to only a single artificial variable $w_{\hat{\imath}}$, $\hat{\imath}$ in G, in the optimal phase-1 basis, where $w_{\hat{\imath}}$ can be removed *only* by pivoting in another artificial variable w_i associated with one of the other redundant constraints in G. To identify all the "essential" members of G (i.e., those having nonzero λ_i in the linear combination above) we need only examine the row of the basic variable $w_{\hat{\imath}}$ in the optimal phase-1 tableau: Any nonbasic artificial variable with a nonzero entry in that row could be brought in to replace $w_{\hat{\imath}}$ in the basis and therefore must be associated with a member of the redundant group G. These redundancy phenomena will be illustrated in the example of Section 5.5.

5.4 ARTIFICIAL VARIABLES IN PHASE 2

Recall that we were originally given the linear programming problem P:

$$\text{Max } z = \mathbf{c}^T\mathbf{x} \tag{5-1}$$

$$\text{subject to} \quad \mathbf{Ax} = \mathbf{b} \tag{5-2}$$

$$\text{and} \quad \mathbf{x} \geq \mathbf{0} \tag{5-3}$$

$$\text{with} \quad \mathbf{b} \geq \mathbf{0} \tag{5-4}$$

We formed the auxiliary problem Q:

$$\text{Max } z' = -\mathbf{1}\cdot\mathbf{w}$$
$$\text{subject to} \quad \mathbf{Ax} + \mathbf{I}_m\mathbf{w} = \mathbf{b}$$
$$\text{and} \quad \mathbf{x}, \mathbf{w} \geq \mathbf{0}$$

In phase 1, problem Q was solved via the simplex method. Now we are considering the case in which the optimal solution to Q, with $z' = 0$, includes some zero-valued artificial variables that cannot be removed from the basis. Let S be the set of subscripts of the artificial basic variables; that is, s is in S if and only if w_s is in the basis. We now observe that the current phase-1 optimum would be a valid BFS for problem P if only P were slightly altered so as to retain in its formulation the artificial variables w_s, $s \in S$. Suppose, then, that we define a modified problem P^* that differs from P only in that, for every w_s appearing in the optimal phase-1 basis, the sth constraint of P^* is

$$a_{s1}x_1 + a_{s2}x_2 + \cdots + a_{sm}x_m + w_s = b_s \tag{5-21}$$

rather than

$$a_{s1}x_1 + a_{s2}x_2 + \cdots + a_{sm}x_m = b_s \tag{5-22}$$

The artificial variables can be thought of as having zero costs in problem P^*.

Obviously, if $\hat{\mathbf{x}}$ is a feasible solution to P, then $(\mathbf{x},\mathbf{w}_{P^*}) = (\hat{\mathbf{x}},\mathbf{0})$, where $\mathbf{w}_{P^*}$ includes only the artificial variables appearing in problem P^*, must be a feasible solution to P^*. The objective values are $z = \mathbf{c}^T\hat{\mathbf{x}}$ in both cases. It follows that *if the optimal solution to P^* happens to have every $w_s = 0$, then its $\mathbf{x}$ component is an optimal solution to P.* (This is easily proved by assuming that the conclusion is false and deducing a contradiction.)

By inventing problem P^* we can move directly from phase 1 to phase 2. Phase 1 ends, to repeat, with a tableau containing the optimal basic feasible solution to problem Q, associated with the optimal basis $\mathbf{B}$; this BFS includes one or more artificial variables w_s that have zero values. Because Q and P^* have the same number of constraints and because every column of $\mathbf{B}$ appears in the constraint set of P^*, the matrix $\mathbf{B}$ is also a basis for P^*.

It now becomes clear that the optimal tableau from phase 1 is converted into the initial phase-2 tableau exactly as was described for the "no-artificial-variables-in-the-basis" case [case (2)] of Section 5.2. First, the nonbasic artificial columns are deleted because their variables do not appear in problem P^*. Because $\mathbf{B}^{-1}$, $\mathbf{b}$, and the columns $\mathbf{a}_j$ and $\mathbf{e}_s$ (associated with the basic variables w_s) are the same in problem P^* as in Q, the tableau entries for $\mathbf{x_B} = \mathbf{B}^{-1}\mathbf{b}$ and $\mathbf{y}_j = \mathbf{B}^{-1}\mathbf{a}_j$ do not have to be changed. The objective of P^*, however, is to maximize $z = \mathbf{c}^T\mathbf{x}$, so $\mathbf{c_B}$ changes as we enter phase 2. It is therefore necessary to recalculate $z = \mathbf{c_B}^T\mathbf{x_B}$ and all $(z_j - c_j) = \mathbf{c_B}^T\mathbf{y}_j - c_j$.

The simplex method is now applied to this initial tableau in order to find the optimal solution to problem P^*. Because $\mathbf{x_B}$ and the $\mathbf{y}_j$ were not altered in the conversion to phase 2, the value of each artificial basic variable $x_{\mathbf{B}i} = w_s$ begins at zero, and its associated elements y_{ij} in all columns $\mathbf{a}_j$ straight across the tableau are also zero (otherwise w_s would have been driven out of the phase-1 optimal basis). Suppose some real variable x_k is selected to enter the basis. All the nonzero elements in the column $\mathbf{y}_k$ are associated with real basic variables, so the simplex exit criterion must choose some real variable $x_{\mathbf{B}r}$ to leave the basis. Consulting the transformation formulas in Chapter 4, we find that the new value $\hat{x}_{\mathbf{B}i}$ of each *artificial* basic variable is the same as the old, namely, zero, because $y_{ik} = 0$. Similarly, its $\hat{y}_{ij}$ for all columns $\mathbf{a}_j$ straight across the tableau have the same zero values as before. It follows that, during a straightforward application of the simplex method to problem P^*, *the artificial variables will never leave the basis and will continue to have zero values.* The optimal solution to P^* will have every $w_s = 0$, so its $\mathbf{x}$ component will constitute an optimal solution to P, as was pointed out earlier in this section.

Incidentally, having proved that the artificial variables never leave the basis and that their tableau rows—the $x_{\mathbf{B}i}$ and all the y_{ij}—consist only of zeros, we need not carry these rows throughout phase 2. They may simply be deleted as the initial phase-2 tableau is formed.

5.5 SUMMARY

The reader may have realized already that it is not always necessary to add m artificial variables in phase 1. The goal is to form an identity matrix that can be used as the initial basis in phase 1. If the ith constraint includes a slack variable, then there is no need to add to it an artificial variable as well—the column $\mathbf{e}_j$ is already present in $\mathbf{A}$. From a theoretical standpoint, it does not matter whether one, two, or m artificial variables are used. The important issue remains whether or not they all can be driven to zero.

Let us summarize the two-phase simplex algorithm for solving the linear programming problem P:

(1) If the matrix $\mathbf{A}$ contains a submatrix $\mathbf{I}_m$, it will serve as an initial basis. There are no redundant constraints and an initial basic feasible solution has been found; proceed to step (4). If $\mathbf{A}$ does not contain an identity submatrix, produce one by adding a sufficient number of nonnegative artificial variables w_i.

(2) *Phase 1.* Use the simplex method to minimize the sum of all the artificial variables. If the minimum value of this sum is positive—or, equivalently, if the optimal value of the phase-1 objective function

$$\text{Max } z' = -\sum w_i$$

is negative—then P has *no feasible solution* and no further work remains to be done. But if the minimum value of the sum is zero, prolong phase 1 (if necessary) in an effort to drive all artificial variables out of the basis.

(3) If all artificial variables can be expelled, then a (possibly degenerate) BFS to problem P has been found. On the other hand, for every artificial variable w_s that cannot be removed from the basis, the sth constraint of problem P is *redundant*. In either case, so long as phase 1 ends with a zero-valued objective, the final tableau is converted into the initial tableau of phase 2 simply by deleting the nonbasic artificial columns and recalculating z and the $(z_j - c_j)$ in accordance with the original objective function (5-1).

(4) *Phase 2.* Apply the simplex method to the modified tableau until the optimal solution to problem P is obtained. Note that an artificial variable that could not be driven out of the basis in phase 1 will remain in the basis with a zero value throughout phase 2.

The two-phase simplex algorithm for solving a linear programming problem is illustrated by the following example.

$$\begin{aligned}
&\text{Max } z = x_1 + x_2 - 2x_3 \\
&\text{subject to} \quad x_1 + x_2 + x_3 = 12 \\
&\qquad 2x_1 + 5x_2 - 6x_3 = 10 \\
&\qquad 7x_1 + 10x_2 - x_3 = 70 \\
&\text{and} \qquad x_i \geq 0 \qquad i = 1, 2, 3
\end{aligned}$$

The constraints may be represented, as usual, by $\mathbf{Ax} = \mathbf{b}$, where $\mathbf{A}$ is 3×3. If the rows of $\mathbf{A}$ are linearly independent, which is not immediately obvious, then only one feasible solution $\mathbf{x} = \mathbf{A}^{-1}\mathbf{b}$ exists, and the objective function becomes superfluous. This points up the fact that sets of simultaneous linear equations can be solved by the simplex method, provided that the variables x_j can be assumed to be nonnegative—and if the x_j cannot, presumably the x_j' can, where

$$x_j' = x_j + 100{,}000$$

Of course, this simplex approach to simultaneous linear equations would be less efficient than finding $\mathbf{A}^{-1}$ by more direct numerical methods.

Returning to the example, we form problem Q:

$$\begin{aligned}
&\text{Max } z' = -w_1 - w_2 - w_3 \\
&\text{subject to } \quad x_1 + x_2 + x_3 + w_1 = 12 \\
&\qquad 2x_1 + 5x_2 - 6x_3 + w_2 = 10 \\
&\qquad 7x_1 + 10x_2 - x_3 + w_3 = 70 \\
&\text{and} \qquad x_i, w_i \geq 0 \qquad \text{for all } i
\end{aligned}$$

If any right-hand-side element b_i had been negative, the ith constraint would have been multiplied through by -1. Phase 1 begins with the BFS consisting of the three artificial variables; the initial basis matrix is $\mathbf{I}_3$. After calculating z' and the $(z_j - c_j)$, as described in Section 5.2, we fill in the tableau (dropping the basis column labels from the far left side):

	x_1	x_2	x_3	w_1	w_2	w_3
$x_{\mathbf{B}1} = w_1 = 12$	1	1	1	1	0	0
$x_{\mathbf{B}2} = w_2 = 10$	2	5	-6	0	1	0
$x_{\mathbf{B}3} = w_3 = 70$	7	10	-1	0	0	1
Max $z' = -92$	-10	-16	$+6$	0	0	0

The entry criterion selects x_2; because

$$\text{Min}\,(^{12}/_1, {}^{10}/_5, {}^{70}/_{10}) = {}^{10}/_5$$

the exit criterion selects w_2. The pivot element is $y_{22} = 5$, and

$$\mathbf{\Phi} = (-0.2, -0.8, -2, 3.2)$$

so the new tableau is

	x_1	x_2	x_3	w_1	w_2	w_3
$x_{\mathbf{B}1} = w_1 = 10$	0.6	0	2.2	1	-0.2	0
$x_{\mathbf{B}2} = x_2 = 2$	0.4	1	-1.2	0	0.2	0
$x_{\mathbf{B}3} = w_3 = 50$	3.0	0	11.0	0	-2.0	1
Max $z' = -60$	-3.6	0	-13.2	0	3.2	0

Now x_3 enters the basis, but either w_1 or w_3 can exit because $10/2.2 = 50/11$. This forewarns us that the next BFS will be degenerate. To break the tie we arbitrarily remove w_1. The pivot element is $y_{13} = 2.2$ and

$$\mathbf{\Phi} = (-6/11, 6/11, -5, 6)$$

so the next tableau is

	x_1	x_2	x_3	w_1	w_2	w_3
$x_{\mathbf{B}1} = x_3 = 50/11$	$3/11$	0	1	$5/11$	$-1/11$	0
$x_{\mathbf{B}2} = x_2 = 82/11$	$8/11$	1	0	$6/11$	$1/11$	0
$x_{\mathbf{B}3} = w_3 = 0$	0	0	0	-5	-1	1
Max $z' = 0$	0	0	0	$+6$	2	0

Phase 1 has ended by finding a feasible solution to the given problem, but the optimal basis contains an artificial variable $x_{\mathbf{B}3} = w_3 = 0$. For every nonbasic real column $\mathbf{a}_j$ (namely, $\mathbf{a}_1$) the corresponding $y_{3j} = 0$, so w_3 cannot be replaced in the basis by any real variable. We conclude therefore that the original constraints are redundant: In particular, Theorem 5.1 assures us that the third constraint is redundant. The final tableau shows, however, that either w_1 or w_2 could replace w_3 in the basis via a simple zero-for-zero pivot, so the first and second constraints are redundant as well. (Of course, this does not imply that all three constraints are superfluous—only that λ_1, λ_2, and λ_3, *all nonzero*, can be found to satisfy

$$\sum_{i=1}^{3} \lambda_i \mathbf{r}_i = \mathbf{0}$$

where $\mathbf{r}_i$ are the rows of our matrix $\mathbf{A}$. Because only one artificial variable remains in the optimal phase-1 basis, only one of the constraints can be said to be superfluous—when any one is deleted, the remaining two are no longer redundant.)

Moving now into phase 2, the objective function is

$$\text{Max } z = x_1 + x_2 - 2x_3$$

Thus $\mathbf{c_B} = (-2, 1, 0)$, where we give the artificial variable a zero cost; actually, the cost is irrelevant because it will be multiplied only by zeros in calculating z and the $(z_j - c_j)$. The necessary computations are

$$z = \mathbf{c_B}^T\mathbf{x_B} = -18/11$$

and
$$z_1 - c_1 = \mathbf{c_B}^T\mathbf{y}_1 - c_1 = 2/11 - 1 = -9/11$$

The reduced costs of the other real columns, which are in the basis, must be zero. The nonbasic artificial columns are dropped, and the initial phase-2 tableau has been constructed:

		x_1	x_2	x_3
$x_{\mathbf{B}1} = x_3 =$	$50/11$	$3/11$	0	1
$x_{\mathbf{B}2} = x_2 =$	$82/11$	$8/11$	1	0
$x_{\mathbf{B}3} = w_3 =$	0	0	0	0
Max $z =$	$-18/11$	$-9/11$	0	0

Here x_1 enters, x_2 exits, the pivot element is $y_{21} = 8/11$, and

$$\mathbf{\Phi} = (-3/8, 3/8, 0, 9/8)$$

and we arrive at the following:

		x_1	x_2	x_3
$x_{\mathbf{B}1} = x_3 =$	$7/4$	0	$-3/8$	1
$x_{\mathbf{B}2} = x_1 =$	$41/4$	1	$11/8$	0
$x_{\mathbf{B}3} = w_3 =$	0	0	0	0
Max $z =$	$27/4$	0	$9/8$	0

All the reduced costs are nonnegative, so this is a maximal solution to the given problem. Moreover, it is unique, because there is no nonbasic column with $z_j - c_j = 0$. Notice, finally, that the row of zeros associated with the artificial basic variable did not have to be carried along in the phase-2 tableaux.

5.6 FURTHER REMARKS ON THE TWO-PHASE METHOD

In the past five sections the theoretical development of the two-phase simplex algorithm has been rigorously presented. In practice a modification of this procedure that tends to shorten computation time is employed. This adjustment concerns the postoptimal pivots of phase 1, which remove as many as possible of the artificial variables from the basis *after* the objective function has been driven to zero. These pivots have various and unpredictable effects on the objective function of the given problem P: Some may raise its value, others may lower it. In the long run we would expect to "break even" on these pivots, so that their net result is to consume computation time without, on the average, moving us closer to the optimum.

It would be desirable, therefore, from the standpoint of computational efficiency, to move directly into phase 2 as soon as the phase 1 objective is driven to zero, even though the basis might contain one or more replaceable artificial variables. To adopt this approach heedlessly, however, without making any changes in the simplex pivoting strategy obviously would introduce a serious difficulty. Whenever the initial phase-2 tableau contained an artificial basic variable $x_{\mathbf{B}i}$ whose y_{ij} were *not* all zero for every nonbasic column $\mathbf{a}_j$, the possibility would exist that the artificial variable might change in value—that is, become positive—during some subsequent pivot. The resulting BFS then would be an infeasible solution to problem P, and there would be no guarantee that feasibility would ever be restored before the "optimum" was reached.

Fortunately, we can eliminate this possibility and insure that after the early termination of phase 1 all artificial variables will remain permanently at their proper values of zero, by employing in phase 2 a slightly modified version of the simplex method. First, no artificial variable can ever be permitted to enter the basis during phase 2. This restriction is quite reasonable: If such a variable were to enter at a positive level it would cause infeasibility, and if it entered with a value of zero it would not improve the objective anyway. Our prohibition against artificial variables is an example of *restricted basis entry*, a general concept that appears in various guises in several different applications of linear programming. It follows that artificial columns may be dropped from the tableau as soon as they become nonbasic, just as they were deleted in the conversion from phase 1 to phase 2.

A second modification of the simplex method for use during phase 2 involves the exit criterion. Suppose that some nonbasic real variable x_k having a negative reduced cost has been chosen to enter the basis. If $y_{ik} \neq 0$ for any artificial variable $x_{\mathbf{B}i}$, then $x_{\mathbf{B}i}$ should be removed from the basis immediately in a zero-for-zero pivot; such a pivot changes none of the values of the basic variables and thus preserves feasibility. If $y_{ik} = 0$ for all artificial $x_{\mathbf{B}i}$, then a real basic variable will necessarily be chosen to exit. Under these circumstances, however, the new values of the artificial $x_{\mathbf{B}i}$ will still be zero, as can be verified from the transformation formula

$$\hat{x}_{\mathbf{B}i} = x_{\mathbf{B}i} - \frac{x_{\mathbf{B}r}}{y_{rk}} y_{ik} \qquad i = 1, \ldots, r-1, r+1, \ldots, m \tag{4-61}$$

Therefore, given that x_k has been selected to enter the basis, the new basic solution will again be feasible provided that the exiting variable is chosen according to the following rules:

(1) If $y_{ik} \neq 0$ for some artificial $x_{\mathbf{B}i}$, then that $x_{\mathbf{B}i}$ leaves the basis (and its column is dropped from the tableau).

(2) If $y_{ik} = 0$ for all artificial $x_{\mathbf{B}i}$, then the usual simplex exit criterion (4-59) is applied, resulting in a real basic variable being chosen to exit.

The simplex method with these modifications will produce in phase 2 a sequence of basic feasible solutions with nondecreasing values of z. No artificial variables will ever have nonzero values; they will, in fact, be disappearing from the problem. When the removal of the artificial variables from the basis was tacked onto the end of phase 1, as described in Section 5.2, the real variables that replaced them were selected in a completely arbitrary manner. By transplanting this process into phase 2 we are able to choose entering variables with favorable reduced costs for our given problem. It must be more promising, in the long run, to get these real variables into the basis, even though they are initially of zero value.

Our modified pivoting procedure is continued, unless the given problem is discovered to have an unbounded solution, until as many of the artificial variables as possible have been removed from the basis *and* the reduced costs of all nonbasic (real) variables are nonnegative. These are precisely the conditions that obtain at the end of the two-phase algorithm as developed rigorously in Section 5.4. The artificial variables that cannot be removed from the optimal basis represent redundant constraints, and in effect, we have optimized the associated problem P^*. Therefore, the $\mathbf{x}$ component of the solution we have found optimizes the original problem P.

5.7 A ONE-PHASE METHOD FOR SOLVING LINEAR PROGRAMS

The two-phase simplex algorithm, as presented above, is credited to Dantzig, Orden, and Wolfe [6], who were all working at the Rand Corporation. An equally well-known procedure for solving LPPs is the so-called "penalties method" of Charnes [2]. Given the linear programming problem P,

$$\text{Max } z = \mathbf{c}^T\mathbf{x} \tag{5-1}$$

$$\text{subject to} \quad \mathbf{A}\mathbf{x} = \mathbf{b} \tag{5-2}$$

$$\text{and} \quad \mathbf{x} \geq \mathbf{0} \tag{5-3}$$

$$\text{with} \quad \mathbf{b} \geq \mathbf{0} \tag{5-4}$$

where $\mathbf{A}$ is an $m \times n$ matrix, the penalties method calls for the formation of an associated problem R:

$$\text{Max } z' = \mathbf{c}^T\mathbf{x} - \mathbf{M}^T\mathbf{w} \tag{5-23}$$

$$\text{subject to} \quad \mathbf{A}\mathbf{x} + \mathbf{I}_m\mathbf{w} = \mathbf{b} \tag{5-24}$$

$$\text{and} \quad \mathbf{x}, \mathbf{w} \geq 0 \tag{5-25}$$

where $\mathbf{w}$ is an m-component vector of artificial variables and $\mathbf{M}$ is an m-component vector containing extremely large positive numbers.

As in the two-phase method, the artificial variables are introduced in order to provide an initial basic feasible solution. The purpose of the huge cost coefficients $-\mathbf{M}$ is to insure that all artificial variables will be driven to zero when problem R is solved via the simplex method. Thus we would like each *penalty cost* component $-M$ to be so unfavorably negative that no basic solution that includes a positive artificial variable can possibly be optimal. For hand computation it is easiest to enter each of these cost coefficients simply as $-M$ and to consider any real number negligible in comparison with any fraction of M. Thus, if the reduced costs of three nonbasic variables are $0.01M - 50$, $-5M + 20$, and $-4M - 20$, the first is taken to be positive, so that its variable is not a candidate for basis entry, while the second is considered "better" (i.e., more negative) than the third. In computer algorithms, however, it is necessary to be specific, and a reasonable value to use for M might be about 10^4 times the largest of the real cost coefficients c_j.

Having constructed problem R, we go on to solve it completely in one phase. Now, for any feasible solution $\hat{\mathbf{x}}$ to problem P it is clear that $(\mathbf{x},\mathbf{w}) = (\hat{\mathbf{x}},\mathbf{0})$ is a feasible solution to R with the same objective value $z = \mathbf{c}^T\hat{\mathbf{x}}$. Therefore, if the optimal solution to R happens to have all artificial variables equal to zero, then its $\mathbf{x}$ component must be an optimal solution to P; this is easily proved by contradiction. On the other hand, because we have "loaded" the cost vector, we may assume that if the optimal solution to problem R has one or more artificial variables with positive values, then problem P has no feasible solution.

Despite the fact that this penalties method solves linear programming problems in one phase rather than two, it is seldom used in computer-solution algorithms. Although it has the minor disadvantage of requiring more time to discover that a problem is infeasible—this is revealed in phase 1 of the two-phase method—its major shortcoming is mechanical. Suppose that problem R had a feasible extreme point in which one of the artificial variables had a value of 0.001. Then, if M were set equal to, say, 5000, this artificial variable would contribute only -5 to the objective, and the extreme point might turn out to be the optimal solution to R in spite of the fact that problem P had feasible solutions. And if M were increased to eliminate this possibility, then a digital computer—even using extended precision—might not retain enough accuracy to allow the $(z_j - c_j)$ to be updated via the usual $\mathbf{\Phi}$-vector method (which would involve the addition and subtraction of very large quantities). Thus the extra labor of recomputing the reduced costs from the definition,

$$z_j - c_j = \mathbf{c_B}^T\mathbf{y}_j - c_j$$

would probably be required at each iteration or every few iterations.

These digital objections apply only to computer algorithms, however. The penalties method, using the symbol M to represent an unspecified, but sufficiently large, number, is certainly a reasonable approach for solution of LPPs by hand; for any given problem it may prove superior to the two-phase method. Exercise 5-14 asks the student to show that, barring ties, the sequence of variables entering and leaving the basis in phase 1 is exactly the same as in the early iterations of Charnes' method of penalties.

5.8 THE REVISED SIMPLEX METHOD

Most of the computational drudgery in the linear programs we have solved so far has been due to the calculation of all the $\mathbf{y}_j$ vectors at every iteration. When the variable x_k is selected for basis entry, the y_{ik} are required for application of the exit criterion, but the y_{ij} for all other columns merely wait to be updated by the pivot. By keeping all the $\mathbf{y}_j$ up to date we are assured of having available the one we need at each pivot.

Remembering that $\mathbf{y}_j = \mathbf{B}^{-1}\mathbf{a}_j$, by definition, suppose we were able to keep track of the basis inverse $\mathbf{B}^{-1}$. At each iteration we could form $\mathbf{c_B}$ from the cost coefficients of the basic variables and compute the row vector $\mathbf{c_B}^T\mathbf{B}^{-1}$. The reduced cost for each nonbasic variable x_j then could be calculated with a single vector multiplication and a subtraction:

$$z_j - c_j = (\mathbf{c_B}^T\mathbf{B}^{-1})\mathbf{a}_j - c_j \tag{5-26}$$

If all of these were nonnegative the current basic solution would be optimal; if not, the entry criterion would choose some variable x_k. We could then compute

$$\mathbf{y}_k = \mathbf{B}^{-1}\mathbf{a}_k \tag{5-27}$$

and $\mathbf{x_B} = \mathbf{B}^{-1}\mathbf{b}$ and apply the exit criterion. Provided that the new basis inverse could be found, we could repeat these steps as many times as necessary, calculating only one $\mathbf{y}$ column at each iteration.

These are precisely the steps executed at each iteration of the *revised simplex method* [21], which, like the *standard simplex method* described in Chapter 4, is an iterative computational algorithm for solving a standard-form linear program, starting with a known basic feasible solution. Both methods work by pivoting from one BFS to another, using the same basis entry and exit criteria; therefore, for any given problem the two methods generate exactly the same sequence of pivots. The only differences between them lie in the computations that are performed at each iteration:

(1) Whereas in a standard simplex pivot the reduced costs are updated directly by means of a transformation formula, in a revised pivot the basis inverse is updated directly and the reduced costs are then calculated via (5-26).

(2) Whereas all $\mathbf{y}$ columns are updated in every standard simplex pivot, only the $\mathbf{y}$ column of the entering variable is computed in a revised pivot.

By now it may have occurred to those students who solved Exercise 4-18(a) that they already know how to update the basis inverse! Consider a linear program for which an identity matrix is available as an initial basic feasible solution, with the initial basic variables being $x_\alpha, x_\beta, \ldots$ (number the variables so that $\alpha, \beta, \ldots$ are consecutive integers). If we begin to solve the problem by the simplex method, in any tableau the $\mathbf{y}$ columns associated with those variables together form the following submatrix:

$$\begin{aligned}[\mathbf{y}_\alpha \mid \mathbf{y}_\beta \mid \cdots] &\equiv [\mathbf{B}^{-1}\mathbf{a}_\alpha \mid \mathbf{B}^{-1}\mathbf{a}_\beta \mid \cdots] \\ &= \mathbf{B}^{-1}[\mathbf{a}_\alpha \mid \mathbf{a}_\beta \mid \cdots] \\ &= \mathbf{B}^{-1}\mathbf{I} = \mathbf{B}^{-1}\end{aligned} \tag{5-28}$$

Evidently these columns must always contain the current basis inverse, and it is clear that the columns of the basis inverse are being updated by the very same $\mathbf{\Phi}$ transformations that update the entire simplex tableau. This justifies the following more general result.

LEMMA 5.1. Whenever the rth column of any invertible $m \times m$ matrix $\mathbf{B}$ is replaced by a column $\mathbf{a}_k$, the columns $\hat{\boldsymbol{\beta}}_j$ of the new matrix inverse can be calculated directly from the columns $\boldsymbol{\beta}_j$ of the old matrix inverse via

$$\hat{\boldsymbol{\beta}}_j = \boldsymbol{\beta}_j + \beta_{rj}\mathbf{\Phi}' \tag{5-29}$$

where β_{rj} is the rth element of $\boldsymbol{\beta}_j$, $\mathbf{y}_k \equiv (y_{1k}, \ldots, y_{mk}) \equiv \mathbf{B}^{-1}\mathbf{a}_k$, and

$$\mathbf{\Phi}' \equiv \left(-\frac{y_{1k}}{y_{rk}}, \ldots, -\frac{y_{r-1,k}}{y_{rk}}, \frac{1}{y_{rk}} - 1, -\frac{y_{r+1,k}}{y_{rk}}, \ldots, -\frac{y_{mk}}{y_{rk}}\right) \tag{5-30}$$

Given any simplex pivot, Lemma 5.1 describes the general procedure for deriving the new basis inverse from the old. In effect, it requires the application of the usual simplex transformations *only* to those tableau columns that correspond to the variables of the original basis (and, of course, only to their $\mathbf{y}_j$, not to their $z_j - c_j$). It is evident from this description that the revised simplex method is not really a new solution technique, but rather an improvement in bookkeeping.

Actually, a modification is available that permits the automatic updating of $\mathbf{c_B}^T\mathbf{B}^{-1}$ as well as $\mathbf{B}^{-1}$. Suppose that an LPP is being solved by the revised simplex method and that, at some stage, $\mathbf{x_B}$ and $\mathbf{B}$ are the current basic feasible solution and basis matrix. Assume that $\mathbf{x_B}$, $\mathbf{B}$, the objective value z, and the basis inverse $\mathbf{B}^{-1}$ all are known and that $\mathbf{c_B}^T\mathbf{B}^{-1}$ has been computed. The current solution uniquely determines an auxiliary matrix

$$\mathbf{B}^* \equiv \begin{bmatrix} \mathbf{B} & \mathbf{0} \\ -\mathbf{c_B}^T & 1 \end{bmatrix} \tag{5-31}$$

Because $\mathbf{B}$ is a basis and therefore has an inverse, we may write

$$(\mathbf{B}^*)^{-1} = \begin{bmatrix} \mathbf{B}^{-1} & \mathbf{0} \\ \mathbf{c_B}^T\mathbf{B}^{-1} & 1 \end{bmatrix} \tag{5-32}$$

The reader may check that $\mathbf{B}^*(\mathbf{B}^*)^{-1} = \mathbf{I}_{m+1}$, the identity matrix of order $(m + 1)$. By assumption, all elements of $(\mathbf{B}^*)^{-1}$ are known.

Knowing $\mathbf{c_B}^T\mathbf{B}^{-1}$, we can compute the reduced cost for every nonbasic variable according to (5-26). Suppose the $(z_j - c_j)$ are scanned and x_k is chosen to enter the basis on the next iteration. The vector $\mathbf{y}_k \equiv \mathbf{B}^{-1}\mathbf{a}_k$ can now be calculated and used by the exit criterion, in conjunction with $\mathbf{x_B}$, to select, say, the rth basic variable $x_{\mathbf{B}r}$ to pivot out. The auxiliary matrix (5-31) associated with the new basic solution will differ from the old $\mathbf{B}^*$ only in that its rth column will be $\mathbf{a}_k^* \equiv (\mathbf{a}_k, -c_k)$ rather than $\mathbf{b}_r^* \equiv (\mathbf{b}_r, -c_{\mathbf{B}r})$. It then follows immediately from Lemma 5.1 that the columns of the new inverse (5-32) may be calculated via $\boldsymbol{\Phi}$ transformations. First, the $\mathbf{y}$ column of the lemma is given by

$$\mathbf{y}_k^* \equiv (\mathbf{B}^*)^{-1}\mathbf{a}_k^* = \begin{bmatrix} \mathbf{B}^{-1} & \mathbf{0} \\ \mathbf{c_B}^T\mathbf{B}^{-1} & 1 \end{bmatrix} \begin{bmatrix} \mathbf{a}_k \\ -c_k \end{bmatrix} = \begin{bmatrix} \mathbf{y}_k \\ z_k - c_k \end{bmatrix} \tag{5-33}$$

[notice in passing that the last row of $(\mathbf{B}^*)^{-1}$ offers a convenient way of calculating the reduced costs of the nonbasic variables in order to determine which of them is to enter the basis]. From $\mathbf{y}_k^*$ we then can form

$$\boldsymbol{\Phi} = \left(-\frac{y_{1k}}{y_{rk}}, \ldots, -\frac{y_{r-1,k}}{y_{rk}}, \frac{1}{y_{rk}} - 1, -\frac{y_{r+1,k}}{y_{rk}}, \ldots, -\frac{y_{mk}}{y_{rk}}, -\frac{(z_k - c_k)}{y_{rk}} \right) \tag{4-75}$$

which is the familiar computational vector of Chapter 4. We now make use of (4-75) to transform the columns of $(\mathbf{B}^*)^{-1}$ in the usual way, yielding the new basis inverse and new value of the vector $\mathbf{c_B}^T\mathbf{B}^{-1}$. Because $\boldsymbol{\Phi}$ is available we also use it to update $\mathbf{x_B}$ and z exactly as in the standard simplex method:

$$\begin{bmatrix} \hat{\mathbf{x}}_\mathbf{B} \\ \hat{z} \end{bmatrix} = \begin{bmatrix} \mathbf{x_B} \\ z \end{bmatrix} + x_{\mathbf{B}r}\boldsymbol{\Phi} \tag{5-34}$$

where hats denote the new postpivot values. These relationships were proved in Chapter 4 [recall (4-60), (4-61), and (4-74)]. Thus, having transformed

$(\mathbf{B}^*)^{-1}$, $\mathbf{x_B}$, and z, we have completed the current pivot and obtained all the information we need to begin the next.

5.9 SUMMARY

Let us outline the steps involved in solving a standard-form linear program by the revised simplex method, assuming that all redundant constraints have been eliminated:

(1) Begin with a basic feasible solution $\mathbf{x_B}$ whose basis inverse is known, and form the matrix $(\mathbf{B}^*)^{-1}$.
(2) For every nonbasic column $\mathbf{a}_j$ multiply the last row of $(\mathbf{B}^*)^{-1}$ by the column $\mathbf{a}_j^* \equiv (\mathbf{a}_j, -c_j)$ to obtain the reduced cost $(z_j - c_j)$. If all reduced costs are nonnegative, then the current solution is optimal; if not, proceed to step (3).
(3) Select the column $\mathbf{a}_k$ with the most negative reduced cost to enter the basis. Perform the multiplication (5-33) to determine $\mathbf{y}_k$ and apply the simplex exit criterion. Identify the variable $x_{\mathbf{B}r}$ that is to leave the basis (unless unboundedness is discovered) and form the $\mathbf{\Phi}$ vector (4-75).
(4) Use this computational vector to transform the columns of $(\mathbf{B}^*)^{-1}$ and to derive the new values of the basic variables and the objective function. Return to step (2).

The reader should note that thus far he has learned four different ways of solving a linear programming problem for which no initial basic feasible solution is available: He may elect either the two-phase or the one-phase strategy, and in each case he may use either standard or revised simplex pivots. To illustrate the two-phase revised simplex approach we shall solve the following example.

$$\begin{aligned}
\text{Maximize } z &= x_1 + 2x_2 + 3x_3 \\
\text{subject to } \quad x_1 + x_2 + x_3 &= 2 \\
3x_1 - 2x_2 + 2x_3 &\le 3 \\
x_1 \qquad + 2x_3 &\ge 3 \\
\text{and} \qquad x_1, x_2, x_3 &\ge 0
\end{aligned}$$

Converting to standard form, the constraints become

$$\begin{aligned}
x_1 + x_2 + x_3 \qquad\qquad &= 2 \\
3x_1 - 2x_2 + 2x_3 + x_4 \qquad &= 3 \\
\text{and} \qquad x_1 \qquad + 2x_3 \qquad - x_5 &= 3
\end{aligned}$$

We must add artificial variables to the first and third constraints in order to produce an initial identity basis for phase 1. Problem Q is then

$$\begin{aligned}
&\text{Maximize } z' = -w_1 - w_3 \\
&\text{subject to} \quad x_1 + x_2 + x_3 + w_1 = 2 \\
&\qquad 3x_1 - 2x_2 + 2x_3 + x_4 = 3 \\
&\qquad x_1 + 2x_3 - x_5 + w_3 = 3 \\
&\text{and} \qquad \text{all } x_i \text{ and } w_i \geq 0
\end{aligned}$$

The initial basic variables are $\mathbf{x_B} = (w_1, x_4, w_3)$. Note that we could not have simply multiplied the third constraint by -1 and used x_5 in the initial basis: Its value would have been -3, which is infeasible.

We now calculate

$$(\mathbf{B}^*)^{-1} = \begin{bmatrix} \mathbf{B}^{-1} & \mathbf{0} \\ \mathbf{c_B}^T\mathbf{B}^{-1} & 1 \end{bmatrix} = \begin{bmatrix} 1 & 0 & 0 & 0 \\ 0 & 1 & 0 & 0 \\ 0 & 0 & 1 & 0 \\ -1 & 0 & -1 & 1 \end{bmatrix} \quad \text{and} \quad \begin{bmatrix} w_1 \\ x_4 \\ w_3 \\ z' \end{bmatrix} = \begin{bmatrix} 2 \\ 3 \\ 3 \\ -5 \end{bmatrix}$$

The reduced costs for nonbasic columns $\mathbf{a}_1$, $\mathbf{a}_2$, $\mathbf{a}_3$, and $\mathbf{a}_5$ are, respectively, -2, -1, -3, and $+1$; these are obtained by multiplying the last row of $(\mathbf{B}^*)^{-1}$ by the various column vectors $\mathbf{a}_j^* = (\mathbf{a}_j, -c_j)$, where the c_j are, of course, the phase-1 costs. The simplex entry criterion selects column $\mathbf{a}_3$ to enter the basis, so

$$\mathbf{y}_3^* = (\mathbf{B}^*)^{-1}\mathbf{a}_3^* = \begin{bmatrix} 1 & 0 & 0 & 0 \\ 0 & 1 & 0 & 0 \\ 0 & 0 & 1 & 0 \\ -1 & 0 & -1 & 1 \end{bmatrix} \begin{bmatrix} 1 \\ 2 \\ 2 \\ 0 \end{bmatrix} = \begin{bmatrix} 1 \\ 2 \\ 2 \\ -3 \end{bmatrix} = \begin{bmatrix} \left.\begin{matrix} \\ \\ \\ \end{matrix}\right\} \mathbf{y}_3 \\ z_3 - c_3 \end{bmatrix}$$

Because $\mathbf{x_B}$ is currently (2, 3, 3), the simplex exit criterion permits the removal of either x_4 or w_3. We choose $x_{\mathbf{B}3} = w_3$ to leave the basis. Applying (4-75), $\mathbf{\Phi} = (-0.5, -1.0, -0.5, 1.5)$, and after the transformations the new values are

$$(\mathbf{B}^*)^{-1} = \begin{bmatrix} 1 & 0 & -0.5 & 0 \\ 0 & 1 & -1.0 & 0 \\ 0 & 0 & 0.5 & 0 \\ -1 & 0 & 0.5 & 1 \end{bmatrix} \quad \text{and} \quad \begin{bmatrix} w_1 \\ x_4 \\ x_3 \\ z' \end{bmatrix} = \begin{bmatrix} 0.5 \\ 0 \\ 1.5 \\ -0.5 \end{bmatrix}$$

This time the reduced costs for $\mathbf{a}_1$, $\mathbf{a}_2$, $\mathbf{a}_5$, and $\mathbf{e}_3$ (the column associated with w_3) are -0.5, -1.0, -0.5, and 1.5. Therefore, $\mathbf{a}_2$ enters the basis,

$$\mathbf{y}_2^* = (\mathbf{B}^*)^{-1}\mathbf{a}_2^* = \begin{bmatrix} 1 \\ -2 \\ 0 \\ -1 \end{bmatrix} = \begin{bmatrix} \left.\begin{matrix} \\ \\ \end{matrix}\right\} \mathbf{y}_2 \\ z_2 - c_2 \end{bmatrix}$$

and $x_{\mathbf{B}1} = w_1$ exits. It follows that $\mathbf{\Phi} = (0, 2, 0, 1)$ and the new values are

$$(\mathbf{B}^*)^{-1} = \begin{bmatrix} 1 & 0 & -0.5 & 0 \\ 2 & 1 & -2.0 & 0 \\ 0 & 0 & 0.5 & 0 \\ 0 & 0 & 0 & 1 \end{bmatrix} \quad \text{and} \quad \begin{bmatrix} x_2 \\ x_4 \\ x_3 \\ z' \end{bmatrix} = \begin{bmatrix} 0.5 \\ 1.0 \\ 1.5 \\ 0 \end{bmatrix}$$

Thus phase 1 terminates with the discovery of a basic feasible solution to the original problem; none of the constraints are redundant.

For phase 2 we return to the first step in our summary of the revised simplex method, the formation of $(\mathbf{B}^*)^{-1}$. Using the current BFS and the original cost coefficients, we may calculate

$$z = \mathbf{c_B}^T\mathbf{x_B} = [2, 0, 3]\cdot(0.5, 1, 1.5) = 5.5$$

and

$$\mathbf{c_B}^T\mathbf{B}^{-1} = [2, 0, 3]\begin{bmatrix} 1 & 0 & -0.5 \\ 2 & 1 & -2.0 \\ 0 & 0 & 0.5 \end{bmatrix} = [2, 0, 0.5]$$

so that phase 2 begins with

$$(\mathbf{B}^*)^{-1} = \begin{bmatrix} 1 & 0 & -0.5 & 0 \\ 2 & 1 & -2 & 0 \\ 0 & 0 & 0.5 & 0 \\ 2 & 0 & 0.5 & 1 \end{bmatrix} \quad \text{and} \quad \begin{bmatrix} x_2 \\ x_4 \\ x_3 \\ z \end{bmatrix} = \begin{bmatrix} 0.5 \\ 1 \\ 1.5 \\ 5.5 \end{bmatrix}$$

The reduced costs for the nonbasic columns $\mathbf{a}_1^* = (1, 3, 1, -1)$ and $\mathbf{a}_5^* = (0, 0, -1, 0)$ are 1.5 and -0.5, so x_5 enters the basis. We have

$$\mathbf{y}_5^* = (\mathbf{B}^*)^{-1}\mathbf{a}_5^* = \begin{bmatrix} 0.5 \\ 2.0 \\ -0.5 \\ -0.5 \end{bmatrix} = \begin{bmatrix} \left.\begin{matrix} \\ \\ \end{matrix}\right\} \mathbf{y}_5 \\ z_5 - c_5 \end{bmatrix}$$

and the second basic variable x_4 must exit. Thus $\mathbf{\Phi} = (-0.25, -0.5, 0.25, 0.25)$ and

$$(\mathbf{B}^*)^{-1} = \begin{bmatrix} 0.5 & -0.25 & 0 & 0 \\ 1.0 & 0.5 & -1 & 0 \\ 0.5 & 0.25 & 0 & 0 \\ 2.5 & 0.25 & 0 & 1 \end{bmatrix} \quad \text{and} \quad \begin{bmatrix} x_2 \\ x_5 \\ x_3 \\ z \end{bmatrix} = \begin{bmatrix} 0.25 \\ 0.5 \\ 1.75 \\ 5.75 \end{bmatrix}$$

We now find that the reduced costs for all nonbasic variables are positive, so the current BFS is the unique optimal solution to the given problem.

5.10 REVISED VERSUS STANDARD SIMPLEX

In Section 5.8 we noted that, although the revised and standard simplex methods generate exactly the same sequence of basic solutions in solving any given linear program, there are significant computational differences between them. It would be interesting to examine these differences further and to compare the two methods from the standpoint of problem-solving efficiency. One obvious way to do this is to count the total number of arithmetic operations required by each in the course of a single pivot; *operation counts* of this sort are frequently used in applied mathematics to measure the relative efficiencies of competing algorithms. Assuming that the calculations are to be performed on a digital computer, the operation count normally includes only multiplications and divisions, which are almost an order of magnitude slower than additions and subtractions.

Let us first consider a single iteration of the standard simplex method for a linear program having m constraints and n variables. After some new variable x_k has been chosen to enter the basis, an average of perhaps $\frac{1}{2}m$ divisions are needed to determine the exiting variable (that is, about half of the y_{ik} will be positive, requiring the quotient $x_{\mathbf{B}i}/y_{ik}$ to be evaluated). Formation of the $\mathbf{\Phi}$ vector than entails $m + 1$ divisions, after which a total of $n - m + 1$ tableau columns must be updated—the original n columns less the m basic columns whose tableau entries are simply unit vectors plus the column containing the values of the $x_{\mathbf{B}i}$ and z. Because $m + 1$ multiplications are required to update each tableau column, the total number of operations per pivot is

$$\tfrac{1}{2}m + (m + 1) + (m + 1)(n - m + 1) = mn - m^2 + \tfrac{3}{2}m + n + 2$$

When the revised simplex method is used, on the other hand, the entering variable cannot be chosen until the reduced costs have been obtained for all of the $n - m$ nonbasic columns; each reduced-cost calculation requires m nontrivial multiplications, as can be seen from (5-33). We then need an additional m^2 multiplications to calculate $\mathbf{y}_k^* = (\mathbf{B}^*)^{-1}\mathbf{a}_k^*$, followed by $\frac{1}{2}m$ divisions to determine the exiting variable, then $m + 1$ divisions to form the $\mathbf{\Phi}$ vector, then $m(m + 1)$ multiplications to transform $(\mathbf{B}^*)^{-1}$, and finally $m + 1$ multiplications to transform $(\mathbf{x_B},z)$, for a grand total of

$$m(n - m) + m^2 + \tfrac{1}{2}m + (m + 1) + m(m + 1) + (m + 1)$$
$$= mn + m^2 + \tfrac{7}{2}m + 2$$

The revised simplex method thus requires about $2m^2 + 2m - n$ more operations per pivot than the standard simplex method. For "square" problems, where m and n are nearly equal, this represents a considerable advantage for the standard approach; thus, when $m = 10$ and $n = 15$, a revised pivot requires 287 operations, compared with only 82 for a standard pivot! The difference between the two methods diminishes as the number of variables increases relative to the number of constraints. Thus, when $m = 10$ and $n = 30$, the standard pivot requires 247 operations and the revised pivot 437. These figures might be considered to reflect a typical real-world linear program; problems for which the revised method actually takes *fewer* operations are extremely rare.

The revised simplex method, however, has other advantages that more than offset this handicap in the long run (although certainly not for every problem). One is its bookkeeping advantage: When a linear program is solved via the revised method, only the m basic columns must be updated at each iteration, whereas a standard simplex pivot transforms all n columns of the tableau. This may be a marginal consideration for hand computations, but in solving large problems by computer each tableau must be written on a disk or a magnetic tape, and the necessary accessing of these peripheral devices is relatively time-consuming.

Another important advantage of the revised method stems from the fact that it calculates all reduced costs and every entering $\mathbf{y}$ column from start to finish via

$$z_j - c_j = (\mathbf{c_B}^T\mathbf{B}^{-1})\mathbf{a}_j - c_j \tag{5-26}$$

and

$$\mathbf{y}_k^* = (\mathbf{B}^*)^{-1}\mathbf{a}_k^* \tag{5-33}$$

These calculations amount to $m(n - m) + m^2 = mn$ multiplications per pivot, which is usually well over half of the total number of operations required. Whenever the $\mathbf{A}$ matrix contains a high proportion of zeros—and linear programs in which 95% of the elements a_{ij} are zero are not uncommon—a great many of the multiplications in (5-26) and (5-33) are trivial and take virtually no time. When the standard simplex method is used, however, all computations involve the $\mathbf{y}$ columns. Although $\mathbf{y}_j = \mathbf{a}_j$ for all j initially, after a few pivot steps many of the zeros have disappeared, causing a much higher percentage of nontrivial multiplications.

This preservation of zeros in the revised simplex method also has the

Table 5-1 IBM 1130 Computation Times in Hours[1]

Number of constraints	Number of variables	Solution time
50	75	0.1–0.5
100	150	0.3–1.3
200	250	0.7–3.0
300	400	2.0–7.0
500	600	6.5–16.5
1000	1200	24–48
1500	2000	50–100

[1] Data is from IBM Manual H20-0562-1, "Linear Programming System/1130, Application Description Manual," June 1969. The times given are for an 1130 computer having 16,000 words of primary memory, a 2.2-microsecond access time, and two disk drives.

advantageous side effect of reducing round-off error. After a tableau column has been subjected to dozens of standard simplex pivots, the round-off error accumulated over all those successive multiplications and divisions becomes quite significant. But because the zero elements persist through every iteration of the revised method, thereby limiting the number of nontrivial multiplications, the round-off error accumulates more slowly and greater accuracy is maintained.

For all these reasons, most computer codes use the revised simplex method, applying it in both phases of the two-phase algorithm. The various codes do differ in several small computational details, tending to be tailored in accordance with the hardware capabilities of the various machines. Somewhat surprisingly, the time required by any one of these codes to solve a given linear programming problem on a given digital computer cannot be predicted with much accuracy, even knowing the dimensions of the **A** matrix *and* the fraction of nonzero elements it contains (a crucial parameter). Two problems that are approximately identical in structure can easily differ in solution time by a factor of 2 or more. Table 5-1 gives the ranges of computation times required to solve linear programs of several different shapes and sizes on an IBM 1130 computer. In these problems the *density* of the **A** matrix (fraction of nonzero a_{ij}) ranges from 1% up to around 10%. In each case, if the user can supply a starting basic feasible solution, the optimum can be obtained in about one-third of the time shown.

It is interesting to compare the solution times given in Table 5-1 with those obtainable on faster and larger machines. In Table 5-2 are listed the exact solution times observed when five specific problems were solved on various computers in the IBM 360 family. Core memory size is given in thousands of bytes, and all solution times are in *minutes*. Again, knowledge of an initial BFS would cut the times down to less than one half of the amounts shown. Notice the difference in solution time between two apparently similar problems, 1 and 4.

Table 5-2 IBM 360 Computation Times in Minutes[1]

Model, memory	Problem 1: 325 const. 452 vars. 2% density	Problem 2: 221 const. 249 vars. 4% density	Problem 3: 226 const. 284 vars. 5% density
360/40, 128K	20.44	29.73	28.22
360/50, 256K	7.36	7.83	8.91
360/65, 256K	2.08	—	3.08

Model, memory	Problem 4: 306 const. 472 vars. 2% density	Problem 5: 499 const. 902 vars. 1% density
360/40, 128K	58.41	113.45
360/50, 256K	20.13	36.33
360/65, 256K	—	10.64

[1] Data from IBM Manual H20-0136-3, "Mathematical Programming System/360, Applications Description," 1968.

EXERCISES

Section 5.2

Exercises 5-1 through 5-3. Solve each of the following linear programs via the two-phase simplex algorithm.

5-1. Max $z = 3x_1 + 2x_2 + 2x_3 + x_4$

$$\begin{aligned} \text{subject to} \quad x_1 - 4x_2 - x_3 + 5x_4 &= 7 \\ 2x_1 + x_2 + 3x_3 + x_4 &= 16 \\ 3x_1 - 2x_2 - x_3 + 2x_4 &= 0 \\ \text{and} \quad x_i \geq 0 \quad &\text{all } i \end{aligned}$$

5-2. Max $z = 2x_1 + x_2 - x_3 + 2x_4$

$$\begin{aligned} \text{subject to} \quad 2x_1 - x_2 - x_3 + x_4 &= 3 \\ -x_1 + 3x_2 + x_3 - x_4 &= 2 \\ x_1 + 3x_2 + 2x_3 + x_4 &= 5 \\ \text{and} \quad x_i \geq 0 \quad &\text{all } i \end{aligned}$$

5-3. Max $z = x_1 + 2x_2 + x_3 - 3x_4$

$$\begin{aligned} \text{subject to} \quad x_1 - x_2 + x_3 - x_4 &= 4 \\ 2x_1 - 4x_2 - 2x_3 + 2x_4 &= -6 \\ x_1 + 2x_2 - 3x_3 - x_4 &= -5 \\ \text{and} \quad x_i \geq 0 \quad &\text{all } i \end{aligned}$$

5-4. How can the simplex method be used to find the inverse, if it exists, of any given square matrix?

5-5. Inasmuch as phase 1 cannot end until all artificial variables have been removed from the basis (or at least driven to zero), it seems wasteful to allow an artificial variable to enter the phase-1 basis just because it happens to have the most negative reduced cost, or to allow *any* variable to enter that will cause a real variable to be removed. Discuss the pros and cons of ignoring the reduced costs and simply choosing at each stage (whenever possible) a pivot in which the entering variable is real and the departing variable, as determined by the usual simplex exit criterion, is artificial. What auxiliary computations would be required? What would be a reasonable way to proceed if at some stage a pivot of the desired type were not possible? Could the termination of phase 1 within a finite number of pivots be guaranteed?

Section 5.5

5-6. Solve via the two-phase simplex algorithm:

$$\begin{aligned} \text{Max } z = 3x_1 + 2x_2 + 2x_3 + x_4 & \\ \text{subject to } \quad x_1 + 3x_2 + x_3 + 2x_4 &= 12 \\ 2x_1 - x_2 + 3x_3 - x_4 &= 7 \\ 3x_1 - 5x_2 + 5x_3 - 4x_4 &= 2 \\ \text{and} \quad x_i \geq 0 \quad \text{all } i & \end{aligned}$$

5-7. Solve via the two-phase simplex algorithm:

$$\begin{aligned} \text{Max } z = 3x_1 + 2x_2 + 2x_3 + x_4 & \\ \text{subject to } \quad 3x_1 + 3x_2 - 2x_3 + 4x_4 &= 6 \\ x_1 + 2x_2 - 2x_3 + 3x_4 &= 5 \\ 4x_1 - x_2 + 4x_3 - 3x_4 &= 2 \\ \text{and} \quad x_i \geq 0 \quad \text{all } i & \end{aligned}$$

5-8. Execute phase 1 of the two-phase simplex algorithm to find a feasible solution to the following set of equations:

$$\begin{aligned} 5x_1 + 3x_2 + x_3 &= 10 \\ x_1 - x_2 &= 1 \\ -x_1 + x_2 + 3x_3 &= 2 \\ 2x_1 - 2x_3 &= 1 \\ 2x_1 - 2x_2 &= 2 \\ \text{and} \quad x_1, x_2, x_3 &\geq 0 \end{aligned}$$

Use information contained in the optimal phase-1 tableau to identify the redundancy relationships that prevail among the various equations.

5-9. Define an *irreducible redundant set* (IRS) to be a redundant set of linear equations

$$a_{i1}x_1 + a_{i2}x_2 + \cdots + a_{in}x_n = b_i \qquad i \in G$$

from which it is impossible to remove a constraint and still preserve redundancy. Two or more IRSs are then said to be mutually *disjoint* if they have no member equations in common. When a linear program is being solved via the two-phase simplex algorithm, does the number of artificial basic variables that must be carried into phase 2 always equal the maximum number of mutually disjoint IRSs that can be found in the constraints? Consider in your answer the following set of three constraints:

$$x_1 + x_2 + \cdots + x_n = 100$$

$$2x_1 + 2x_2 + \cdots + 2x_n = 200$$

and $$3x_1 + 3x_2 + \cdots + 3x_n = 300$$

If your answer was "no," can you suggest another, more valid relationship between artificial variables in phase 2 and redundant constraints?

Section 5.7

5-10. Solve the problem of Exercise 5-1 using the one-phase, or "penalties," method.

5-11. Solve the problem of Exercise 5-2 using the method of penalties.

5-12. Solve the problem of Exercise 5-6 using the method of penalties.

5-13. Discuss and compare the ways in which infeasibility and unboundedness are discovered by (a) the two-phase simplex algorithm and (b) the one-phase method of penalties.

5-14. Show that the sequence of bases generated by the method of penalties in solving any given linear program must begin with the precise sequence of bases that would be generated in phase 1 of the two-phase algorithm, provided that no ties occur in applying the simplex entry and exit criteria (or, if they do occur, that they are always resolved in the same way).

Section 5.9

Exercises 5-15 through 5-18. Solve each of the following linear programs using (a) the two-phase revised simplex algorithm and (b) the one-phase revised simplex algorithm.

5-15. Max $z = 3x_1 + 2x_2 + 2x_3 + x_4$

subject to

$$\begin{aligned} x_1 - 4x_2 - x_3 + 5x_4 &= 7 \\ 2x_1 + x_2 + 3x_3 + x_4 &= 16 \\ 3x_1 - 2x_2 - x_3 + 2x_4 &= 0 \end{aligned}$$

and $x_i \geq 0$ all i

5-16. Max $z = 2x_1 + x_2 - x_3 + 2x_4$

subject to
$$\begin{aligned} 2x_1 - x_2 - x_3 + x_4 &= 3 \\ -x_1 + 3x_2 + x_3 - x_4 &= 2 \\ x_1 + 3x_2 + 2x_3 + x_4 &= 5 \end{aligned}$$
and $x_i \geq 0 \quad$ all i

5-17. Max $z = x_1 + 2x_2 + x_3 - 3x_4$

subject to
$$\begin{aligned} 2x_1 + x_2 - x_3 - x_4 &= 3 \\ 3x_1 + 6x_2 + 2x_3 - x_4 &= 18 \\ -x_1 + 4x_2 + 4x_3 + x_4 &= 12 \end{aligned}$$
and $x_i \geq 0 \quad$ all i

5-18. Max $z = x_1 + x_2 + x_3 - x_4$

subject to
$$\begin{aligned} x_1 - 2x_2 + x_4 &= 1 \\ 2x_1 + x_3 - 2x_4 &= 15 \\ x_2 - 2x_3 - x_4 &= 1 \end{aligned}$$
and $x_i \geq 0 \quad$ all i

5-19. Much of the computation time required by the revised simplex method is spent in calculating the reduced costs $(z_j - c_j)$ of the $n - m$ nonbasic variables. Discuss the possibility of obtaining greater overall efficiency by devoting less effort to the choice of the entering variable. Might it be practical simply to choose the first variable that is found to have a negative reduced cost (in a maximization problem)? What about labeling or "remembering" those whose reduced costs are negative at one iteration and considering them for basis entry at the *next*?

Section 5.10

5-20. How many nontrivial multiplications and divisions are performed in a single revised simplex pivot when the constraint matrix **A** is $m \times n$ and has density α? Compare this number to the operation count of a standard simplex pivot when α is close to zero, and comment on how the relative values of m and n affect this comparison. Assuming that the expected number of pivots required to solve a given linear program is roughly proportional to m and independent of n (a very crude approximation of reality), derive a rule of thumb showing how total solution time—i.e., the total number of multiplications and divisions required to solve the problem—varies with m and n when the revised simplex method is used and when $\alpha = 1$. What would the rule of thumb be when α is close to zero?

5-21. Compute the number of multiplications and divisions required to solve the simultaneous equation system $\mathbf{Ax} = \mathbf{b}$ by *gaussian elimination*,[1] where **A** is an

[1] Gaussian elimination proceeds as follows. Solve the first equation for x_1, producing

$$x_1 = \frac{b_1}{a_{11}} - \frac{a_{12}x_2}{a_{11}} - \cdots - \frac{a_{1n}x_n}{a_{11}}$$

and substitute this expression for x_1 into each of the remaining $n - 1$ equations. Solve the second equation, which now includes only $n - 1$ variables, for x_2 and substitute into the remaining $n - 2$ equations. Continue in this manner until the last equation has been solved for x_n. Then substitute this final value of x_n into the first $n - 1$ equations, yielding the value of x_{n-1}. Similarly, substitute that value of x_{n-1} into the first $n - 2$ equations, and continue until the value of x_1 has been obtained.

$n \times n$ matrix, $\mathbf{A}^{-1}$ is assumed to exist, and all elements a_{ij} are different from zero. Now count the operations that would be required to solve the system via the following steps:

(a) Add n artificial variables to form an initial BFS and construct a standard simplex tableau, omitting the row that contains the reduced costs and the value of the objective.

(b) Choose any nonbasic real variable x_k to enter and any artificial basic variable $x_{\mathbf{B}i}$ to leave, provided only that $y_{ik} \neq 0$ (we shall not worry about preserving the nonnegativity of the basic variables; in fact, the eventual solution $\mathbf{x} = \mathbf{A}^{-1}\mathbf{b}$, in general, will not be nonnegative).

(c) Use $\mathbf{y}_k$ to form the computational $\mathbf{\Phi}$ vector, with the last component omitted, and perform a standard simplex pivot, deleting from the tableau the column of the exiting basic variable.

(d) Repeat steps (b) and (c) until exactly n pivots have been performed, at which time n columns will remain in the tableau and the basis will consist of the n real variables, as desired.

Are there any values of n for which this simplex approach is superior to gaussian elimination? How would your answer be affected if a substantial fraction of the a_{ij} were zero?

6

ILLUSTRATIVE CASE PROBLEMS

6.1 LINEAR PROBLEMS IN A NONLINEAR WORLD

This chapter will be devoted entirely to examples illustrating when and how linear programming techniques can be used in operations research. The problems described below are drawn from various management decision-making contexts, with several involving actual plant operations. Most are sufficiently general to be of interest to the practicing engineer as well, who is usually concerned either with minimizing the overall cost of some system subject to constraints on its performance or with optimizing system performance subject to constraints on available resources.

The reader will notice that none of the examples are what might be called "hard" engineering problems, that is, those limited strictly to the interplay of physical forces. The reason for their absence is basically that the physical world is not really very linear. For example, one does not double the strength of a dam or building by pouring twice as much concrete, nor does a rocket accelerate twice as rapidly when the thrust of its engines is doubled. On the other hand, true linearity often does appear in the economic world simply because of artificial rules and routines imposed by human beings for the purpose of regulating their affairs. Two pairs of shoes do not really cost twice as much to produce as one, but they do cost twice as much to buy (although a hundred pairs would probably not cost a hundred times as much). Other important linear relationships are based upon such constant ratios as the cost per pound of raw materials, the hourly wages of workers, the output per minute of an assembly line, the rate of return on invested dollars, and the carrying cost per unit in inventory.

All of the cases discussed below are fictitious and some rather lighthearted, but each illustrates a fairly common problem-type. Regardless of his own professional discipline, the reader is urged to study this chapter as carefully as he has the others. Its most important purposes are to show how real-world problems can be recognized and expressed in mathematical terms and to indicate what the important considerations are in deciding whether or not a dubious relationship can be treated as linear. Modeling is more an art than a science and is best learned by example: The more experience the student has, the more skillful he can become.

6.2 RESOURCE ALLOCATION

We begin with the production problem introduced in Chapter 1. The American Eagle Munitions Company manufactures three finished products: bullets, hand grenades, and artillery shells. The amounts of raw materials required for 1 ton of each, along with the profit margins that the company can make, are given in Table 6-1. Current inventory includes 100 tons of gunpowder and 150 tons each of lead and steel. Using only this inventory, how much of each product should American Eagle manufacture in order to maximize its profit?

Table 6-1 Data for American Eagle Problem

	Gun-powder	Lead	Steel	Profit
1 ton bullets	0.1 ton	0.7 ton	0.2 ton	$300
1 ton hand grenades	0.5		0.5	$600
1 ton artillery shells	0.3	0.5	0.2	$900

The formulation is quite simple. Let x_1, x_2, and x_3 represent the numbers of tons of bullets, grenades, and artillery shells to be produced. Because each product requires gunpowder and only a finite amount of it is available, we have a scarce-resource constraint:

$$0.1x_1 + 0.5x_2 + 0.3x_3 \leq 100$$

Inequality is permitted because the company is not compelled to use up all of its gunpowder. Similarly, the lead and steel constraints are

$$0.7x_1 \qquad\qquad + 0.5x_3 \leq 150$$

and

$$0.2x_1 + 0.5x_2 + 0.2x_3 \leq 150$$

The variables are nonnegative, and the objective function must be

$$\text{Max } z = 3x_1 + 6x_2 + 9x_3$$

where costs are in hundreds of dollars. The problem is evidently a linear program.

The constraints may be converted to standard form by the addition of slack variables:

$$\begin{aligned} 0.1x_1 + 0.5x_2 + 0.3x_3 + x_4 \qquad\qquad &= 100 \\ 0.7x_1 \qquad\quad + 0.5x_3 \qquad + x_5 \qquad &= 150 \\ 0.2x_1 + 0.5x_2 + 0.2x_3 \qquad\qquad + x_6 &= 150 \end{aligned}$$

An identity basis is immediately available, and the problem may be solved by the standard simplex method, which leads to the following optimal tableau:

	x_1	x_2	x_3	x_4	x_5	x_6
$x_{\mathbf{B}1} = x_2 = 20$	-0.64	1	0	2	-1.2	0
$x_{\mathbf{B}2} = x_3 = 300$	1.40	0	1	0	2.0	0
$x_{\mathbf{B}3} = x_6 = 80$	0.24	0	0	-1	0.2	1
Max $z = 2820$	5.76	0	0	12	10.8	0

Note that no bullets are made and that, because x_6 is positive, the steel in inventory is not entirely used up.

The student was asked to show in Exercise 4-18(b) that the reduced costs of the slack variables in the optimal primal tableau must equal the optimal values of the dual variables. This can be demonstrated easily in two steps:

(1) The optimal solution to the dual problem is

$$\mathbf{u}_0^T = \mathbf{c_B}^T\mathbf{B}^{-1} \tag{4-50}$$

as was shown in the proof of the optimality theorem (Theorem 4.4).

(2) The reduced cost of the jth slack variable in the optimal tableau is

$$z_j - c_j = \mathbf{c_B}^T\mathbf{B}^{-1}\mathbf{e}_j - 0 = u_{0j}$$

where $\mathbf{e}_j$ is a unit vector and u_{0j} is the jth component of $\mathbf{u}_0$, Q.E.D.

In the problem above the slack variable for gunpowder, x_4, has an optimal reduced cost of 12. Therefore the marginal value of an extra ton of it (assuming that this increment would not carry us onto a different segment of the piecewise linear curve of Figure 3–3) must be \$1200: It would be profitable for American Eagle to buy more gunpowder and alter its production plan if the price were below \$1200 per ton. On the other hand, additional steel would have no value at all, a conclusion that might have been reached also by merely observing that the optimal production plan does not exhaust the available steel supply.

Placing limits upon production capacity leads to a simple variation of the basic resource-allocation problem we have been discussing. For example, the speed of the production line or the lack of great demand might dictate that no more than 200 tons of artillery shells could be produced. We would then have an extra constraint

$$x_3 \leq 200 \tag{6-1}$$

called an *upper-bound* constraint; it is linear, so the problem would still be a linear program. Although it is perfectly legitimate to add (6-1) to the constraint set and solve routinely via the simplex method, there also exist special computational procedures that speed up the solution process by treating upper-bound constraints somewhat differently (just as nonnegativity constraints, which establish *lower* bounds, are also treated differently by the simplex method). These upper-bound procedures are, in fact, included in most modern computer codes for solving linear programs. The student need not concern himself with the gory computational details, but it is worth bearing in mind that upper-bound constraints can be handled more efficiently than ordinary linear constraints and have considerably less effect on overall solution time.

Let us turn our attention now to the objective function that was used in determining American Eagle's best production scheme. In the real world a manufacturer's total profit is not a simple linear function of the number of items or tons produced but tends instead to vary with production levels in a complicated nonlinear fashion. In particular, profit margins tend to decline at high production levels: Workers must be paid overtime wages, less efficient

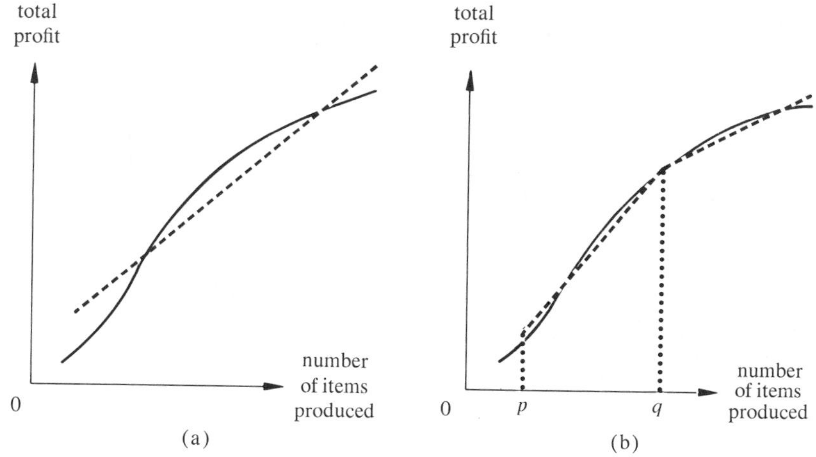

Figure 6-1

equipment must be activated, machines can overheat, and so on. To fit a single straight line to the *entire* profit curve, as shown for one product in Figure 6-1(a), obviously would introduce a serious error in the problem formulation.

On the other hand, it is often reasonable to treat profit as a linear function of the number of items produced *over some given range*—typically, that range within which the plant operates at roughly peak efficiency. For example, if the plant manager can tell from resource constraints, production capacity, or other considerations that the number of items to be produced will be somewhere between p and q in Figure 6-1(b), then he should have no qualms about fitting a straight line to that nearly linear segment of the profit curve. The slope of the line can be taken as the profit margin per item, and the production problem will be a linear program.

But suppose that it is quite possible for the production level to exceed q, so that a single straight line simply does not fit the curve very well. The plant manager can still approximate the profit curve to any desired degree of accuracy by means of a *piecewise linear function*, as illustrated by the broken line in Figure 6-1(b). Provided that successive pieces have smaller and smaller slopes, linear programming can still be used. Consider the example of Figure 6-1(b). Instead of using, say, x_1 to represent the number of items of type 1 to be produced let

$$x_{11} = \text{number of type-1 items produced at profit } c_{11} \text{ apiece}$$

and $$x_{12} = \text{number of type-1 items produced at profit } c_{12} \text{ apiece}$$

where c_{11} and c_{12} are the slopes of the first and second segments of the piecewise linear curve, with $c_{11} > c_{12}$. The profit from type-1 items is then represented in the objective function by $c_{11}x_{11} + c_{12}x_{12}$ instead of by c_1x_1. Items in both profit categories require exactly the same inputs, so that each resource constraint

$$a_{i1}x_1 + a_{i2}x_2 + a_{i3}x_3 \leq b_i$$

in the original formulation now becomes

$$a_{i1}x_{11} + a_{i1}x_{12} + a_{i2}x_2 + a_{i3}x_3 \leq b_i$$

(assuming that analogous transformations are not required for type-2 and type-3 items). Finally, because only q items can be produced at the higher profit margin, the upper-bound constraint

$$x_{11} \leq q$$

is needed too. We have again formulated a linear program.

There is another constraint implied by the problem's conditions that we did not state explicitly, namely, that no item is to be sold at the lower profit c_{12} until q items have been sold at c_{11}. This could be represented algebraically as

$$(x_{11} - q)x_{12} = 0 \tag{6-2}$$

That is, either x_{11} is at its upper bound or $x_{12} = 0$. To have included the constraint (6-2) would have rendered the problem nonlinear, but fortunately that was not necessary. Unit increases in x_{11} and x_{12} consume the same amounts of each raw material; therefore, because $c_{11} > c_{12}$, the objective function "prefers" to maximize x_{11} and to leave x_{12} at zero. Its ability to do so is limited ultimately by the upper-bound constraint on x_{11}, and the overall effect is to guarantee that (6-2) will hold. Notice, incidentally, that our transformation would *not* work if $c_{11} < c_{12}$, that is, if profits on the first few items produced were lower than profits on later items. The objective then would prefer to increase x_{12}, and our linear programming formulation might well lead to an optimal solution with $x_{11} = 0$ and $x_{12} > 0$, a physical impossibility. To prevent this, a constraint such as (6-2) would have to be included, and the problem would no longer be linear.

6.3 THE DIET PROBLEM

A hospital dietitian is planning a breakfast menu that may include any (nonnegative) amounts of n different foods, such as corn flakes, eggs, prunes, etc. Each patient requires at least the minimum amounts b_i, $i = 1, \ldots, m$, of various nutrients (protein, riboflavin, and so on). If 1 ounce of the jth food costs c_j cents and contains a_{ij} ounces of the ith nutrient, what menu will satisfy all the dietary requirements at minimum cost?

This well-known problem-type is essentially the opposite of the resource allocation problem discussed in Section 6.2, which sought the production scheme satisfying the scarce-resource constraints at maximum profit. Whereas those resource constraints were of the form

$$a_{i1}x_1 + a_{i2}x_2 + \cdots + a_{in}x_n \leq b_i$$

the constraints of the diet problem are

$$a_{i1}x_1 + a_{i2}x_2 + \cdots + a_{in}x_n \geq b_i$$

with x_j here representing the amount of the jth food in the diet of each patient. Instead of maximizing

$$z = c_1x_1 + c_2x_2 + \cdots + c_nx_n$$

as in the resource-allocation problem, the dietitian seeks to minimize it. Thus the diet problem is another straightforward linear program. The optimal value of its ith dual variable must represent the cost saving per patient that would be afforded by a 1-ounce (unit) decrease in the minimum requirement for the ith nutrient; from another point of view, it represents also the maximum price that the dietitian would just be willing to pay for an ounce of the ith nutrient in pure form.

6.4 MACHINE SCHEDULING

The problem of scheduling jobs on several different production lines or machines arises quite frequently in plant management and also lends itself to solution via linear programming. Consider a metal foundry that converts swords into plowshares of four different sizes: large, extra large, super, and jumbo. Weekly demands for these four types are d_1, d_2, d_3, and d_4, respectively. The conversion operation can be performed on any one of the three machines of various ages and capabilities; in 1 hour the ith machine can produce a_{ij} plowshares of type j at an operating cost of c_i dollars, $c_i > 0$. If no machine may run for more than 40 hours per week, how should machine time be allocated in order to satisfy the weekly demands at minimum cost? Assume that start-up times and down times between jobs are negligible.

Again the formulation is not difficult. Let x_{ij} be the number of hours per week scheduled on machine i for producing plowshares of type j; an alternate choice of variable would have been the number of type-j plowshares produced weekly by the ith machine. There are two types of constraints, corresponding to the requirements that demands be satisfied and that available machine time not be exceeded. The former are expressed by the following:

$$\begin{aligned}
a_{11}x_{11} \qquad\qquad &+ a_{21}x_{21} \qquad\qquad &+ a_{31}x_{31} \qquad\qquad &= d_1 \\
a_{12}x_{12} \qquad\qquad &+ a_{22}x_{22} \qquad\qquad &+ a_{32}x_{32} \qquad\qquad &= d_2 \\
a_{13}x_{13} \qquad\qquad &+ a_{23}x_{23} \qquad\qquad &+ a_{33}x_{33} \qquad\qquad &= d_3 \\
a_{14}x_{14} \qquad\qquad &+ a_{24}x_{24} \qquad\qquad &+ a_{34}x_{34} &= d_4
\end{aligned}$$

where equality signs are used because it cannot be optimal to produce more plowshares than are demanded. The machine-time constraints are

$$\begin{aligned}
x_{11} + x_{12} + x_{13} + x_{14} \qquad\qquad\qquad\qquad\qquad\qquad\qquad\qquad &\leq 40 \\
x_{21} + x_{22} + x_{23} + x_{24} \qquad\qquad\qquad\qquad &\leq 40 \\
x_{31} + x_{32} + x_{33} + x_{34} &\leq 40
\end{aligned}$$

Subject to these constraints, the overall cost is to be minimized:

$$\text{Min } z = \sum_{i=1}^{3} c_i \sum_{j=1}^{4} x_{ij}$$

The variables are nonnegative, everything is linear, and we have another linear program. Notice that, of the 12 variables, at most 7 can be greater than zero in the optimal solution; moreover, if any of the machines is scheduled for less than 40 hours, its slack variable also will have to appear in the final basis, thereby further reducing the number of positive x_{ij}.

A simple variation on the scheduling problem occurs when machine operating costs depend on the product as well as on the machine. The objective then becomes

$$\text{Min } z = \sum_i \sum_j c_{ij}x_{ij}$$

with the constraints remaining unchanged. This sort of dependence tends to occur whenever the products differ among themselves in size, composition, or complexity; usually the cost difference is due to an extra laborer or a piece of auxiliary equipment.

A somewhat more complicated version of the problem is obtained when the machines are permitted to run overtime at higher cost. Now we must resort to the same sort of maneuver that was used in the case of the piecewise linear profit function in Section 6.2. Suppose for simplicity that only machine 1 can run overtime. Its hourly operating cost is c_1 dollars for the first 40 hours (independent of product) and c_1' afterward, with $c_1' > c_1$. Let x_{1j} represent the number of regular hours and x_{1j}' the number of overtime hours per week scheduled on machine 1 for the manufacture of product j, $j = 1, \ldots, 4$. The objective function then becomes

$$\text{Min } z = \sum_{i=1}^{3} c_i \sum_{j=1}^{4} x_{ij} + c_1' \sum_{j=1}^{4} x_{1j}'$$

and the demand constraints are altered to

$$a_{1j}(x_{1j} + x_{1j}') + a_{2j}x_{2j} + a_{3j}x_{3j} = d_j \qquad j = 1, \ldots, 4$$

while the time constraints remain just as they were. As before, we have omitted a legitimate restriction, namely, that nothing is produced on machine 1 during overtime hours unless its 40 regular hours have been completely scheduled. This might be expressed as

$$(x_{11} + x_{12} + x_{13} + x_{14} - 40)\cdot(x_{11}' + x_{12}' + x_{13}' + x_{14}') = 0 \qquad \text{(6-3)}$$

which is obviously analogous to (6-2). Again the optimum-seeking objective function will "prefer" to schedule regular hours whenever possible, so that (6-3) need not be explicitly stated.

There is another aspect of machine-scheduling problems that is more difficult to handle. Before a machine or production line is ready to accept a new production job, in general, it must be "warmed up," cleaned, and/or adjusted. The initial delay after the machinery is turned on is known as the *start-up time* and may be quite protracted, especially if ovens or heating elements have to be brought to high temperatures. Subsequent delays between production jobs, called *setup times*, also are required for changing the stream of raw-material inputs, adjusting tolerances and gear ratios, and so on. Because these delays are "yes-or-no" affairs and do not vary with the duration of the production jobs that they precede, they cannot be modeled with linear equations; only if the plant manager is willing to ignore them can linear programming techniques be used.

To see why these delays are essentially nonlinear, let us develop the algebraic expressions that represent them. Assuming that start-up and setup times can be treated as equal (the problem is even more complicated if this is not so), we are postulating that if machine i is scheduled to manufacture product j for any positive number of hours, then a delay of t_{ij} hours must be incurred, for $i = 1, 2, 3$ and $j = 1, 2, 3, 4$. Usually some or all of the t_{ij} will be equal. Let us define a yes-or-no or *binary variable* y_{ij} as follows, for all i and j:

$$y_{ij} = \begin{cases} 0 & \text{if } x_{ij} = 0 \\ 1 & \text{if } x_{ij} > 0 \end{cases} \tag{6-4}$$

Putting (6-4) into words, y_{ij} has a value of zero (no setup delay) if product j is not scheduled at all on machine i or a value of one if it is; no other value of y_{ij} is physically meaningful. Observe now that if the logical conditions (6-4) could be enforced by means of linear algebraic constraints, we would have a linear program: The time constraint for the ith machine (with no overtime permitted) would be

$$\sum_{j=1}^{4} (x_{ij} + t_{ij}y_{ij}) \leq 40 \qquad i = 1, 2, 3$$

and the objective function would become

$$\text{Min } z = \sum_{i=1}^{3} c_i \sum_{j=1}^{4} (x_{ij} + t_{ij}y_{ij})$$

The demand constraints, of course, would not be affected.

In order to enforce (6-4), however, it is necessary to use either a nonlinear constraint or an *integer variable*. In the former approach the following pair of constraints is added for each y_{ij}:

$$x_{ij}(y_{ij} - 1) = 0 \tag{6-5}$$

and

$$y_{ij} \geq 0 \tag{6-6}$$

If x_{ij} is positive, (6-5) requires $y_{ij} = 1$; if $x_{ij} = 0$ the objective function is free to push y_{ij} down as far as possible, namely, to $y_{ij} = 0$. The second way of expressing the conditions of (6-4) introduces a new kind of variable:

$$y_{ij} \geq 0 \tag{6-6}$$

$$x_{ij} - 40y_{ij} \leq 0 \tag{6-7}$$

and

$$y_{ij} \text{ an integer} \tag{6-8}$$

Here, if $x_{ij} = 0$, then y_{ij} is not constrained by (6-7) and, as before, the objective of minimizing total cost guarantees $y_{ij} = 0$. But if x_{ij} is positive, y_{ij} is forced by (6-7) to be positive as well and, because of (6-8), it must jump at least to 1.

Because x_{ij} cannot be greater than 40, y_{ij} will never be obliged to exceed 1 [if the upper bound on x_{ij} were unknown, the same effect could be achieved by using an arbitrarily large number in place of the 40 in equation (6-7)]. Once again (6-4) is satisfied.

Unfortunately, neither of these approaches is amenable to linear programming. In general, problems involving yes-or-no variables—which usually are associated with some sort of *fixed charge*—must be solved by integer programming or combinatorial methods, both of which are beyond the scope of this book.

6.5 THE TRANSPORTATION PROBLEM

The Consolidated Cocoa Company, a wholly owned subsidiary of American Eagle Munitions, processes cocoa beans into chocolate at two different plantations. The chocolate is then shipped every week to warehouses in three major cities for retail distribution. Shipping costs per pound are listed in Table 6-2. Weekly demands for chocolate in the three cities are 3000, 2500, and 1500 pounds, respectively, and maximum production capacity is 5000 pounds per week at plantation A and 2500 pounds per week at plantation B. Assuming that all production costs are the same at the two plantations, what production/shipping strategy should CoCoCoaCo adopt to minimize its shipping costs?

This classical application of linear programming is known as a *transportation problem*. Its essential elements are displayed in Figure 6-2, in what is called a *transportation network*, or a *network flow diagram*. The long straight lines or *arcs* in the diagram represent shipping routes along which commodities may travel from *origins* to *destinations*—or, in our problem, from plantations to cities. The label on each arc gives its per-unit shipping cost. Origins are labeled with their respective *supplies*, that is, their maximum weekly production capacities, and destinations with their *demands*.

With the aid of the network flow diagram the problem formulation becomes trivial. Let x_{Ai} and x_{Bi} be the numbers of pounds shipped each week from plantations A and B to city i, for $i = 1, 2, 3$. We assume that negative or backward flow (from a city to a plantation) is not permitted; so that

$$x_{Ai}, x_{Bi} \geq 0 \qquad i = 1, 2, 3$$

Table 6-2 Data for Consolidated Cocoa Problem

	City 1	City 2	City 3
from plantation A to	20¢	15¢	10¢
from plantation B to	16¢	12¢	10¢

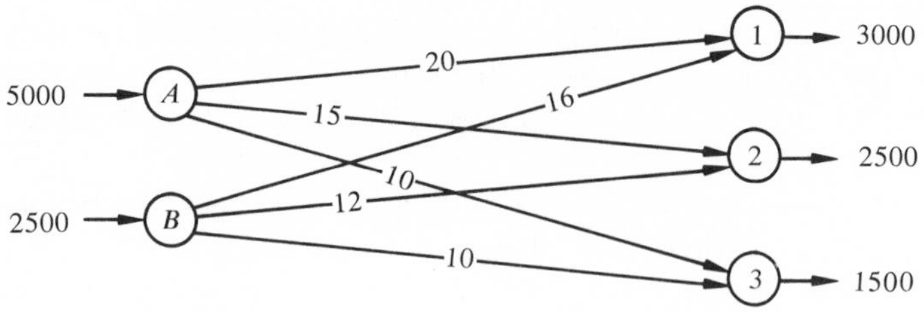

Figure 6-2

The limited production capacities impose the following supply constraints:

$$x_{A1} + x_{A2} + x_{A3} \leq 5000$$

and

$$x_{B1} + x_{B2} + x_{B3} \leq 2500$$

The demand constraints are

$$\begin{aligned} x_{A1} + x_{B1} &= 3000 \\ x_{A2} + x_{B2} &= 2500 \end{aligned}$$

and

$$x_{A3} + x_{B3} = 1500$$

where equality signs are used because overfulfillment of demands cannot be optimal (this follows from the fact that all transportation costs are nonnegative). Subject to these restrictions, the company's objective is to minimize its overall shipping cost:

$$\text{Min } z = 20x_{A1} + 15x_{A2} + 10x_{A3} + 16x_{B1} + 12x_{B2} + 10x_{B3} \qquad (6\text{-}9)$$

After adding slack variables s_1 and s_2 to the two supply constraints, we must still use three artificial variables to obtain an initial identity basis. The constraints in standard form for phase 1 are then

$$\begin{aligned} x_{A1} + x_{A2} + x_{A3} + s_1 &= 5000 \\ x_{B1} + x_{B2} + x_{B3} + s_2 &= 2500 \\ x_{A1} + x_{B1} + w_3 &= 3000 \\ x_{A2} + x_{B2} + w_4 &= 2500 \\ x_{A3} + x_{B3} + w_5 &= 1500 \end{aligned}$$

The objective in phase 1 is to minimize the sum of the artificial variables,

$$\text{Min } z' = w_3 + w_4 + w_5$$

so the initial tableau is as follows:

	x_{A1}	x_{A2}	x_{A3}	x_{B1}	x_{B2}	x_{B3}	s_1	s_2	w_3	w_4	w_5
$x_{B1} = s_1 = 5000$	1	1	1	0	0	0	1	0	0	0	0
$x_{B2} = s_2 = 2500$	0	0	0	1	1	1	0	1	0	0	0
$x_{B3} = w_3 = 3000$	1	0	0	1	0	0	0	0	1	0	0
$x_{B4} = w_4 = 2500$	0	1	0	0	1	0	0	0	0	1	0
$x_{B5} = w_5 = 1500$	0	0	1	0	0	1	0	0	0	0	1
Min $z' = 7000$	1	1	1	1	1	1	0	0	0	0	0

To break the six-way tie we arbitrarily choose x_{A1} to enter the basis; w_3 exits, $\mathbf{\Phi} = (-1, 0, 0, 0, 0, 0, -1)$, and the new tableau is

	x_{A1}	x_{A2}	x_{A3}	x_{B1}	x_{B2}	x_{B3}	s_1	s_2	w_3	w_4	w_5
$s_1 = 2000$	0	1	1	−1	0	0	1	0	−1	0	0
$s_2 = 2500$	0	0	0	1	1	1	0	1	0	0	0
$x_{A1} = 3000$	1	0	0	1	0	0	0	0	1	0	0
$w_4 = 2500$	0	1	0	0	1	0	0	0	0	1	0
$w_5 = 1500$	0	0	1	0	0	1	0	0	0	0	1
Min $z' = 4000$	0	1	1	0	1	1	0	0	−1	0	0

The usual labels x_{Bi} have been dropped from the far left-hand column of the tableau in order to avoid confusion with the variables of this problem. We continue pivoting, always breaking ties so as to hasten the removal of artificial variables from the basis, until we arrive at the optimal phase-1 tableau:

	x_{A1}	x_{A2}	x_{A3}	x_{B1}	x_{B2}	x_{B3}	s_1	s_2	w_3	w_4	w_5
$x_{A2} = 2500$	0	1	0	0	1	0	0	0	0	1	0
$s_2 = 500$	0	0	0	0	0	0	1	1	−1	−1	−1
$x_{A1} = 2500$	1	0	1	0	−1	0	1	0	0	−1	0
$x_{B1} = 500$	0	0	−1	1	1	0	−1	0	1	1	0
$x_{B3} = 1500$	0	0	1	0	0	1	0	0	0	0	1
Min $z' = 0$	0	0	0	0	0	0	0	0	−1	−1	−1

At this point, the last row of the tableau must be recalculated, based on the original objective function (6-9). Using $\mathbf{c_B}^T = [15, 0, 20, 16, 10]$, we obtain the following initial values of the phase-2 objective function and reduced costs:

Min z = 110,500	0	0	4	0	−1	0	4	0

where the artificial variables have been deleted. Because we are minimizing, both x_{A3} and s_1 are candidates for basis entry. Choosing x_{A3}, we arrive after two more pivots at the following optimal tableau:

		x_{A1}	x_{A2}	x_{A3}	x_{B1}	x_{B2}	x_{B3}	s_1	s_2
x_{A2} =	2,500	0	1	0	0	1	0	0	0
s_1 =	500	0	0	0	0	0	0	1	1
x_{A1} =	500	1	0	0	0	−1	−1	0	−1
x_{B1} =	2,500	0	0	0	1	1	1	0	1
x_{A3} =	1,500	0	0	1	0	0	1	0	0
Min z =	102,500	0	0	0	0	−1	−4	0	−4

The optimal solution to our transportation problem is illustrated on the network flow diagram in Figure 6-3; this time the labels on the arcs are the *flows*, that is, the amounts to be shipped each week along the various routes. Minimum overall shipping cost is $1025 weekly.

The student doubtless has noticed the rather surprising property shared by all of the above tableaux: Every y_{ij} value at every iteration was −1, 0, or +1. As a result, the computation was extremely simple: The pivot element was always +1, and not a single multiplication or division was ever required! Moreover, the persistence of 1's and 0's in the **y** columns does not appear to

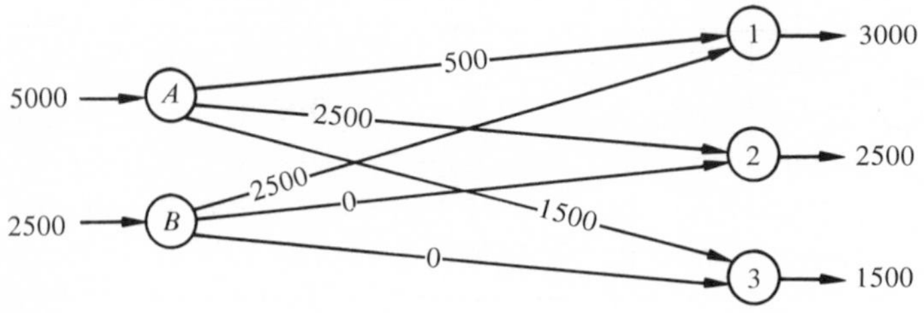

Figure 6-3

have depended in any way upon the actual numerical values of the supplies, demands, and shipping costs. Although different sets of numbers would lead to different sequences of pivots, there is every reason to suspect that the property observed in our example would continue to prevail.

It should occur to the student at this point that the simplex method, with all its tedious updating of **y** columns, just might not be a very efficient procedure for solving transportation problems. Those recurring arrays of 0s and 1s, and particularly the repetitive pattern of the initial tableau, constitute a sort of special structure in the following sense: A problem is said to have *special structure* if it displays strong or interesting structural properties that are not characteristic of the general class to which it belongs. One of the major concerns of the applied mathematician is the development of computational algorithms that take the fullest possible advantage of whatever structure a problem may possess. Obviously, the transportation problem is a highly specialized form of linear program. Because the simplex method exploits only the linearity of its constraints and objective function, we should not be surprised to discover that it can be solved much more efficiently by other means. The best known of the "transportation algorithms" will be discussed in detail in Chapter 8.

Several modifications of the basic transportation problem are also of interest. For various physical reasons upper bounds, or *capacities*, might be placed on the flows permitted from origins to destinations, leading to constraints of the form

$$x_{A2} \leq 2000$$

When these simple linear constraints are present, the network is said to have *capacitated arcs*. Another variation that preserves linearity is the assessment of a per-unit storage (or spoilage) cost for undelivered supply; that is, if the total amount of commodity shipped away from origin A in the optimal solution is 8 units while the supply at A is 10 units, then 2 units must be stored at a cost of c_A cents apiece. One straightforward way to incorporate this storage penalty into the problem formulation would be to include origin A's slack variable in the objective function, with a cost coefficient of c_A. In general, storage costs are most appropriate when the origins are warehouses or similar facilities with goods actually sitting in them. Those goods, if undelivered, can cost money in several different ways: They represent tied-up capital, they consume space, they may have to be inspected or guarded, and so on. In the problem we solved earlier, storage costs probably would not have been applicable because the "supplies" were maximum production capacities—presumably CoCoCoaCo simply would not produce any more chocolate than called for by the optimal shipping plan.

There is another extremely important variant of the transportation problem that should be mentioned in this section. Our basic formulation specifies a strictly linear objective function; that is, the total cost C of shipping x_{A1} units from origin A to destination 1 must be

$$C = c_{A1}x_{A1} \qquad \text{for any } x_{A1} \geq 0 \tag{6-10}$$

where c_{A1} is the cost per unit. Unfortunately, however, unless shipping is via some common carrier that charges by the pound (and offers no quantity discounts), (6-10) is not likely to be a very realistic cost function. A somewhat more common situation is represented by

$$C = \begin{cases} 0 & \text{if } x_{A1} = 0 \\ d_{A1} + c_{A1}x_{A1} & \text{if } x_{A1} > 0 \end{cases} \tag{6-11}$$

As before, c_{A1} is a per-unit shipping cost, but here d_{A1} is an additional *fixed charge* that must be paid if the route from A to 1 is used at all. Typically, d_{A1} might include the cost of leasing a truck, hiring a driver, and paying for gas and oil, and c_{A1} would cover the per-unit loading, unloading, and insurance costs. The trouble with (6-11), however, is that because of the fixed charge it is not a continuous linear function over all nonnegative x_{A1}. The same situation appeared in a different guise in Section 6.4, where the fixed charge was assessed for the production delay preceding each manufacturing job. In order to formulate that problem algebraically, it was necessary to define a binary variable

$$y_{ij} = \begin{cases} 0 & \text{if } x_{ij} = 0 \\ 1 & \text{if } x_{ij} > 0 \end{cases} \tag{6-4}$$

and then to use it in either a nonlinear or an integer constraint; as a result, linear programming techniques could not be applied. We reach the same frustrating conclusion in the case of the fixed-charge transportation problem: Unless the fixed charges are small enough to allow approximated cost functions of the form (6-10) to be employed, more sophisticated (and far less efficient) mathematical methods are required for solution.

6.6 JOB TRAINING

The Luna Cheese Company now has a working force of 400 experienced men producing its famous green cheeses. The company would like to increase this number to 600 over the next six weeks. Each pair of newly hired employees must be personally trained for two weeks by one experienced worker, during which time none of them actually produces anything. One man can make one cheese in one week, and the company's backlog of orders requires that exactly D_i

cheeses be made during the ith week, $i = 1, \ldots, 6$; because over-aging destroys the delicate flavor, cheese may not be produced ahead of time and stored. Assuming that trainees are paid the same wages as workers, how should Luna schedule its training program in order to minimize its labor costs over the six-week period?

Let x_i be the number of men hired at the start of the ith week, $i = 1, \ldots, 5$ (a trainee hired any later than week 5 would not be ready to work after the sixth week). Because trainees hired in week 1 will be paid for six weeks during the period of interest and those hired in week 2 will be paid for five, and so on, Luna's objective function is simply

$$\text{Min } z = 6x_1 + 5x_2 + 4x_3 + 3x_4 + 2x_5 \qquad (6\text{-}12)$$

A manpower constraint must be written for each week to insure that the required cheese can be produced; in general, the number of experienced workers available for cheesemaking is 400 plus whatever new men have finished their training, less the number of workers currently engaged in teaching. Therefore the constraints for the six weeks are

$$\begin{aligned}
400 - \tfrac{1}{2}x_1 &\geq D_1 \\
400 - \tfrac{1}{2}x_1 - \tfrac{1}{2}x_2 &\geq D_2 \\
400 + x_1 - \tfrac{1}{2}x_2 - \tfrac{1}{2}x_3 &\geq D_3 \\
400 + x_1 + x_2 - \tfrac{1}{2}x_3 - \tfrac{1}{2}x_4 &\geq D_4 \\
400 + x_1 + x_2 + x_3 - \tfrac{1}{2}x_4 - \tfrac{1}{2}x_5 &\geq D_5 \\
400 + x_1 + x_2 + x_3 + x_4 - \tfrac{1}{2}x_5 &\geq D_6
\end{aligned}$$

Inequalities must be used because it is possible that the optimal training schedule will call for some workers to remain idle during one or more weeks. Finally, because a total of 200 trainees must be hired,

$$x_1 + x_2 + x_3 + x_4 + x_5 = 200$$

The variables are nonnegative, and again we have a linear program.

Suppose we vary the problem by allowing cheeses to be stored or *inventoried* for later delivery, at a cost of \$10 per cheese per week. Because the objective now will be to minimize the total labor and inventory costs, we need to know also that Luna's workers are paid \$100 per week [note that in (6-12) we were simply minimizing the total number of weeks' wages paid to the new employees during the training period]. Let y_i represent the number of cheeses left in inventory just after the demand D_i has been satisfied at the end of the ith week, $i = 1, \ldots, 5$; for simplicity we stipulate that the inventory level is zero initially and must be zero after the sixth week. The objective function is then

$$\text{Min } z = 600x_1 + 500x_2 + 400x_3 + 300x_4 + 200x_5 + 10 \sum_{i=1}^{5} y_i$$

In order to formulate the constraints properly we must introduce a third set of variables: Let p_i be the number of cheeses produced in the ith week, $i = 1, \ldots, 6$. The number of cheeses made in any week must equal the sum of the number demanded plus the net change in inventory level. This relationship can be expressed in a set of "material balance," or "bookkeeping," equations:

$$\begin{aligned} p_1 &= D_1 + y_1 \\ p_i &= D_i + y_i - y_{i-1} \qquad i = 2, \ldots, 5 \\ p_6 &= D_6 - y_5 \end{aligned} \tag{6-13}$$

Additional constraints are required to insure that enough cheesemakers are available each week to produce the scheduled number of cheeses:

$$\begin{aligned} 400 - \tfrac{1}{2}x_1 &\geq p_1 \\ 400 - \tfrac{1}{2}x_1 - \tfrac{1}{2}x_2 &\geq p_2 \\ 400 + x_1 - \tfrac{1}{2}x_2 - \tfrac{1}{2}x_3 &\geq p_3 \\ 400 + x_1 + x_2 - \tfrac{1}{2}x_3 - \tfrac{1}{2}x_4 &\geq p_4 \\ 400 + x_1 + x_2 + x_3 - \tfrac{1}{2}x_4 - \tfrac{1}{2}x_5 &\geq p_5 \\ 400 + x_1 + x_2 + x_3 + x_4 - \tfrac{1}{2}x_5 &\geq p_6 \end{aligned} \tag{6-14}$$

These are almost identical to the manpower constraints used earlier, except that here the number of cheeses made each week is not necessarily the same as the number demanded. As before, workers may be idle. The remaining constraints are

$$x_1 + x_2 + x_3 + x_4 + x_5 = 200$$

$$x_i \geq 0 \qquad i = 1, \ldots, 5$$

$$y_i \geq 0 \qquad i = 1, \ldots, 5$$

and

$$p_i \geq 0 \qquad i = 1, \ldots, 6$$

and the problem is still linear.

At this point it may be tempting to revise the formulation by using equations (6-13) to substitute for the variables p_i in (6-14). This procedure would appear to eliminate six variables and six constraints from the problem, thereby lessening significantly the computation time required for its solution. However, we must be careful not to neglect the nonnegativity restrictions $p_i \geq 0$. These are constraints, too, and if the variables p_i are to be eliminated, the expressions substituted for them still must be required to be nonnegative:

$$D_1 + y_1 \geq 0$$

$$D_i + y_i - y_{i-1} \geq 0 \qquad i = 2, \ldots, 5$$

and

$$D_6 - y_5 \geq 0$$

Thus, six constraints really would not be saved after all; the net result of eliminating the p_i would be to decrease the number of variables by six while leaving the number of constraints unchanged. The latter is the critical parameter, so the expected savings in computation time would be no more than a few percent.

6.7 AN INVESTMENT PROBLEM

Uwanimus University plans to build a new physics laboratory over the next 12-month period, and the builders have determined that the cash outlays listed in Table 6-3, which are payable on the first day of each month, will be needed to finance the construction. Uwanimus has $20,000 in cash now and will raise the balance of the money by selling its shares of American Eagle Munitions Company, which pay dividends at the rate of 0.5% per month. In order to avoid driving down the price of the stock, the University has decided not to liquidate more than $100,000 worth in any one month. Whenever the cash on hand is insufficient to meet the construction outlay, the difference must be borrowed at an interest rate of 1% per month. Assuming that dividends from unsold stock are committed to other projects and cannot be used directly to finance the laboratory, how should Uwanimus schedule its stock sales in order to minimize the total costs of borrowing money *and* of losing dividends through premature liquidation? For simplicity, all transactions are considered to take place on the first day of the month.

Table 6-3 Data for Uwanimus University Problem

Month	1	2	3	4	5	6
Dollars ($000 omitted)	140	120	80	40	30	150
Month	7	8	9	10	11	12
Dollars ($000 omitted)	90	40	40	50	130	120

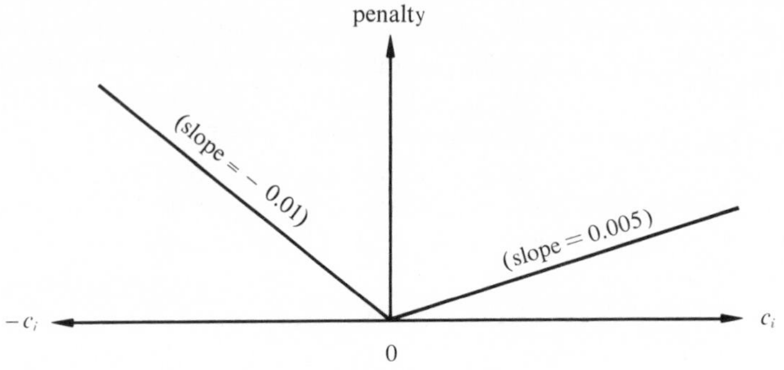

Figure 6-4

For this problem the formulation will not be entirely straightforward. Let s_i be the value of the stock sold at the start of the ith month, $i = 1, \ldots, 12$, and let c_i be the *net cash position* (positive if cash on hand, negative if in debt) immediately after the ith outlay. All variables will be measured in units of \$1000. In general, the net cash position obeys a simple recursion rule:

$$c_i = c_{i-1} + s_i - D_i \qquad i = 1, \ldots, 12 \tag{6-15}$$

where D_i stands for the ith construction outlay and $c_0 = 20$. The only other constraints in the problem are

$$0 \leq s_i \leq 100 \qquad i = 1, \ldots, 12$$

As for the objective function, the University wants to minimize the sum of its penalty costs, which are defined as follows:

$$\text{penalty in the } i\text{th month} = \begin{cases} 0.005c_i & \text{if } c_i > 0 \\ 0 & \text{if } c_i = 0 \\ -0.01c_i & \text{if } c_i < 0 \end{cases}$$

The penalties measure dividend losses for positive c_i and interest payments for negative c_i. This nonlinear function, sketched in Figure 6-4, brings up the question: How can it be represented within a linear programming format?

The trick is to define for each c_i a new pair of nonnegative variables:

$$\begin{aligned} & c_i \equiv u_i - v_i \\ \text{where} \qquad & u_i, v_i \geq 0 \qquad i = 1, \ldots, 12 \end{aligned} \tag{6-16}$$

By means of this maneuver we can express the penalty costs as legitimate linear functions,

$$i\text{th penalty} = 0.005u_i + 0.01v_i \tag{6-17}$$

as should be clear from the following considerations. For any given value of c_i, the values of u_i and v_i that minimize the function (6-17) subject to (6-16) are as follows:

$$\text{If } c_i > 0, \quad u_i = c_i \text{ and } v_i = 0$$
$$\text{If } c_i = 0, \quad u_i = 0 \text{ and } v_i = 0$$
$$\text{If } c_i < 0, \quad u_i = 0 \text{ and } v_i = -c_i$$

Observe that when these values are substituted into (6-17), the precisely correct penalty is obtained in all three cases. Therefore the University's financial problem finally can be formulated as a linear program:

$$\text{Min } z = \sum_{i=1}^{12} (0.005u_1 + 0.01v_i)$$

$$\begin{aligned}
\text{subject to} \quad & u_1 - v_1 = 20 + s_1 - D_1 & \\
& u_i - v_i = u_{i-1} - v_{i-1} + s_i - D_i & i = 2, \ldots, 12 \\
& s_i \leq 100 & i = 1, \ldots, 12 \\
\text{and} \quad & s_i, u_i, v_i \geq 0 & i = 1, \ldots, 12
\end{aligned}$$

When the cash outlays D_i given in the problem statement are used, the optimum amounts of stock to sell in months 1 through 12, respectively, are found to be (in \$1000 units)

100, 100, 100 60, 80, 100 90, 40, 40 100, 100, 100

If these numbers are compared to the table of monthly outlays, it can be seen that the best financial strategy lies simply in preferring advance liquidation over borrowing (except at the beginning) whenever one or the other is necessary. Using the liquidation schedule above, the University will pay \$800 in interest and will suffer a \$600 loss of dividends, for a total of \$1400 in finance costs.[1]

[1] The sequential constraints (6-15) constitute another sort of special structure that is not well exploited by linear programming techniques. The best and most direct way to solve the problem of this section is via *dynamic programming*, a general computational approach that is basically quite simple, but can be applied to a great many problem types. One rather good text on this subject is G. Nemhauser's *Introduction to Dynamic Programming*, John Wiley & Sons, New York, 1966.

6.8 A MINIMAX STRATEGY

Sunshine Mining, Inc. is building a dock-side loading facility for hoisting its iron ore onto river barges and has budgeted \$20,000 for insurance against various types of catastrophe. In the event of fire the facility would be completely destroyed, for a loss of \$1,000,000, and the insurance company would pay out \$100 for each dollar's worth of fire insurance. A flood would cause an estimated \$200,000 damage and Sunshine would collect \$500 for each of its flood-insurance dollars. Finally, if a barge collided with the facility, the damage would total about \$500,000 and Sunshine would collect \$50 per dollar. Rather than trying to estimate the probability of each type of disaster, Sunshine's managers have decided to allocate the \$20,000 in such a way as to minimize the loss if the *most costly* (in terms of net loss after insurance recovery) of the three disasters occurs. How much should Sunshine spend on each type of insurance in order to accomplish its objective of minimizing the maximum loss it can possibly suffer?

Let f, w, and c represent the amounts of money in \$1000 units spent on fire, flood (water), and collision insurance in that order. We then have the following constrained optimization problem:

$$\begin{aligned} &\text{Min}\,\{\text{Max}\,(1000 - 100f,\ 200 - 500w,\ 500 - 50c)\} \\ &\text{subject to}\quad f + w + c \leq 20 \\ &\text{and}\qquad\qquad f, w, c \geq 0 \end{aligned}$$

That nonlinear objective function looks forbidding, but fortunately we have another trick at our disposal. Suppose we define a new variable z to be Sunshine's maximum possible loss under any given insurance plan, where an insurance plan consists of a set of feasible values of f, w, and c. Then the minimum value of z, which is what we are seeking, can be obtained by solving the following linear program:

$$\begin{aligned} &\text{Min } z \\ \text{subject to}\qquad & z \geq 1000 - 100f \\ & z \geq 200 - 500w \\ & z \geq 500 - 50c \\ & f + w + c \leq 20 \\ \text{and}\qquad & f, w, c \geq 0 \end{aligned}$$

For any feasible insurance plan, the first three constraints require z to be *at least as large* as the maximum loss, while the objective function guarantees that it will be *no larger*. Thus, z must exactly equal the maximum possible loss, and Sunshine's problem can be solved by finding the insurance plan that minimizes it. Applying the simplex method, we eventually discover that \$9875 should be spent on fire insurance, \$375 on flood, and \$9750 on collision, for a maximum possible net loss of \$12,500.

In the foregoing example Sunshine's managers followed what is known to game theorists as a *minimax strategy*. Such a strategy is fundamentally conservative: In effect, it assumes that the worst will happen and proceeds to guard against it insofar as available resources allow. The minimax approach may often be adopted by intelligent adversaries engaged in various forms of conflict (arms races, retail price wars, or political campaigns), particularly when each has enough information about the other to locate and attack his weakest point. On the other hand, a straightforward analysis of probabilities and expectations seems more appropriate when one's adversary is blind chance or Nature (as in Sunshine's insurance problem), essentially because leaving a critical point undefended does not guarantee or even increase the likelihood that it will be attacked.

6.9 PRODUCTION AND INVENTORY PLANNING

In our final example we consider the very common problem faced by any manufacturing facility that must schedule its production and maintain inventory in order to satisfy nonuniform customer demands. A number of variations will be discussed, and several of the algebraic maneuvers introduced earlier will reappear. We begin by stating the problem in its simplest form. A company that manufactures one product at a single plant is scheduled to ship out a total of D_i product units to its customers at the end of the ith time period, $i = 1, 2, 3, \ldots$. Management has decreed that *back orders* are prohibited; in other words, all orders must be filled on time. In any one period, up to M units can be manufactured during regular working hours at a basic per-unit cost of c_0 dollars (over and above the plant's fixed costs); additional units may be produced on an overtime basis for $c_0 + c$ dollars apiece, with overtime production not permitted to exceed N units in any period. If the cost of carrying inventory is taken to be r dollars per unit per period, what is the most economical production schedule, and what inventory levels will it call for in each period?

Before going further, we should point out that the problem as stated above may be a vast oversimplification of the real-world situation that gave rise to it. Production costs are never truly linear, for example, and the use of a linear approximation is justified only if the unknown production levels can be expected to fall within a region where the approximation is fairly accurate. The inventory carrying charge is linear insofar as it includes the cost of tied-up capital, but the component of it that reflects the cost of storage space may be a quite different sort of function. Finally, very few businesses enjoy the luxury of knowing what their future demands or sales will be. Even those that produce only in response to individual orders usually must make guesses about the orders ahead of time in order to decide how much of each raw material to buy. Unless future demands can be forecast with great precision, to treat them as ordinary constants, as done above, is grossly improper and introduces serious inaccuracies into the formulation. Demands that are even moderately uncertain in most cases should be modeled as random variables, although that approach unfortunately requires statistical solution methods that are much more complicated and less efficient than linear programming. We shall have more to say on this subject later on.

Returning to our basic production/inventory problem, let x_i and y_i be the numbers of units produced in the ith period during regular hours and overtime, respectively, and let I_i be the number of units in inventory immediately after the demand D_i is satisfied at the end of the ith period. Then for each $i = 1, 2, \ldots,$ we may write the following set of constraints:

$$I_i = I_{i-1} + x_i + y_i - D_i$$

$$x_i \leq M$$

$$y_i \leq N$$

and $$x_i, y_i, I_i \geq 0$$

where I_0 is the amount of inventory on hand initially. The total cost associated with any production schedule $\{x_i, y_i\}$ is given by

$$c_0 \sum_i x_i + (c_0 + c) \sum_i y_i + r \sum_i I_i \tag{6-18}$$

and this expression is to be minimized. In the long run, however, each unit demanded will be made either during regular hours at a cost of c_0 or during overtime at a cost of $c_0 + c$. Therefore, we may cancel out the c_0 charged for every unit and concentrate instead on minimizing the *extra* costs associated with overtime production. The objective function then can be represented more simply:

$$\text{Min } c \sum_i y_i + r \sum_i I_i \tag{6-19}$$

This completes the formulation of what turned out to be a linear program; for an N-period problem there would be $3N$ constraints, of which $2N$ would merely impose upper bounds, and $3N$ nonnegative variables.

A great many variations of this problem that arise in practice can be modeled within the linear framework; some of them will be quite familiar to the reader by now. In order to keep the presentation clear we shall discuss them individually.

Multiple Plants

In one of the simplest variations the company manufactures its products at two (or more) plants instead of one and may ship to any customer out of either. Let the following variables be defined for plant A and for period i:

x_{Ai} = number of units produced during regular hours

y_{Ai} = number of units of overtime production

d_{Ai} = number of units shipped out to customers

and I_{Ai} = number of units remaining in inventory at the end of the period

and let x_{Bi}, y_{Bi}, d_{Bi}, and I_{Bi} be defined analogously for plant B. The inventory level at each plant obeys the same sequential relationship as I_i did in the basic problem:

$$I_{Ai} = I_{A,i-1} + x_{Ai} + y_{Ai} - d_{Ai}$$

and

$$I_{Bi} = I_{B,i-1} + x_{Bi} + y_{Bi} - d_{Bi}$$

In addition, the units shipped must satisfy total customer demand in every period:

$$d_{Ai} + d_{Bi} = D_i$$

The remaining constraints are the various upper bounds on production levels. Production costs in general will differ at the two plants, so the objective function must be analogous to (6-18) rather than (6-19); using obvious notation, we have the following:

$$\text{Min } c_{A0} \sum_i x_{Ai} + (c_{A0} + c_A) \sum_i y_{Ai} + c_{B0} \sum_i x_{Bi} + (c_{B0} + c_B) \sum_i y_{Bi} + r \sum_i I_{Ai} + r \sum_i I_{Bi}$$

where the inventory carrying cost is assumed to be the same at both plants.

Limited Storage Capacity

Another very common variation on the basic production/inventory problem—we are returning to the single-plant formulation—arises whenever the space available for storing inventory is limited. In most cases the plant has set aside a storage area that has a capacity of, say, A product units. Simple upper-bound constraints of the form

$$I_i \leq A \qquad \text{for all } i$$

then must be added to the linear programming formulation.

Under certain circumstances, however, the company might be planning to lease (or build with borrowed capital) a new storage facility just large enough to house the expected peak inventory, where inventory levels will depend upon the production schedule that is yet to be determined. This situation could arise in a business in which customer orders varied consistently from month to month or from season to season, but repeated the same 12-month pattern every year. Such a company would want to consider the possibility of scheduling more overtime production than necessary during certain months in order to reduce the maximum inventory level required. If the cost of the storage facility were a linear function of its capacity—a dubious assumption, to be sure—the basic linear program could be modified to take it into account. The problem need be formulated only for a single demand cycle, with the optimal production and inventory schedule presumably being repeated in every succeeding cycle (note that an extra constraint is required to insure that the cycle begins and ends with the same number of units in inventory). Let z be the unknown maximum inventory level, and let R be the cost per unit of capacity of leasing and maintaining the storage facility for one demand cycle. Then the production schedule that minimizes overall cost is obtained by adding the constraints

$$I_i \leq z \qquad \text{for all } i$$

to the basic linear program and including the term Rz in the objective function (6-19), which now represents total cost per demand cycle. Because Rz is being minimized, z takes on the value of the largest inventory level—this, of course, is the same device that was used in the Sunshine Mining problem.

Multiple Products

It is somewhat more usual for a plant to have inventory problems in several products rather than in one. When this occurs, the sequence of inventory levels in each product will be governed by the familiar recursion rule

$$I_{ki} = I_{k,i-1} + x_{ki} + y_{ki} - D_{ki}$$

where I_{ki} is the number of units of product k in inventory at the end of the ith period, and so on. If the products are made on different machines or assembly lines, the upper limits on production are usually stated separately; that is, for product k

$$x_{ki} \leq M_k$$

and $$y_{ki} \leq N_k \qquad \text{for all } i$$

But when several different products are made on the same assembly line, production constraints are *linked*:

$$\sum_k x_{ki} \leq M$$

and $$\sum_k y_{ki} \leq N \qquad \text{for all } i$$

For expressing inventory storage capacity, *linking constraints* are almost always appropriate (except in special circumstances, as when different products must be stored in different vats or rooms). The simplest version is

$$\sum_k I_{ki} \leq A \qquad \text{for all } i$$

where A is the capacity in units of the inventory storage area. This, however, is not adequate when products differ in size, a situation that calls for a limit on total volume rather than on the total number of units. Letting v_k be the volume of each unit of product k and V be the total available storage volume, the capacity constraints become

$$\sum_k v_k I_{ki} \leq V \qquad \text{for all } i$$

Minimum Inventory Level

Let us return to the single-product environment and consider again the matter of demand uncertainty. We said earlier that the assumption of known demands is legitimate only when extremely accurate forecasts are available; this condition might obtain in businesses whose products are made in accordance with customer specifications (custom-tailored garments, for example) or in businesses with a great many individual customers (so that the "law of averages" can be relied upon to some extent). Our linear programming formulation, however, supposes that future demands are known *with absolute certainty*, not just approximately. The most important consequence of this treatment is that it is likely to yield an optimal production schedule that plans on exactly zero inventory in certain periods. Obviously such planning would be disrupted in the real world whenever demand in any of those periods proved to be even slightly greater than expected; too few product units would be available to fill all the orders and the infeasible condition of negative inventory would result. If the plant manager tried to make up the shortage by producing more units than had been planned, he might find that his "optimal" schedule as determined via linear programming already called for maximum production levels in the next few periods.

The best way of avoiding all these headaches—provided that the plant is committed to using a linear programming model—is to plan for a minimum inventory level, or *safety stock*, S in every period,

$$I_i \geq S \qquad \text{for all } i \tag{6-20}$$

In addition to preserving the linearity of the formulation, so that the problem can still be solved by the simplex algorithm, a properly chosen minimum inventory level provides a guarantee that the plant will be able to satisfy all or nearly all of its customer orders on time, even if demand forecasts turn out to be slightly inaccurate. The most important advantage of this simple method for dealing with uncertainty is that it does not require management to determine complicated probability distributions for all future demands; only a single parameter, the safety stock, needs to be specified. One reasonable procedure for doing so is to solve the linear program several times, using different values of S in (6-20). Then, knowing the minimum overall cost implied by various safety-stock levels, the plant manager can use his judgment to choose a suitable level in light of recent sales history, available capital, and so on. The cost of all the computation involved is not likely to be prohibitive because the simplex method is so efficient.[1]

Before going on, we shall repeat an earlier warning that linear programming techniques should *not* be employed when substantial uncertainty is present in the demand forecasts. In the real world this will quite often be the case: *a majority of inventory problems arising in practice are dealt with by statistical methods* rather than via linear programming.

Back Orders

In almost every manufacturing business, even those whose demand forecasts are very accurate, it sometimes happens that a customer order cannot be satisfied immediately because the product units requested are not available in inventory. Such an event is called a *stock-out*, and the unfilled order, also known as a *back order*, can be thought of as "negative inventory." Like positive inventory levels, back orders cost money, for a variety of reasons: Time is wasted in responding to complaints and making alternative arrangements, penalty charges may have to

[1] The general technique of solving a sequence of linear programs that differ only in the value of a single parameter, in this case the safety stock S, is called *parametric programming* (to be discussed in Section 7.7). The overall procedure is especially efficient because each optimal solution can be used as a starting basic solution for the next problem.

be paid or price discounts offered, the customer has a certain probability of taking his business elsewhere, and so on. These costs, however, are not infinite and in practice may be rather modest. Provided that management is able to quantify them, an overall *back-order penalty cost* can be calculated and introduced explicitly into the objective function of the production/inventory problem. It then follows that the optimal production schedule is the one that minimizes the sum of production, inventory, *and back-order* costs. Thus, for example, the company will want to consider the possibility of planning for negative inventories in certain periods in order to avoid higher inventory levels or overtime production in others.

Suppose management has decided that negative inventories are permissible and that it costs b dollars to have one unit back-ordered for one period. Then the penalty cost associated with any inventory level is given by the following function:

$$\text{Penalty cost in period } i = \begin{cases} rI_i & \text{if } I_i > 0 \\ 0 & \text{if } I_i = 0 \\ -bI_i & \text{if } I_i < 0 \end{cases}$$

We encountered a similar situation in the Uwanimus University problem (recall Figure 6-4), and the trick we used then also will serve us here. Substitute for each I_i in the set of constraints a new pair of nonnegative variables

$$I_i \equiv \alpha_i - \beta_i$$

where

$$\alpha_i, \beta_i \geq 0 \qquad \text{for all } i$$

Then remove the term

$$r \sum_i I_i$$

from the original objective function (6-19) and replace it with

$$r \sum_i \alpha_i + b \sum_i \beta_i$$

This is to be minimized, so the correct penalty cost will be assessed for any inventory level, and the linearity of the formulation is preserved.

Employment Levels

Another option sometimes available to management, particularly in labor-intensive industries, is the adjustment of maximum production capacities through the hiring and firing of workers. It may be economical, for example, to increase the work force during certain periods of heavy demand in order to avoid having to build up large and expensive inventories beforehand. At other times idle employees may have to be laid off. In general, the size of the work force cannot be adjusted free of charge; it costs money both to hire and to fire workers (personnel expenses and preliminary training in the former case, severance pay in the latter). But provided that the relationship between number of men working and number of units produced can be treated as linear, these new considerations all can be incorporated into our linear programming format.

Suppose that in one period a single worker can produce up to p units during regular working hours and up to q additional units during overtime. If n_i represents the number of employees during the ith period, the production-capacity constraints $x_i \leq M$ and $y_i \leq N$ are replaced by

$$x_i \leq pn_i$$

and

$$y_i \leq qn_i \qquad \text{for all } i$$

(recall that x_i and y_i represent the total numbers of units produced in the ith period during regular time and overtime, respectively). Because the work force is now variable, labor costs must be represented explicitly in the objective function. Let

w = regular wages per period paid to each employee

c_0 = material cost (includes everything except labor and fixed overhead) of producing one unit at any time

c = labor cost of producing one unit *during overtime*

Then the total of all manufacturing costs incurred in the ith period is

$$wn_i + c_0(x_i + y_i) + cy_i \tag{6-21}$$

where, as usual, the plant's overhead expenses have been ignored. Note that the labor costs of all units produced during regular working hours are covered by the "wages" term wn_i.

Other terms representing the penalties paid by the company for increasing or decreasing its work force must appear also in the objective function. Let the costs of hiring and of firing one employee be h and f dollars, respectively; then the "employment cost" incurred in the ith period is

$$h(n_i - n_{i-1}) \qquad \text{if } n_i > n_{i-1}$$

$$0 \qquad \text{if } n_i = n_{i-1}$$

or

$$f(n_{i-1} - n_i) \qquad \text{if } n_i < n_{i-1}$$

By now the student can probably see how to proceed: Add the set of equations

$$n_i - n_{i-1} = \gamma_i - \delta_i$$

where $\gamma_i, \delta_i \geq 0 \quad \text{for all } i$

to the constraint set and include in the objective function the terms

$$h \sum_i \gamma_i + f \sum_i \delta_i$$

which will correctly represent the employment costs. Assuming for simplicity that back orders are forbidden, the only remaining costs are the inventory carrying charges. Hence the optimal production/inventory/employment schedule can be obtained by solving the following linear program:

$$\text{Min} \sum_i (wn_i + cy_i + h\gamma_i + f\delta_i + rI_i)$$

subject to

$$\left.\begin{aligned} I_i &= I_{i-1} + x_i + y_i - D_i \\ x_i &\leq pn_i \\ y_i &\leq qn_i \\ n_i - n_{i-1} &= \gamma_i - \delta_i \\ \text{and} \quad x_i, y_i, I_i, n_i, \gamma_i, \delta_i &\geq 0 \end{aligned}\right\} \text{for all } i$$

The term $c_0(x_i + y_i)$ that appears in (6-21) was left out of the objective function because the material cost c_0 will inevitably be incurred for every unit demanded, regardless of the schedule chosen.

Multistage Manufacturing

As our final variation on the basic problem, we shall consider a simple example of *multistage processing* (see Figure 6-5). Suppose that a certain manufacturing process consists of two successive production operations and that after undergoing operation 1 each product unit must be temporarily stored in semifinished form in inventory 1, perhaps to allow a chemical reaction to take place. After operation 2, all finished units not needed to fill customer orders immediately are kept in inventory 2. For simplicity we suppose further that the manufacturing

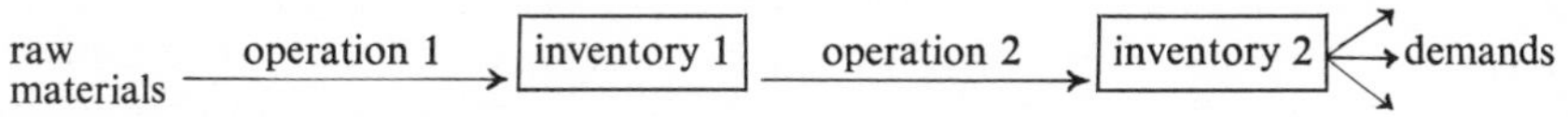

Figure 6-5

cycle is timed so that any product unit undergoing operation 2 during period i must have been sitting in inventory 1 at the end of period $i - 1$ (although it may have been stored there much earlier). The formulation will not be difficult; for each period i let

x_{1i} = number of product units undergoing operation 1 during regular hours

y_{1i} = number of units undergoing operation 1 during overtime

I_{1i} = number of units in inventory 1 at the end of the period

with x_{2i}, y_{2i}, and I_{2i} defined analogously for operation 2 and inventory 2. Then the equations governing inventory balances are

$$\left.\begin{aligned} I_{1i} &= I_{1,i-1} + x_{1i} + y_{1i} - x_{2i} - y_{2i} \\ x_{2i} &+ y_{2i} \le I_{1,i-1} \\ I_{2i} &= I_{2,i-1} + x_{2i} + y_{2i} - D_i \\ \text{and} \qquad I_{1i}&, I_{2i} \ge 0 \end{aligned}\right\} \text{for all } i$$

where we have assumed that no back orders are permitted and no safety stocks are required. If the plant has a fixed inventory capacity and a constant work force, the only remaining constraints are the upper bounds on the various production and inventory levels, which may or may not be linked. Finally, the objective function includes terms for overtime production costs and inventory carrying charges; note that, in general, the carrying cost per unit in inventory 1 will differ from that of inventory 2.

EXERCISES

General

Exercises 6-1 through 6-3. Formulate each of the following problems as a linear program.

6-1. A machine shop produces four types of ball bearings. Each week customers demand r_j bearings of type j, $j = 1, \ldots, 4$. Three adjustable machines are available for production, each of which can be operated up to 2400 minutes per week. In 1 minute the ith machine can produce a_{ij} ball bearings of type j at a cost of c_i dollars (that is, machine operating costs are independent of ball-bearing type); setup times and costs are assumed to be negligible. The machine shop's objective is to produce the required output within the time available and at a minimum cost. Formulate this problem twice, using different sets of variables:

(a) Let x_{ij} be the number of minutes per week during which machine i is producing bearings of type j.

(b) Let x_{ij} be the number of type-j bearings produced each week by machine i.

6-2. *The caterer's problem*: This is a famous problem from the annals of operations research literature. A professional caterer will require r_j fresh napkins on each of N successive days, $j = 1, \ldots, N$. These may be newly purchased or laundered after previous use. Suppose the laundry has two types of service, "fast" and "slow." The fast service costs f cents per napkin and requires m days; the slow service costs s and takes n days. Naturally, $m < n$ and $f > s$. If new napkins cost c cents apiece and the caterer starts with none, how can he meet his requirements at a minimum total cost? Assume the napkins have no resale value.

6-3. A college has five men's and five women's dormitories, in which m_i males and f_i females, $i = 1, \ldots, 5$, are initially housed. All ten buildings have a maximum capacity of C students, so that

$$m_i \leq C \quad \text{and} \quad f_i \leq C \qquad i = 1, \ldots, 5$$

At midsemester the administration decides to integrate the sexes and wishes to integrate the dormitories in such a way that the total population of every dorm will be between 40% and 60% male. Letting d_{ij} be the straight-line distance from male dorm i to female dorm j, how should students be shifted so as to minimize the total moving distance for all students? Note that we have not bothered to give the distances between any two men's dormitories. Why are they irrelevant? That is, why is it never optimal to shift a man from one male dormitory to another?

6-4. Let a penalty function of the following form be associated with each of the n unrestricted variables x_j:

$$p(x_j) = \begin{cases} x_j & \text{if } x_j \geq 0 \\ 0 & \text{if } x_j \leq 0 \end{cases}$$

Can linear programming techniques be used to minimize

$$\sum_{j=1}^{n} p(x_j)$$

subject to a set of linear constraints on the x_j?

6-5. Can the following optimization problem be formulated as a linear program?

$$\begin{aligned} &\text{Max}\,\{\text{Max}\,(\mathbf{c}_1{}^T\mathbf{x}, \quad \mathbf{c}_2{}^T\mathbf{x})\} \\ &\text{subject to} \quad \mathbf{A}\mathbf{x} \leq \mathbf{b} \\ &\text{and} \qquad\qquad \mathbf{x} \geq \mathbf{0} \end{aligned}$$

where the value of the objective function for any feasible solution $\hat{\mathbf{x}}$ is either $\mathbf{c}_1{}^T\hat{\mathbf{x}}$ or $\mathbf{c}_2{}^T\hat{\mathbf{x}}$, whichever is greater. If not, how can it be solved?

6-6. (a) Can the following problem be formulated as a single linear program?

$$\text{Min } z = \sum_{j=1}^{n} |x_j|$$

$$\text{subject to } \sum_{j=1}^{n} a_{ij}x_j \geq b_i \qquad i = 1, \ldots, m$$

$$\text{with } x_j \text{ unrestricted} \qquad j = 1, \ldots, n$$

where $|x_j|$ denotes the absolute value of x_j. Would your answer be the same if the objective function were to be maximized rather than minimized?

(b) Suppose the objective function were

$$\text{Min } z = \sum_{j=1}^{n} c_j|x_j|$$

What conditions would the coefficients c_j have to satisfy in order to allow the problem to be formulated as a single linear program?

(c) Suppose the objective function were

$$\text{Max } z = |\mathbf{c}^T\mathbf{x}|$$

where $\mathbf{c} > \mathbf{0}$. Could the problem be solved by linear programming methods? If so, how many individual linear programs would have to be solved in order to obtain the desired optimal solution?

(d) Repeat part (c) for the objective function

$$\text{Max } z = \sum_{j=1}^{n} c_j|x_j|$$

where $c_j > 0, j = 1, \ldots, n$.

6-7. Can the following problem be solved by linear programming techniques?

$$\text{Min } z = x_1^2$$

$$\text{subject to } \sum_{j=1}^{n} a_{ij}x_j \geq b_i \qquad i = 1, \ldots, m$$

$$\text{with } x_j \text{ unrestricted} \qquad j = 1, \ldots, n$$

Would your answer be the same if the objective function were

$$\text{Min } z = \sum_{j=1}^{n} x_j^2$$

What if it were

$$\text{Min } z = \left(\sum_{j=1}^{n} x_j\right)^2$$

7

THE DUAL SIMPLEX METHOD AND POSTOPTIMALITY PROBLEMS

7.1 INTRODUCTION AND MOTIVATION

It often happens that, after formulating and solving a linear programming problem, an engineer or operations research analyst wants also to perform certain additional computations, either to obtain more information about the problem just solved or to determine the optimal solutions of one or more other LPPs that are nearly identical to it. The need for this sort of "postoptimal" problem-solving may arise for any of several reasons:

(1) Sometimes an error is belatedly discovered in the original formulation—a wrong value used for one of the cost coefficients, perhaps, or a variable inadvertently omitted—and the analyst is then confronted with a new problem that differs only slightly from the one he has just solved.

(2) Another possibility is that the computational work may have to be performed on the basis of *incomplete information*. If the analyst must write his report before he knows, for example, whether a certain constraint will actually hold, the best he can do is to cover both alternatives by solving the linear program twice, once with and once without the questionable constraint. In such a case it would presumably be desirable for him to use whichever optimal solution he obtained first as a starting point for finding

the other. To take a second example, suppose the values of a few of the cost coefficients in some linear program are not known with certainty, so that the analyst must use estimates for them in order to solve the problem. He may then wish to perform a *sensitivity analysis* to determine, for each unknown cost coefficient c_j, either the range of c_j values over which his solution remains optimal or the rate of change of the optimal objective value with respect to changes in c_j. If the optimal solution proves to be highly sensitive to the values used for the unknown coefficients, he may recommend spending some time and money to obtain better estimates of them.

(3) A sensitivity analysis is useful also when, for one reason or another, the decision-maker considers choosing a policy that differs slightly from the optimal linear programming solution and wants to determine how much that deviation from the optimum will cost him.

(4) A variation on the theme of "incomplete information" is the situation in which the various elements of the linear program are known with certainty, but are to some extent within the power of the decision-maker to control. In such a case the analyst might be asked, for example, to solve an LPP both before and then after the addition of a certain constraint. This would enable him to calculate the overall impact of that constraint upon the optimal objective value and thereby determine whether or not it is worthwhile to pay the price of eliminating it. Generalizing this idea, it might even be desirable to solve a sequence of several linear programs in which, say, the right-hand side b_i of one of the constraints is progressively increased. Using these results, the optimal value of the objective then could be plotted as a function of b_i, allowing the most profitable level of investment in the ith resource (i.e., the most profitable value of b_i) to be determined.

We have just described several types of *postoptimality problems* and the reasons that an operations research analyst might have for trying to solve them. Generally speaking, they arise when relatively slight changes are made in the parameters or the structure of a given linear program *after* its optimal solution has been found. With any postoptimality problem the important question is always "Can the information contained in the optimal simplex tableau[1] be used efficiently to help solve the new LPP?" If a great many of the a_{ij}, b_i, and c_j of the original problem have been changed or if numerous variables and constraints have been added and deleted, the answer is likely to be "No; start again from scratch." But for small alterations the optimal tableau usually can be adjusted and used as the starting point for a further sequence of pivots.

Before we go on to discuss the various postoptimal solution procedures, however, a major digression is in order. Thus far the reader has learned how to

[1] In this and similar contexts the word "tableau" should be interpreted more generally to refer *either* to the standard simplex tableau of Section 4.8 *or* to the auxiliary matrix inverse (5-32) of the revised simplex method, depending on which type of pivoting is being used.

generate simplex pivots only under the highly restrictive condition that all basic variables must have nonnegative values at every stage. As we shall see later on, though, there are several important types of postoptimality problems in which the necessary modifications to the optimal tableau may introduce negative values for some of the variables. For example, suppose that, after obtaining an optimal solution with basis matrix $\mathbf{B}$ for some linear program, we discover that incorrect values were assigned to one or two components of the right-hand-side vector $\mathbf{b}$. Because the errors did not affect the columns of the activity matrix, $\mathbf{B}$ is still a legitimate and probably near-optimal basis for the problem we want to solve, and it would clearly be to our advantage to make whatever corrections are necessary in the "optimal" tableau so that we can continue pivoting from $\mathbf{B}$. Suppose further that when the values of the basic variables are recalculated via $\mathbf{x_B} = \mathbf{B}^{-1}\mathbf{b}$, one or more of them turn out to be negative. We would find ourselves—in our present state of knowledge—unable to proceed further because the pivoting methods we have studied all operate upon *basic feasible solutions* and make no provision for dealing with and getting rid of negative basic variables.

Fortunately, we need not be stymied by the appearance of negative variables. There is another pivoting procedure, called the *dual simplex method*, which is ideally suited for application to the problem just described. Although it is closely related to the simplex algorithms of Chapters 4 and 5, we have placed it in this chapter because in practice it is most often used in various types of postoptimality analysis. The next three sections will be devoted to the development of this dual simplex method, after which we shall be ready for a detailed discussion of postoptimality problems.

7.2 THE ASSOCIATED DUAL SOLUTION

We begin by introducing the notion of the *associated dual solution*, which is of crucial importance both in understanding the dual simplex method and in establishing it mathematically. Recall that in the proof of the optimality theorem (Theorem 4.4) we showed that if $\mathbf{B}$ is an optimal basis for the primal problem

$$\begin{aligned} &\text{Max } z = \mathbf{c}^T\mathbf{x} \\ &\text{subject to} \quad \mathbf{A}\mathbf{x} = \mathbf{b} \\ &\text{and} \qquad\qquad \mathbf{x} \geq \mathbf{0} \end{aligned}$$

then $\mathbf{u}_0{}^T = \mathbf{c_B}{}^T\mathbf{B}^{-1}$ is an optimal solution to the dual problem

$$\begin{aligned} &\text{Min } z' = \mathbf{b}^T\mathbf{u} \\ &\text{subject to} \quad \mathbf{A}^T\mathbf{u} \geq \mathbf{c} \end{aligned}$$

Let us generalize this idea in the following way: Given any basic solution to the primal problem, with basis matrix $\mathbf{B}$ and cost vector $\mathbf{c_B}$, the set of values

$$\hat{\mathbf{u}}^T = \mathbf{c_B}^T\mathbf{B}^{-1} \tag{7-1}$$

will be said to constitute the *associated solution* to the dual problem. Observe that a primal solution and its associated dual solution have the same objective value in their respective problems because

$$z = \mathbf{c_B}^T\mathbf{x_B} = \mathbf{c_B}^T\mathbf{B}^{-1}\mathbf{b} = \hat{\mathbf{u}}^T\mathbf{b} = z' \tag{7-2}$$

In particular, (7-2) holds for the *optimal* primal BFS; in this case we saw that the associated dual solution is optimal as well.

When a primal basic feasible solution is *not* optimal, however, neither is its associated dual solution, a fact that can be demonstrated in two different ways. First, we can remark simply that the value of the primal objective function z at a nonoptimal BFS is less than the optimal value of z. This implies that the associated dual solution has a lower (i.e., better) objective value than the dual optimum; thus the associated dual solution is "superoptimal" and must be infeasible. Second, we can obtain more direct evidence of dual infeasibility by evaluating the jth primal reduced cost

$$z_j - c_j = \mathbf{c_B}^T\mathbf{B}^{-1}\mathbf{a}_j - c_j = \hat{\mathbf{u}}^T\mathbf{a}_j - c_j \tag{7-3}$$

and comparing it with the jth dual constraint

$$\mathbf{a}_j^T\mathbf{u} \geq c_j$$

where $\mathbf{a}_j^T$ is the jth row of $\mathbf{A}^T$. The latter is equivalent to

$$\mathbf{u}^T\mathbf{a}_j - c_j \geq 0 \tag{7-4}$$

and it is evident from (7-3) and (7-4) that *the jth primal reduced cost is negative* —a condition that prevents the primal BFS from being optimal—*if and only if the associated dual solution $\hat{\mathbf{u}}^T$ violates the jth dual constraint.* In fact, the magnitude of the negative $(z_j - c_j)$ is just equal to the amount by which the jth dual constraint is being violated. Note that when the primal is a *minimization* problem, the dual constraints are of the form

$$\mathbf{a}_j^T\mathbf{u} \leq c_j$$

so that *positive* reduced costs correspond to unsatisfied dual constraints.

Incidentally, the simplex method can be given a rather interesting interpretation in light of this discussion. When a variable x_k enters the basis during a maximizing simplex pivot, its reduced cost is brought from a negative level up to zero (recall that $z_j - c_j = 0$ for all basic variables x_j). Thus, the pivot can be viewed as a mechanism for searching the dual constraints, finding one that is being violated, and generating a new associated dual solution that satisfies it! In this sense the simplex method is a stepwise procedure that reduces "dual infeasibility" while increasing the value of the primal objective until, at the final pivot, primal optimality and dual feasibility are simultaneously attained.

The foregoing properties of the associated dual solution can be illustrated using the sample problem that was solved in Section 4.8. Its representation in standard form was

$$\begin{aligned}
&\text{Max } z = 2x_1 + 4x_2 + 3x_3 + x_4 \\
&\text{subject to} \quad 3x_1 + x_2 + x_3 + 4x_4 + x_5 = 12 \\
&\qquad x_1 - 3x_2 + 2x_3 + 3x_4 + x_6 = 7 \\
&\qquad 2x_1 + x_2 + 3x_3 - x_4 + x_7 = 10 \\
&\text{and} \qquad x_i \geq 0 \quad i = 1, \ldots, 7
\end{aligned}$$

The dual of this problem is as follows:

$$\begin{aligned}
&\text{Min } z' = 12u_1 + 7u_2 + 10u_3 \\
&\text{subject to} \quad 3u_1 + u_2 + 2u_3 \geq 2 \\
&\qquad u_1 - 3u_2 + u_3 \geq 4 \\
&\qquad u_1 + 2u_2 + 3u_3 \geq 3 \\
&\qquad 4u_1 + 3u_2 - u_3 \geq 1 \\
&\qquad u_1 \geq 0 \\
&\qquad u_2 \geq 0 \\
&\text{and} \qquad u_3 \geq 0
\end{aligned}$$

where the three constraints $u_i \geq 0$ are associated with the slack variables x_5, x_6, and x_7. Let us skip to the second (primal) basic feasible solution, which was $\mathbf{x_B} = (x_5, x_6, x_2) = (2, 37, 10)$, with $\mathbf{c_B}^T = [0, 0, 4]$ and $z = \mathbf{c_B}^T\mathbf{x_B} = 40$. The initial basic columns $\mathbf{a}_5$, $\mathbf{a}_6$, and $\mathbf{a}_7$ formed an identity matrix, and we recall [see equation (5-28)] that the current basis inverse can be read directly from the last three columns of the tableau:

$$\mathbf{B}^{-1} = \begin{bmatrix} 1 & 0 & -1 \\ 0 & 1 & 3 \\ 0 & 0 & 1 \end{bmatrix}$$

This can be checked by writing out the basis matrix

$$\mathbf{B} = [\mathbf{a}_5 \,\vdots\, \mathbf{a}_6 \,\vdots\, \mathbf{a}_2] = \begin{bmatrix} 1 & 0 & 1 \\ 0 & 1 & -3 \\ 0 & 0 & 1 \end{bmatrix}$$

and then verifying that $\mathbf{B}^{-1}\mathbf{B} = \mathbf{I}_3$.

The dual solution associated with the current primal basis is

$$\hat{\mathbf{u}}^T = \mathbf{c}_\mathbf{B}{}^T\mathbf{B}^{-1} = [0, 0, 4]$$

which yields a dual objective value of 40; this equals the current value of the primal objective function, as expected. The reduced cost in the fourth column of the primal tableau is -5, which implies both that $\mathbf{x}_\mathbf{B}$ cannot be an optimal (maximal) solution to the primal problem and that $\hat{\mathbf{u}}^T$ violates the fourth constraint of the dual. Moreover, the value -5 should represent the exact amount by which that constraint is being violated, and so it does:

$$\hat{\mathbf{u}}^T\mathbf{a}_4 - c_4 = [0, 0, 4]\cdot(4, 3, -1) - 1 = -5$$

Note that all other primal reduced costs are nonnegative and that $\hat{\mathbf{u}}^T$ satisfies all the other dual constraints.

In the third and final primal tableau, all the reduced costs are nonnegative, so we know that optimality has been attained. The optimal solution is $\mathbf{x}_\mathbf{B} = (x_4, x_6, x_2) = (0.4, 37.0, 10.4)$, with $\mathbf{c}_\mathbf{B}{}^T = [1, 0, 4]$, $z = \mathbf{c}_\mathbf{B}{}^T\mathbf{x}_\mathbf{B} = 42.0$, and

$$\mathbf{B}^{-1} = \begin{bmatrix} 0.2 & 0 & -0.2 \\ 0.0 & 1 & 3.0 \\ 0.2 & 0 & 0.8 \end{bmatrix}$$

It follows that the new associated dual solution

$$\hat{\mathbf{u}}^T = \mathbf{c}_\mathbf{B}{}^T\mathbf{B}^{-1} = [1, 0, 3]$$

also must be optimal. The reader may verify that $\hat{\mathbf{u}}^T$ satisfies all the dual constraints and has the same objective value, 42, as the primal optimum.

7.3 THE DUAL SIMPLEX METHOD

Insofar as its mode of operation is concerned, the dual simplex method is quite similar to the pivoting procedure described in Chapter 4, which in this context should be called the *primal* simplex method.[1] Both methods solve the standard-form linear programming problem by generating a sequence of pivots that proceeds from one basic solution to another until the optimum is reached. Where they differ is in the algebraic criteria they use to determine the variables entering and leaving the basis at each iteration: in this respect, as we shall see below, the dual simplex method is essentially the opposite of the primal.

It is important to bear in mind that in solving an LPP via a pivoting algorithm we are not committed to using any particular basis entry and exit criteria. Given a current basis, an entering variable x_k, and an exiting variable $x_{\mathbf{B}r}$, a simplex pivot is a straightforward and legitimate computational procedure that leads to a new basis. It can be executed for any nonbasic x_k and any $x_{\mathbf{B}r}$, subject only to the restriction that the pivot element y_{rk} be nonzero [of course, unless our simplex exit criterion (4-59) is used, the variables, in general, will not be restricted to nonnegative values]. Thus, from any one basis a great many pivots are possible, and we are free to devise whatever algebraic rules or criteria we like for choosing from them.

Once we have decided to use a pivoting procedure (and to pivot one variable at a time), the selection of the entry and exit criteria completely specifies an algorithm for solving linear programming problems. The algorithm will be said to be *efficient* if, in comparison with other possible approaches, it obtains optimal solutions with relatively little computational labor. We say also that an algorithm is *automatic* if, given any basis, its criteria always lead to unique choices; in that case it can be coded and applied to any problem by a "mindless" computer.

For example, one rather poor, but perfectly valid, algorithm for solving a standard-form LPP would be to begin with any basic solution, not necessarily nonnegative, and then carry out pivots by choosing at random the variables to enter and leave the basis, with the proviso that the pivot element be nonzero. After each iteration a test would be performed to see whether $\mathbf{x_B} \geq \mathbf{0}$ *and* all $(z_j - c_j) \geq 0$. If so, $\mathbf{x_B}$ would be a maximizing solution; if not, the process would continue. This, of course, usually would take an absurdly long time.

Far more efficient algorithms than this have been devised, and the best of them seems to be the *simplex*, or *primal simplex*, *method* presented in Chapter 4. Unlike the previous "technique," the primal simplex method is automatic, provided that some simple tie-breaking rules are furnished. Its strategy is to begin with one basic *feasible* solution $\mathbf{x_B} \geq \mathbf{0}$ and to pivot to another, keeping

[1] In this connection, remember that the pivoting procedure of Chapter 4 may also be called the *standard* simplex method when necessary to distinguish it from the revised simplex method. The standard/revised dichotomy refers to bookkeeping differences, the primal/dual dichotomy to differences in the criteria for choosing pivots (discussion to follow).

all variables nonnegative and improving (or maintaining) the value of the objective z at each iteration. When all reduced costs are nonnegative, the maximum solution has been obtained.

The strategy of seeking a solution with all $(z_j - c_j) \geq 0$, subject to a side condition $\mathbf{x_B} \geq \mathbf{0}$ that is maintained at every step, suggests a directly opposite approach to solving linear programs. Why not begin with a superoptimal basic solution, having all $(z_j - c_j) \geq 0$, but one or more $x_{\mathbf{B}i} < 0$, and attempt to find a sequence of pivots that maintains the nonnegativity of the reduced costs and leads eventually to a basic *feasible* solution? We ought to be able to develop pivoting rules that bring the objective z along a monotonically "worsening" path from its superoptimal starting point to its optimal feasible value, just as z monotonically improves in the primal simplex method.

These expectations are realized, in fact, in the well-known *dual simplex method*, an algorithm constructed by C. E. Lemke [19]. Its name refers to the fact that *dual feasibility* is preserved during its operation, just as primal feasibility is satisfied by all basic solutions arising in the primal simplex method. More precisely, by "preserving dual feasibility" we mean that when the problem

$$\begin{aligned} &\text{Max } z = \mathbf{c}^T\mathbf{x} \\ &\text{subject to} \quad \mathbf{A}\mathbf{x} = \mathbf{b} \\ &\text{and} \qquad\qquad \mathbf{x} \geq \mathbf{0} \end{aligned}$$

is being solved by the dual simplex method, the condition

$$z_j - c_j \geq 0 \qquad j = 1, \ldots, n$$

is enforced at every step. It follows that for each basic solution $\mathbf{x_B}$ the associated dual solution

$$\hat{\mathbf{u}}^T = \mathbf{c_B}^T\mathbf{B}^{-1} \tag{7-1}$$

satisfies every constraint of the dual problem and thus can be said to be "dual feasible"; this was demonstrated in Section 7.2. Note that, in the case of a primal *minimization* problem, the associated dual solution (7-1) satisfies all the dual constraints if and only if $(z_j - c_j) \leq 0, j = 1, \ldots, n$.

The dual simplex method begins with a basic solution to the primal maximization (or minimization) problem having all reduced costs greater (less) than or equal to zero. Unfortunately, most problems have no obvious initial basic solution with this dual feasibility property, and for these the dual simplex method is not particularly useful. Computational procedures have been invented, in fact, which can get the dual simplex method started, but, in general, they require considerably more labor than the more straightforward two-phase primal simplex method. There is one particular type of linear program, however, to which the dual simplex method may be applied quite conveniently because an initial basic solution with the dual feasibility property is immediately available. We refer to LPPs of the following format:

$$\begin{aligned} &\text{Min } z = \mathbf{c}^T\mathbf{x} \\ &\text{subject to} \quad \mathbf{Ax} \geq \mathbf{b} \\ &\text{and} \quad \mathbf{x} \geq \mathbf{0} \\ &\text{with} \quad \mathbf{c} \geq \mathbf{0} \end{aligned}$$

where $\mathbf{A}$ is, as usual, an $m \times n$ matrix. The constraints are converted to standard form by the addition of a vector $\mathbf{x}_s$ of nonnegative surplus variables:

$$\mathbf{Ax} - \mathbf{I}_m\mathbf{x}_s = \mathbf{b}$$

where $\mathbf{I}_m$ is an $m \times m$ identity matrix. Consider the initial basic solution

$$\mathbf{x_B} = \mathbf{x}_s = -\mathbf{b}$$

with basis matrix $\mathbf{B} = -\mathbf{I}_m$. Because $\mathbf{c_B}^T = \mathbf{0}$, the reduced costs are

$$z_j - c_j = \mathbf{c_B}^T\mathbf{B}^{-1}\mathbf{a}_j - c_j = -c_j \leq 0 \qquad j = 1, \ldots, n$$

which is precisely the dual feasibility requirement for a minimization problem. Therefore, $\mathbf{x_B}$ can be used to initiate the dual simplex method. Note that if $\mathbf{b} \leq \mathbf{0}$, then $\mathbf{x_B}$ is already a feasible solution to the primal problem and immediately must be optimal; this follows from the minimization version of Theorem 4.4 and can be checked easily by inspection of the given problem. But if even one component of $\mathbf{b}$ is positive, the basic solution $\mathbf{x_B}$ is not feasible and work remains to be done.

From now on let us assume we are solving a *minimization* linear program in standard form:

$$\begin{aligned} &\text{Min } z = \mathbf{c}^T\mathbf{x} \\ &\text{subject to} \quad \mathbf{Ax} = \mathbf{b} \\ &\text{and} \quad \mathbf{x} \geq \mathbf{0} \end{aligned} \tag{7-5}$$

All of our results and rules will have obvious parallels for the case of maximization. Before proceeding to develop the computational steps of the dual simplex method, we make the observation that *if we can find a basic solution* $\mathbf{x_B}$ *with all reduced costs nonpositive* (*dual feasibility*), *then the primal problem does not have an unbounded minimum.* This follows from the existence theorem [3.7(b)]; by hypothesis we have a feasible solution $\hat{\mathbf{u}}^T = \mathbf{c_B}^T\mathbf{B}^{-1}$ to the dual problem

$$\begin{aligned} &\text{Max } z' = \mathbf{u}^T\mathbf{b} \\ &\text{subject to} \quad \mathbf{A}^T\mathbf{u} \le \mathbf{c} \\ &\text{with} \quad \mathbf{u} \text{ unrestricted} \end{aligned} \tag{7-6}$$

so the primal cannot be unbounded.

Suppose, then, that we currently have a basis $\mathbf{B}$ and basic solution $\mathbf{x_B} = \mathbf{B}^{-1}\mathbf{b}$ for the given primal problem (7-5), with all $(z_j - c_j) \le 0$, but with one or more $x_{\mathbf{B}i} < 0$. Because $\hat{\mathbf{u}}^T = \mathbf{c_B}^T\mathbf{B}^{-1}$ is a feasible solution to the dual problem (7-6), its objective value must be less than or equal to the optimal value of z'. By the duality theorem, the latter is the same as the optimal value of the primal objective, so for the current basic solution

$$z = \mathbf{c_B}^T\mathbf{x_B} = \mathbf{c_B}^T\mathbf{B}^{-1}\mathbf{b} = \hat{\mathbf{u}}^T\mathbf{b} \le z'_{\max} = z_{\min}$$

It follows that the primal objective function currently has an optimal or superoptimal value; that is, z is less than or equal to its minimum *feasible* value $z_{\min}$.

In general $z < z_{\min}$; we therefore want to develop a sequence of pivots that will increase the primal objective as rapidly as possible. Recall the transformation formula giving the change in z during a simplex pivot: When x_k replaces $x_{\mathbf{B}r}$ in the basis, the new value (denoted, as usual, by a hat) is

$$\hat{z} = z - \frac{x_{\mathbf{B}r}}{y_{rk}}(z_k - c_k) \tag{4-74}$$

Because our goal is to evolve a nonnegative basic solution (maintaining dual feasibility), we want to select a negative variable $x_{\mathbf{B}r}$ to be driven to zero and removed from the basis. Thus, from (4-74), because $(z_k - c_k) \le 0$ for any entering variable x_k, the new objective value $\hat{z}$ will be greater than the old, as desired, *only if we choose an entering variable with* $y_{rk} < 0$.

At this point a question comes up: How shall we proceed if, having chosen $x_{\mathbf{B}r} < 0$ to leave the basis, we cannot find any nonbasic variable x_j such that $y_{rj} < 0$? Taking a cue from the unboundedness theorem (4.5), we might conjecture that this condition is the signal for infeasibility of the primal problem. That conjecture would be accurate.

THEOREM 7.1. Suppose some $x_{\mathbf{B}r}$ is negative in a dual simplex tableau, where all $z_j - c_j \leq 0$ (minimization). If $y_{rj} \geq 0$ for all nonbasic variables x_j, then the primal problem has no feasible solution.

The dual of the given minimization problem can be represented as

$$\text{Max } z' = \mathbf{u}^T\mathbf{b}$$
$$\text{subject to} \quad \mathbf{u}^T\mathbf{a}_j \leq c_j \qquad j = 1, \ldots, n$$
$$\text{with} \quad \mathbf{u} \text{ unrestricted}$$

where $\mathbf{a}_j$ denotes the jth column of $\mathbf{A}$. We are given that $\hat{\mathbf{u}}^T = \mathbf{c_B}^T\mathbf{B}^{-1}$ is a feasible solution to the dual program, where $\mathbf{B}^{-1}$ and $\mathbf{c_B}$ are the current primal basis inverse and basic cost vector. Consider the vector

$$\mathbf{u}_0{}^T = \hat{\mathbf{u}}^T - \theta\boldsymbol{\beta}_r \tag{7-7}$$

where $\boldsymbol{\beta}_r$ denotes the rth row of $\mathbf{B}^{-1}$. If θ is any positive real number, $\mathbf{u}_0{}^T$ must be a feasible solution to the dual problem because

$$\begin{aligned} \mathbf{u}_0{}^T\mathbf{a}_j &= \hat{\mathbf{u}}^T\mathbf{a}_j - \theta\boldsymbol{\beta}_r\mathbf{a}_j \\ &= \hat{\mathbf{u}}^T\mathbf{a}_j - \theta y_{rj} \\ &\leq \hat{\mathbf{u}}^T\mathbf{a}_j \leq c_j \end{aligned}$$

The first inequality follows from the nonnegativity of θ and y_{rj} (it should be remarked that if x_j is in the basis, then y_{rj} is simply an element of the identity matrix and is therefore nonnegative), and the second holds because $\hat{\mathbf{u}}^T$ is a feasible solution to the dual.

The dual objective value of this feasible solution $\mathbf{u}_0{}^T$ is

$$z' = \mathbf{u}_0{}^T\mathbf{b} = \hat{\mathbf{u}}^T\mathbf{b} - \theta\boldsymbol{\beta}_r\mathbf{b} = \hat{\mathbf{u}}^T\mathbf{b} - \theta x_{\mathbf{B}r}$$

But by our hypothesis $x_{\mathbf{B}r}$ is negative; therefore as θ approaches infinity in (7-7), z' increases without bound while $\mathbf{u}_0{}^T$ remains feasible. It follows that the dual problem is unbounded and, from the existence theorem [3.7(b)], that the primal has no feasible solution, Q.E.D.

Returning to the development of the dual simplex method, suppose we have chosen $x_{\mathbf{B}r} < 0$ to leave the basis, where our criterion for making this choice is simply that $x_{\mathbf{B}r}$ is the most negative of the variables. We now want to select a variable x_k to enter the basis in such a way that dual feasibility will be maintained. Recall that the simplex transformation formula for the reduced costs is

$$\hat{z}_j - c_j = (z_j - c_j) - \frac{y_{rj}}{y_{rk}}(z_k - c_k) \tag{4-70}$$

It has been established that y_{rk} must be less than zero if z is to increase toward $z_{\min}$, as desired. Because $(z_k - c_k) \leq 0$ no matter which variable enters, it is clear from (4-70) that for any nonbasic variable x_j whose y_{rj} is *positive* or *zero*, the reduced cost after the pivot always will be

$$\hat{z}_j - c_j \leq z_j - c_j \leq 0$$

But for those variables x_j with *negative* y_{rj}, we can ensure $(\hat{z}_j - c_j) \leq 0$ only by requiring that

$$z_j - c_j \leq \frac{y_{rj}}{y_{rk}}(z_k - c_k)$$

or, dividing through by the negative quantity y_{rj},

$$\frac{z_j - c_j}{y_{rj}} \geq \frac{z_k - c_k}{y_{rk}}$$

This establishes our *dual simplex entry criterion*: Given that $x_{\mathbf{B}r}$ is to pivot out of the basis, the variable x_k must enter, where

$$\frac{z_k - c_k}{y_{rk}} = \min_j \left(\frac{z_j - c_j}{y_{rj}},\quad y_{rj} < 0\right) \tag{7-8}$$

If this criterion cannot be applied because no y_{rj} is negative, then Theorem 7.1 guarantees that the given problem is infeasible.

Having determined $x_{\mathbf{B}r}$ and x_k, it remains only to perform the pivot. The computations required are precisely the same as those of a primal simplex pivot and thus are given by the transformation formulas developed in Chapter 4. As we have seen, the objective z must increase [unless the reduced cost $(z_k - c_k)$ of the entering variable is zero] toward its minimum feasible value, while the reduced costs remain nonpositive. Notice also that because the exiting variable $x_{\mathbf{B}r}$ and the pivot element y_{rk} are both negative, the newly entering variable

$$\hat{\mathbf{x}}_{\mathbf{B}r} = \hat{x}_k = \frac{x_{\mathbf{B}r}}{y_{rk}} \tag{4-60}$$

assumes a positive value. Thus, although the values of the other $x_{\mathbf{B}i}$ will change unpredictably, the pivot is known to be substituting one positive basic variable for a negative one, thereby taking "on the average" a step toward primal feasibility.

After the pivot, the new basic variables may all be nonnegative. If so, primal and dual feasibility have been obtained, and by Theorem 4.4 (minimization version) the optimal solution has been found. If not, another pivot will be required.

7.4 SUMMARY

The dual simplex method, using standard pivots, solves the *minimization* problem (7-5) by executing the following steps:

(1) Begin with a basic solution $\mathbf{x_B}$ to the primal problem (7-5) with all $(z_j - c_j) \leq 0$. If none exists, the problem has an *unbounded* minimum (unless both primal and dual have no feasible solutions, but that is highly unlikely).

(2) If the current $\mathbf{x_B}$ is nonnegative, it is *optimal* and the algorithm terminates. But if one or more of the $x_{\mathbf{B}i}$ are negative, go on to step 3.

(3) *Exit criterion.* Choose the variable with the most negative value to leave the basis; that is, $x_{\mathbf{B}r}$ exits where

$$x_{\mathbf{B}r} = \min_i \,(x_{\mathbf{B}i})$$

(4) *Entry criterion.* If y_{rj} is nonnegative for all nonbasic variables x_j, the problem has *no feasible solution.* But if at least one $y_{rj} < 0$, then choose x_k to enter the basis, where

$$\frac{z_k - c_k}{y_{rk}} = \min_j \left(\frac{z_j - c_j}{y_{rj}}, \quad y_{rj} < 0\right)$$

(5) Perform the pivot, calculating the new values of the variables, the objective function, the $\mathbf{y}$ columns, and the reduced costs, according to the simplex transformation formulas. Return to step 2.

When the dual simplex method is used to solve a *maximization* problem, the above steps must be modified slightly. The value of the objective function is initially greater than its optimal feasible value and is reduced rather than increased at each iteration, the dual feasibility condition

$$z_j - c_j \geq 0 \qquad j = 1, \ldots, n$$

being preserved throughout. The ultimate goal is to eliminate primal infeasibility, so the exit criterion is the same as in step 3. It follows from the transformation formula

$$\hat{z} = z - \frac{x_{\mathbf{B}r}}{y_{rk}}(z_k - c_k) \tag{4-74}$$

that the pivot element y_{rk} again must be negative if z is to decrease toward its optimal value. If we now examine the reduced-cost transformation

$$\hat{z}_j - c_j = (z_j - c_j) - \frac{y_{rj}}{y_{rk}}(z_k - c_k) \tag{4-70}$$

we can see that

$$\hat{z}_j - c_j \geq z_j - c_j \geq 0$$

whenever $y_{rj} \geq 0$. But when $y_{rj} < 0$, the new value of the jth reduced cost will be nonnegative if and only if

$$\frac{z_j - c_j}{y_{rj}} \leq \frac{z_k - c_k}{y_{rk}}$$

Therefore the *entry criterion for maximization* is as follows: Given that $x_{\mathbf{B}r}$ is to leave the basis, the variable x_k enters, where

$$\frac{z_k - c_k}{y_{rk}} = \max_j \left(\frac{z_j - c_j}{y_{rj}}, \quad y_{rj} < 0 \right) \tag{7-9}$$

Note that the quotients compared in (7-9) are nonpositive, so that the one having the smallest magnitude will be selected; similarly, the primal simplex exit criterion (4-59) and the dual simplex minimization entry criterion (7-8) also choose the "smallest-magnitude" quotient.

Let us use the dual simplex method with standard pivots to solve the following linear programming problem:

$$\text{Min } z = 3x_1 + 6x_2 + x_3$$

$$\begin{aligned} \text{subject to} \quad x_1 + x_2 + x_3 &\geq 6 \\ x_1 - 5x_2 - x_3 &\geq 4 \\ x_1 + 5x_2 + x_3 &\geq 24 \end{aligned}$$

$$\text{and} \qquad x_1, x_2, x_3 \geq 0$$

First, the constraints are converted to standard form by the addition of surplus variables:

$$\begin{aligned} x_1 + \ x_2 + x_3 - x_4 \qquad\qquad &= 6 \\ x_1 - 5x_2 - x_3 \qquad - x_5 \qquad &= 4 \\ x_1 + 5x_2 + x_3 \qquad\qquad - x_6 &= 24 \\ \text{with} \quad x_j \geq 0 \quad j = 1, \ldots, 6 \end{aligned}$$

The surplus variables form an initial basis with $\mathbf{B}^{-1} = -\mathbf{I}_3$. For this basis $\mathbf{c_B} = \mathbf{0}$, so the reduced cost of each variable x_j is given by

$$z_j - c_j = \mathbf{c_B}^T\mathbf{B}^{-1}\mathbf{a}_j - c_j = -c_j$$

and the value of the objective function is

$$z = \mathbf{c_B}^T\mathbf{x_B} = 0$$

The $\mathbf{y}$ columns also may be written directly because

$$\mathbf{y}_j \equiv \mathbf{B}^{-1}\mathbf{a}_j = -\mathbf{a}_j \qquad j = 1, \ldots, 6$$

Without further ado, we can construct the initial simplex tableau according to the usual format:

	x_1	x_2	x_3	x_4	x_5	x_6
$x_{B1} = x_4 = -6$	-1	-1	-1	1	0	0
$x_{B2} = x_5 = -4$	-1	5	1	0	1	0
$x_{B3} = x_6 = -24$	-1	-5	-1	0	0	1
Min $z =$ 0	-3	-6	-1	0	0	0

All reduced costs are nonpositive, so this basic solution satisfies the dual feasibility condition; thus we know already that the problem cannot have an unbounded minimum. The current $\mathbf{x_B}$ is not yet feasible, however, and we select $x_{B3} = x_6$ to leave the basis. Applying the entry criterion,

$$\operatorname*{Min}_j \left(\frac{z_j - c_j}{y_{3j}}, \quad y_{3j} < 0\right) = \min\left(\frac{-3}{-1}, \frac{-6}{-5}, \frac{-1}{-1}\right) = \frac{-1}{-1} = \frac{z_3 - c_3}{y_{33}}$$

so the variable x_3 enters the basis. We form the computational vector $\mathbf{\Phi}$ about the pivot element $y_{33} = -1$:

$$\mathbf{\Phi} = (-1, 1, -2, -1)$$

The new tableau is then

		x_1	x_2	x_3	x_4	x_5	x_6
$x_{B1} = x_4 =$	18	0	4	0	1	0	-1
$x_{B2} = x_5 =$	-28	-2	0	0	0	1	1
$x_{B3} = x_3 =$	24	1	5	1	0	0	-1
Min $z =$	24	-2	-1	0	0	0	-1

The value of z has increased to 24 and dual feasibility has been maintained, but because $x_5 < 0$, another pivot is required. The exiting variable must be $x_{B2} = x_5$, so x_1 enters. The pivot element is $y_{21} = -2$ and $\mathbf{\Phi} = (0, -1.5, 0.5, -1.0)$, leading to the next tableau:

		x_1	x_2	x_3	x_4	x_5	x_6
$x_{B1} = x_4 =$	18	0	4	0	1	0	-1.0
$x_{B2} = x_1 =$	14	1	0	0	0	-0.5	-0.5
$x_{B3} = x_3 =$	10	0	5	1	0	0.5	-0.5
Min $z =$	52	0	-1	0	0	-1.0	-2.0

Because this basic solution is feasible, we have obtained the optimal solution to the given problem; the minimum value of the objective function is 52.

7.5 POSTOPTIMALITY PROBLEMS

With the dual simplex method added to our arsenal, we are now ready for a full-scale attack on postoptimality problems. As we noted in Section 7.1, these problems are concerned with exploiting the information contained in the optimal simplex tableau of some previously solved linear program, for the purpose of solving one or more other LPPs that differ from it only slightly. There are several different types of postoptimality problems, but they can be grouped rather arbitrarily into three general categories:

(1) *Problems involving discrete parameter changes.* These arise when the numerical values of a few of the a_{ij}, the b_i, or the c_j are altered—often in

order to correct keypunch errors, misplaced decimals, or other defects in the original formulation.

(2) *Problems involving continuous parameter changes.* This category includes problems in which one or more of the parameters are perturbed or varied in some given direction, either in order to see how much perturbation can be tolerated before the solution originally obtained ceases to be optimal (*sensitivity analysis*) or in order to determine the sequence of basic solutions that become optimal as the perturbation is extended further and further (*parametric programming*).

(3) *Problems involving structural changes.* These arise when the analyst reformulates the problem by adding or deleting constraints or variables, presumably in order to reflect an alternate operating mode or a fundamental change in the real-world system that his model represents.

Of course, it is possible to imagine a great variety of more complicated problems involving different combinations of structural and parametric changes.

In the next few sections we shall discuss seven or eight basic types of postoptimality problems, covering all three of the above categories. Our attention will be restricted to the simpler perturbations of the original linear program; the more complicated variations will be left to the exercises and to the reader's ingenuity. For most problems, as we shall see, figuring out how to modify the optimal tableau and proceed to the new solution is a fairly straightforward exercise in linear algebra. It should be borne in mind, however, that when extensive modifications are required, the postoptimal computations may well become so tedious that it is more efficient simply to re-solve the new LPP from the beginning.

Henceforth we shall assume that an optimal basis matrix $\mathbf{B}$ and basic solution $\mathbf{x_B} = \mathbf{B}^{-1}\mathbf{b}$ have been obtained for the linear programming problem P:

$$\begin{aligned} &\text{Max } z = \mathbf{c}^T\mathbf{x} \\ &\text{subject to} \quad \mathbf{Ax} = \mathbf{b} \\ &\text{and} \qquad\quad \mathbf{x} \geq \mathbf{0} \end{aligned}$$

where $\mathbf{A}$ is $m \times n$. We shall assume further that $\mathbf{B}^{-1}$ is explicitly available, either because the revised simplex method was used or because it can be read directly from the slack variable columns of the final standard simplex tableau.

7.6 PROBLEMS INVOLVING DISCRETE PARAMETER CHANGES

Of all postoptimality problems in this category, the easiest to handle are those involving changes in the cost vector $\mathbf{c}$. Suppose that, after obtaining the optimal solution to problem P, the analyst wants to substitute a new cost vector $\mathbf{c}^*$ and solve the resulting linear program

$$\begin{aligned} &\text{Max } z^* = \mathbf{c}^{*T}\mathbf{x} \\ &\text{subject to} \quad \mathbf{Ax} = \mathbf{b} \\ &\text{and} \qquad\qquad \mathbf{x} \geq \mathbf{0} \end{aligned}$$

(incidentally, z^* is read as "z star," and so on). The constraint set is not affected at all, so the formerly optimal basic solution $\mathbf{x_B} = \mathbf{B}^{-1}\mathbf{b}$ remains at least feasible. If standard simplex pivots are used, the tableau columns $\mathbf{y}_j = \mathbf{B}^{-1}\mathbf{a}_j$ also are unaffected, and it is necessary only to recalculate the objective function and the reduced costs. Their new values are given by

$$z^* = \mathbf{c}_\mathbf{B}^{*T}\mathbf{x_B} \tag{7-10}$$

and

$$z_j^* - c_j^* = \begin{cases} 0 & \text{if } x_j \text{ basic} \\ \mathbf{c}_\mathbf{B}^{*T}\mathbf{y}_j - c_j^* & \text{if } x_j \text{ nonbasic} \end{cases} \tag{7-11}$$

where we adopt the policy of labeling all postoptimally adjusted values with asterisks. When the revised simplex method is used, the vector $\mathbf{c_B}^T\mathbf{B}^{-1}$ in the bottom row of the auxiliary matrix inverse (5-32) is recalculated instead, and the new values of the reduced costs are then obtained via

$$z_j^* - c_j^* = (\mathbf{c}_\mathbf{B}^{*T}\mathbf{B}^{-1})\mathbf{a}_j - c_j^*$$

In either case, if all $(z_j^* - c_j^*)$ are nonnegative, the current solution $\mathbf{x_B}$ is already optimal for the new problem; if not, pivoting proceeds via the primal simplex method until the optimum is found. Note that when all of the altered cost coefficients are associated with nonbasic variables (i.e., when $c_j^* = c_j$ for all basic x_j), the postoptimal adjustments are even simpler because z^* and the z_j^* do not have to be recalculated.

Although the computations (7-11) are quite straightforward, they do involve a fairly large number of multiplications. Unless $\mathbf{c}^*$ differs from $\mathbf{c}$ in almost every component, it is much more efficient to recalculate the $(z_j^* - c_j^*)$ as follows. Let the *change* in the jth cost coefficient be $\Delta c_j \equiv c_j^* - c_j$, $j = 1, \ldots, n$. This implies $\mathbf{c}_\mathbf{B}^* = \mathbf{c_B} + \Delta\mathbf{c_B}$, where the components of $\Delta\mathbf{c_B}$ are the respective changes in the basic cost coefficients. The adjusted value of the jth reduced cost, then, is given by

$$\begin{aligned} z_j^* - c_j^* = \mathbf{c}_\mathbf{B}^{*T}\mathbf{y}_j - c_j^* &= \mathbf{c_B}^T\mathbf{y}_j + (\Delta\mathbf{c_B})^T\mathbf{y}_j - c_j - \Delta c_j \\ &= (z_j - c_j) + \sum_{i=1}^{m} (\Delta c_{\mathbf{B}i})y_{ij} - \Delta c_j \end{aligned} \tag{7-12}$$

(Of course, if x_j is in the basis, then $z_j^* - c_j^* = 0$.) Notice that for any nonbasic x_j the transformation (7-12) requires only one multiplication per *changed* basic cost coefficient; by comparison, (7-11) requires m multiplications regardless of the number of changes. A similar transformation is available for recalculating the objective value:

$$z^* = \mathbf{c}_\mathbf{B}^{*T}\mathbf{x}_\mathbf{B} = \mathbf{c}_\mathbf{B}{}^T\mathbf{x}_\mathbf{B} + (\Delta\mathbf{c}_\mathbf{B})^T\mathbf{x}_\mathbf{B} = z + \sum_{i=1}^{m} (\Delta c_{\mathbf{B}i})x_{\mathbf{B}i} \tag{7-13}$$

Postoptimality problems created by discrete changes in the right-hand-side vector $\mathbf{b}$ also are quite straightforward. Suppose that the m-component vector $\mathbf{b}^*$ is substituted for $\mathbf{b}$ in problem P and that no other changes are made, so that $x_{\mathbf{B}1}, \ldots, x_{\mathbf{B}m}$ continue to constitute a set of basic variables. Because the cost coefficients and the vectors $\mathbf{y}_j \equiv \mathbf{B}^{-1}\mathbf{a}_j$ all remain as they were, so do the reduced costs; this means that the adjusted solution still will have the dual feasibility property. In general, however, when the new values of the basic variables are recalculated via

$$\mathbf{x}_\mathbf{B}^* = \mathbf{B}^{-1}\mathbf{b}^* \tag{7-14}$$

some of them will turn out to be negative—if not, $\mathbf{x}_\mathbf{B}^*$ is immediately optimal—and it will be necessary to apply the dual simplex method in order to solve the problem. Note that, because a dual feasible solution already is known to exist, the new problem cannot have an unbounded optimum; on the other hand, it may be infeasible.

As was true of adjusting the reduced costs, obtaining the new values of the basic variables by means of a transformation formula usually will be more efficient than calculating them directly. Define $\Delta\mathbf{b} \equiv \mathbf{b}^* - \mathbf{b}$; then, using the identity (2-1),

$$\mathbf{x}_\mathbf{B}^* = \mathbf{B}^{-1}\mathbf{b}^* = \mathbf{B}^{-1}\mathbf{b} + \mathbf{B}^{-1}(\Delta\mathbf{b}) = \mathbf{x}_\mathbf{B} + \sum_{i=1}^{m} (\Delta b_i)\boldsymbol{\beta}_i \tag{7-15}$$

where $\boldsymbol{\beta}_i$ is the ith column of $\mathbf{B}^{-1}$. When only one or two of the Δb_i are different from zero, as is frequently the case, (7-15) requires only m or $2m$ multiplications, which is far fewer for large problems than the m^2 required by (7-14).

There remain to be considered those postoptimality problems that involve discrete changes in the activity coefficients a_{ij}. In order to simplify the discussion we shall require that all changed coefficients lie in the same column of the $\mathbf{A}$ matrix; that is, having solved problem P, we shall substitute the column $\mathbf{a}_q^*$ for $\mathbf{a}_q$ and then seek the optimal solution of the resulting linear program. It is easily seen that if x_q is nonbasic, the problem is quite simple: The current solution

$\mathbf{x_B} = \mathbf{B}^{-1}\mathbf{b}$ remains feasible, and it is necessary only to recalculate $\mathbf{y}_q^* = \mathbf{B}^{-1}\mathbf{a}_q^*$ and $(z_q^* - c_q) = \mathbf{c_B}^T\mathbf{y}_q^* - c_q$. If $(z_q^* - c_q)$ is nonnegative, $\mathbf{x_B}$ is already optimal; if not, pivoting proceeds via the primal simplex method. Note also that a transformation formula analogous to (7-15) can be written for $\mathbf{y}_q^*$.

Things become much more complicated when x_q is a basic variable, because the changes in $\mathbf{a}_q$ may disrupt both primal and dual feasibility, may destroy the basis (i.e., if $\mathbf{a}_q^*$ is a linear combination of the other basic columns), and may even introduce redundancy in the constraint set. This type of postoptimality problem should not even be attempted if the original linear program were solved by the revised simplex method; the postoptimal computations to be described below are based on the $\mathbf{y}$ columns, and calculating them all via $\mathbf{y}_j = \mathbf{B}^{-1}\mathbf{a}_j$ simply would take too much time. We therefore suppose that the optimal solution to problem P is available in a standard simplex tableau. The best way to proceed, then, is to try to pivot $x_{\mathbf{B}r} = x_q$ out of the basis under the most favorable circumstances possible *before* the substitution of $\mathbf{a}_q^*$ for $\mathbf{a}_q$ takes place. Accordingly, the tableau columns are scanned, one by one, in the hope of finding a nonbasic variable x_k with the property that

$$\frac{x_{\mathbf{B}r}}{y_{rk}} = \min_i \left(\frac{x_{\mathbf{B}i}}{y_{ik}}, \quad y_{ik} > 0\right) \tag{7-16}$$

This, of course, is the basis exit criterion of the primal simplex method. What we are trying to find is a pivot that will remove $x_{\mathbf{B}r} = x_q$ from the basis and produce another primal feasible solution (of course, that solution, in general, would *not* be dual feasible). Such a pivot would leave us with a "post-post-optimality" problem in which only a single nonbasic column $\mathbf{a}_q$ would have to be adjusted. This situation already has been discussed: Using the new basis inverse and basic cost vector, the values $\mathbf{y}_q^* = \mathbf{B}^{-1}\mathbf{a}_q^*$ and $(z_q^* - c_q) = \mathbf{c_B}^T\mathbf{y}_q^* - c_q$ would be calculated and entered in the tableau, and, regardless of the sign of $(z_q^* - c_q)$, the adjusted linear program then could be solved by the primal simplex method.

If no nonbasic variable x_k can be found such that (7-16) holds, the next best course of action is to use the *dual* simplex entry criterion (7-9) to select the variable that will replace $x_{\mathbf{B}r} = x_q$. Assuming that this criterion can be applied (i.e., that some y_{rj} is negative), the ensuing pivot will produce a new basic solution $\mathbf{x}_B$ which is again dual feasible (though not primal feasible). The new $\mathbf{B}^{-1}$ and $\mathbf{c}_B$ then can be used to calculate $\mathbf{y}_q^*$ and $(z_q^* - c_q)$, which replace $\mathbf{y}_q$ and $(z_q - c_q)$ in the tableau as usual. If $(z_q^* - c_q)$ turns out to be nonnegative, then the current $\mathbf{x_B}$ is a dual feasible solution to the adjusted linear program, which therefore can be solved directly by the dual simplex method. If $(z_q^* - c_q) < 0$, however, the current solution is neither primal nor dual feasible, and it is usually best to abandon the postoptimal computations and simply solve the new LPP from the beginning. In the latter case, one might be tempted to try a couple of improvised pivots in the hope of arriving at a basic solution with all $(z_j - c_j) \geq 0$; from that point it would be possible to proceed directly to the optimum via the dual simplex method. This informal approach, however, is highly uncertain and, therefore, is not recommended; while it might work out successfully for any given problem, it is more likely to lead into a morass of negative basic variables and negative reduced costs from which there is no clearly marked escape.

In the rare combination of circumstances that

(1) No nonbasic variable x_k having the property (7-16) can be found
(2) $y_{rj} \geq 0$ for all nonbasic x_j, so that the dual simplex entry criterion cannot be employed

the adjusted linear program should simply be re-solved from the beginning. Note that it does *not* follow from condition (2) that Theorem 7.1, maximization version, applies to the original problem.

Our discussion of the postoptimality problems that arise when a column of the **A** matrix is changed will be illustrated with a group of examples based on the following linear program P:

$$\text{Max } z = 3x_1 + 4x_2 + 2x_3 + 4x_4 + 2x_5 + 5x_6$$

$$\begin{aligned} \text{subject to} \quad x_1 + x_2 + x_3 + x_4 + 2x_6 &= 10 \\ 2x_2 + 2x_4 + x_5 + 3x_6 &= 6 \\ x_1 + 4x_2 + 2x_3 + 5x_4 + x_5 + 9x_6 &= 25 \end{aligned}$$

$$\text{and} \qquad x_i \geq 0 \qquad i = 1, \ldots, 6$$

Suppose this problem has been solved by means of the two-phase simplex algorithm, using standard pivots; the optimal tableau is

	x_1	x_2	x_3	x_4	x_5	x_6
$x_{B1} = x_1 = 1$	1	0	0	-1	0	-2
$x_{B2} = x_3 = 9$	0	1	1	2	0	4
$x_{B3} = x_5 = 6$	0	2	0	2	1	3
Max $z = 33$	0	2	0	1	0	3

and the associated basis inverse is[1]

$$\mathbf{B}^{-1} = [\mathbf{a}_1 \mid \mathbf{a}_3 \mid \mathbf{a}_5]^{-1} = \begin{bmatrix} 1 & 1 & 0 \\ 0 & 0 & 1 \\ 1 & 2 & 1 \end{bmatrix}^{-1} = \begin{bmatrix} 2 & 1 & -1 \\ -1 & -1 & 1 \\ 0 & 1 & 0 \end{bmatrix}$$

We shall consider four different postoptimality problems, each stemming from a discrete change in one of the activity coefficients a_{ij}.

First, let $a_{24}^* = 1$ in problem P. Because x_4 is nonbasic, we need compute only

$$\mathbf{y}_4^* = \mathbf{B}^{-1}\mathbf{a}_4^* = \begin{bmatrix} 2 & 1 & -1 \\ -1 & -1 & 1 \\ 0 & 1 & 0 \end{bmatrix} \begin{bmatrix} 1 \\ 1 \\ 5 \end{bmatrix} = \begin{bmatrix} -2 \\ 3 \\ 1 \end{bmatrix}$$

and $$z_4^* - c_4 = \mathbf{c_B}^T\mathbf{y}_4^* - c_4 = [3, 2, 2]\cdot(-2, 3, 1) - 4 = -2$$

The new reduced cost is negative, so the solution $\mathbf{x_B}$ is not optimal for the adjusted problem, and we must continue via the primal simplex method. The entering variable is necessarily x_4, and $x_{\mathbf{B}2} = x_3$ is chosen to exit. The resulting BFS is $x_1 = 7$, $x_4 = 3$, $x_5 = 3$, and $z = 39$, which turns out to be the optimum.

Next, let $a_{25}^* = 2$ in problem P (with a_{24} restored to its original value of 2). In this case the alteration involves a basic column, and before substituting $\mathbf{a}_5^*$ for $\mathbf{a}_5$ we first must attempt to remove $x_{\mathbf{B}3} = x_5$ from the basis. The very first nonbasic variable we examine, x_2, has the property we are seeking:

$$\operatorname*{Min}_i \left(\frac{x_{\mathbf{B}i}}{y_{i2}},\ y_{i2} > 0\right) = \min\left(\frac{9}{1}, \frac{6}{2}\right) = \frac{6}{2} = \frac{x_{\mathbf{B}3}}{y_{32}}$$

Therefore, we can pivot x_2 into the basis to replace x_5, knowing that the new basic solution will be (primal) feasible:

	x_1	x_2	x_3	x_4	x_5	x_6
$x_{\mathbf{B}1} = x_1 = 1$	1	0	0	−1	0	−2
$x_{\mathbf{B}2} = x_3 = 6$	0	0	1	1	−0.5	2.5
$x_{\mathbf{B}3} = x_2 = 3$	0	1	0	1	0.5	1.5
Max $z = 27$	0	0	0	−1	−1	0

[1] Note that we must compute the basis inverse separately because it was not preserved in the final tableau; the artificial columns that would have contained it were dropped at the end of phase 1. This belated calculation of $\mathbf{B}^{-1}$ would have been extremely time-consuming if

We now have a postoptimality problem that is essentially the same as the one discussed previously. Using the current basis inverse and basic cost vector,

$$\mathbf{B}^{-1} = [\mathbf{a}_1 \mid \mathbf{a}_3 \mid \mathbf{a}_2]^{-1} = \begin{bmatrix} 1 & 1 & 1 \\ 0 & 0 & 2 \\ 1 & 2 & 4 \end{bmatrix}^{-1} = \begin{bmatrix} 2 & 1 & -1 \\ -1 & -1.5 & 1 \\ 0 & 0.5 & 0 \end{bmatrix}$$

and $$\mathbf{c_B}^T = [3, 2, 4]$$

we calculate the new tableau entries

$$\mathbf{y}_5^* = \mathbf{B}^{-1}\mathbf{a}_5^* = \begin{bmatrix} 2 & 1 & -1 \\ -1 & -1.5 & 1 \\ 0 & 0.5 & 0 \end{bmatrix} \begin{bmatrix} 0 \\ 2 \\ 1 \end{bmatrix} = \begin{bmatrix} 1 \\ -2 \\ 1 \end{bmatrix}$$

and $$z_5^* - c_5 = \mathbf{c_B}^T\mathbf{y}_5^* - c_5 = [3, 2, 4]\cdot(1, -2, 1) - 2 = 1$$

and proceed via the primal simplex method.

Suppose now that $a_{31}^* = -1$ in problem P. Again the alteration involves a basic column, this time $\mathbf{a}_1$, so we wish to remove $x_{\mathbf{B}1} = x_1$ from the basis. Testing the primal simplex exit criterion on each of the nonbasic variables in turn, we discover that none of them would drive x_1 out of the basis; in fact, the exiting variable would be x_5 in all three cases. We must settle, therefore, for the alternate strategy of using the dual simplex entry criterion (7-9) to select the variable that will replace x_1:

$$\operatorname*{Max}_j \left(\frac{z_j - c_j}{y_{1j}},\ y_{1j} < 0\right) = \max\left(\frac{1}{-1}, \frac{3}{-2}\right) = \frac{1}{-1} = \frac{z_4 - c_4}{y_{14}}$$

Thus x_4 enters the basis, $\mathbf{\Phi} = (-2, 2, 2, 1)$, and the new tableau is

	x_1	x_2	x_3	x_4	x_5	x_6
$x_{\mathbf{B}1} = x_4 = -1$	-1	0	0	1	0	2
$x_{\mathbf{B}2} = x_3 = 11$	2	1	1	0	0	0
$x_{\mathbf{B}3} = x_5 = 8$	2	2	0	0	1	-1
Max z = 34	1	2	0	0	0	1

the number of constraints had not been so small. When an LPP of even moderate size is being solved and postoptimality problems are anticipated, either the artificial columns should be carried along "on the side" throughout phase 2 or revised simplex pivoting should be used, in order that the optimal basis inverse will be available at the end without extra computation.

Note that dual feasibility has been preserved. As in the previous example, we now compute the current basis inverse

$$\mathbf{B}^{-1} = [\mathbf{a}_4 \mid \mathbf{a}_3 \mid \mathbf{a}_5]^{-1} = \begin{bmatrix} 1 & 1 & 0 \\ 2 & 0 & 1 \\ 5 & 2 & 1 \end{bmatrix}^{-1} = \begin{bmatrix} -2 & -1 & 1 \\ 3 & 1 & -1 \\ 4 & 3 & -2 \end{bmatrix}$$

and use it to obtain the new tableau entries for the adjusted problem:

$$\mathbf{y}_1^* = \mathbf{B}^{-1}\mathbf{a}_1^* = \begin{bmatrix} -2 & -1 & 1 \\ 3 & 1 & -1 \\ 4 & 3 & -2 \end{bmatrix} \begin{bmatrix} 1 \\ 0 \\ -1 \end{bmatrix} = \begin{bmatrix} -3 \\ 4 \\ 6 \end{bmatrix}$$

and $$z_1^* - c_1 = \mathbf{c_B}^T\mathbf{y}_1^* - c_1 = [4, 2, 2]\cdot(-3, 4, 6) - 3 = 5$$

We are pleased to find that the reduced cost $(z_1^* - c_1)$ is nonnegative; this means that $\mathbf{x_B} = (x_4, x_3, x_5) = (-1, 11, 8)$ is a dual feasible solution to the adjusted linear program, which can therefore be solved from this point via the dual simplex method. If $(z_1^* - c_1)$ had turned out to be negative, $\mathbf{x_B}$ would have been neither primal nor dual feasible, and we would have had to re-solve the adjusted LPP from the beginning.

As our final example, let $a_{33}^* = 4$ in problem P. We noted earlier that $x_{\mathbf{B}2} = x_3$ could not be driven out of the basis in a primal simplex pivot. We now find that it cannot be removed in a dual pivot either, because all the y_{2j} are nonnegative. Under the circumstances, the recommended course of action is to return to the beginning, substitute a_{33}^* for a_{33}, and solve the problem all over again.

7.7 PROBLEMS INVOLVING CONTINUOUS PARAMETER CHANGES

In this section we shall be concerned principally with questions of the following form: How far can a certain numerical parameter or set of parameters of problem P—typically the cost vector $\mathbf{c}$ or the right-hand-side vector $\mathbf{b}$—be perturbed in a certain specified direction before $\mathbf{x_B}$ ceases to be optimal? In other words, how "sensitive" is the optimal solution to perturbation in those parameters? The process of investigating any question of this sort (whether in linear programming or in some other branch of operations research) is known, appropriately enough, as *sensitivity analysis*. A natural extension of the above question is also of interest: In some cases one may wish to determine not only the critical amount of perturbation at which $\mathbf{x_B}$ just stops being optimal, but also the various basic solutions that become optimal, one after the other, as the perturbation is extended further and further beyond that initial critical point. This information can be obtained by means of an iterative computational approach called *parametric programming*, an example of which will be presented below.

Most applications of sensitivity analysis and parametric programming involve either a perturbation of the cost vector or a perturbation of the right-hand-side vector. Of these, the former is rather more common and will be considered first. We begin with a precise statement of the problem we are proposing to solve: Let the cost vector $\mathbf{c}$ of problem P be replaced by

$$\mathbf{c}^* = \mathbf{c} + \theta\mathbf{s} \tag{7-17}$$

where $\mathbf{s}$ is a specified n-component vector and θ is a nonnegative scalar parameter, and determine the maximum value of θ for which $\mathbf{x_B}$ is still an optimal solution. Observe that we are restricting ourselves to cases in which the general cost vector $\mathbf{c}^*$ is a linear function of a single parameter; every $\mathbf{c}^*$ satisfying (7-17) lies on a straight line in euclidean n space. Multiparameter or nonlinear sensitivity analyses are perfectly possible in theory, but the mass of computational work involved is usually prohibitive; in any case, expressions more complicated than (7-17) will not be considered here. Note that when only a single cost coefficient is being perturbed, as is frequently the case, $\mathbf{s}$ is simply a unit vector.

Returning to the sensitivity analysis of (7-17), our computations will be based on the optimal solution to problem P, which, by assumption, already has been obtained. Its values are $\mathbf{x_B} = \mathbf{B}^{-1}\mathbf{b}$ and $z = \mathbf{c_B}^T\mathbf{x_B}$. For problem P, of course, the cost vector is $\mathbf{c}$ and $\theta = 0$. As θ is increased, the feasible region and the values of the variables $\mathbf{x_B}$ remain unaffected, but the reduced costs change; for any value of θ the jth reduced cost is given by

$$\begin{aligned} z_j^* - c_j^* &= \mathbf{c_B^*}^T\mathbf{y}_j - c_j^* = (\mathbf{c_B} + \theta\mathbf{s_B})^T\mathbf{y}_j - c_j - \theta s_j \\ &= z_j - c_j + \theta(\mathbf{s_B}^T\mathbf{y}_j - s_j) \end{aligned} \tag{7-18}$$

where $\mathbf{s_B}$ contains those components of $\mathbf{s}$ that correspond, respectively, to the components of $\mathbf{c}$ in $\mathbf{c_B}$. Because we are dealing with a maximization problem, the current solution $\mathbf{x_B}$ will remain optimal only as long as all the reduced costs $(z_j^* - c_j^*)$ are nonnegative. Now, whenever $(\mathbf{s_B}^T\mathbf{y}_j - s_j) \geq 0$, we have

$$z_j^* - c_j^* \geq z_j - c_j \geq 0$$

for all positive values of θ. Thus, if $(\mathbf{s_B}^T\mathbf{y}_j - s_j) \geq 0$ for *all* variables x_j, $\mathbf{x_B}$ will remain optimal no matter how large θ becomes. If some $(\mathbf{s_B}^T\mathbf{y}_j - s_j)$ is negative, however, the reduced cost $(z_j^* - c_j^*)$ is nonnegative if and only if

$$\theta \leq \frac{-(z_j - c_j)}{\mathbf{s_B}^T\mathbf{y}_j - s_j} \tag{7-19}$$

It follows that $\mathbf{x_B}$ ceases to be optimal as soon as θ is large enough to violate (7-19) for some variable x_j, that is, as soon as θ exceeds the *critical value*

$$\theta^{(1)} = \min_j \left(\frac{-(z_j - c_j)}{\mathbf{s_B}^T \mathbf{y}_j - s_j}, \quad \mathbf{s_B}^T \mathbf{y}_j - s_j < 0 \right) \tag{7-20}$$

Actually, this statement is not quite true; if $\mathbf{x_B}$ is degenerate, it might remain optimal even after one of the reduced costs $(z_j^* - c_j^*)$ has become negative, a point that was touched upon in Exercise 4-21(a). Notice, finally, that $\theta^{(1)}$ can be zero if one or more $(z_j - c_j) = 0$, that is, if the optimal solution to problem P was not unique.

Some sensitivity analyses are performed solely in order to determine $\theta^{(1)}$. Quite often, however, the analyst will be interested also in identifying the new solution that becomes optimal as θ passes through $\theta^{(1)}$. Let x_k be the nonbasic variable that yields the minimum quotient in (7-20); that is,

$$\theta^{(1)} = \frac{-(z_k - c_k)}{\mathbf{s_B}^T \mathbf{y}_k - s_k}$$

When $\theta = \theta^{(1)}$, the reduced cost $(z_k^* - c_k^*)$ is zero and the adjusted linear program

$$\begin{aligned} &\text{Max } z = \mathbf{c}^{*T}\mathbf{x} = (\mathbf{c} + \theta^{(1)}\mathbf{s})^T\mathbf{x} \\ &\text{subject to} \quad \mathbf{A}\mathbf{x} = \mathbf{b} \\ &\text{and} \qquad\qquad \mathbf{x} \geq \mathbf{0} \end{aligned} \tag{7-21}$$

in general, will have at least two optimal extreme points: the original optimum —call it $\mathbf{x}^{(1)}$—and a new extreme point $\mathbf{x}^{(2)}$, which results when x_k is pivoted into the basis (of course, if x_k enters in a zero-for-zero pivot, then $\mathbf{x}^{(1)} = \mathbf{x}^{(2)}$). When θ is increased beyond $\theta^{(1)}$, the reduced cost $(z_k^* - c_k^*)$ becomes negative and, in general, $\mathbf{x}^{(2)}$ replaces $\mathbf{x}^{(1)}$ as the unique optimum.[1]

[1] It will be true usually, but not always, that the new optimal solution (optimal for $\theta > \theta^{(1)}$) can be obtained from the old via a single pivot. As an example of the exceptional situation, suppose problem P calls for maximizing $z = x_1 + x_2 + x_3$ over the unit cube in E^3; the point $(x_1, x_2, x_3) = (1, 1, 1)$ is obviously the unique optimal solution. Now let $\mathbf{s}^T = [0, -1, -1]$, so that

$$\mathbf{c}^* = \mathbf{c} + \theta\mathbf{s} = \begin{bmatrix} 1 \\ 1 \\ 1 \end{bmatrix} + \theta \begin{bmatrix} 0 \\ -1 \\ -1 \end{bmatrix} = \begin{bmatrix} 1 \\ 1 - \theta \\ 1 - \theta \end{bmatrix}$$

When $0 \leq \theta < 1$, we have $\mathbf{c}^* > \mathbf{0}$, and $(1, 1, 1)$ continues to be the unique optimum. At $\theta = 1$, $\mathbf{c}^{*T} = [1, 0, 0]$ and the four extreme points having $x_1 = 1$ are all optimal. But when $\theta > 1$, the unique optimal solution is $(1, 0, 0)$, which cannot be reached by a single pivot from $(1, 1, 1)$.

Having obtained $\mathbf{x}^{(2)}$, the analyst may want to continue his investigation by considering still larger values of θ. The next question that arises is, how much *further* can θ be increased beyond $\theta^{(1)}$ before the new solution $\mathbf{x}^{(2)}$, in its turn, ceases to be optimal? The answer to this can be obtained by means of a second sensitivity analysis, which will yield a second critical value and a second adjusted linear program having a new alternate optimum $\mathbf{x}^{(3)}$. If the cycle is repeated enough times—assuming for simplicity that an unbounded optimum is never encountered—a basic solution eventually will be found that remains optimal as θ approaches infinity. The overall procedure, in which sensitivity analysis alternates with simplex pivoting, is known as *parametric programming*; if carried to its conclusion, it will generate a sequence of critical values $\theta^{(1)}, \theta^{(2)}, \ldots, \theta^{(p)}$ and a sequence of solutions $\mathbf{x} = \mathbf{x}^{(1)}, \mathbf{x} = \mathbf{x}^{(2)}, \ldots, \mathbf{x} = \mathbf{x}^{(p+1)}$ such that

$$\begin{array}{lll} \mathbf{x}^{(1)} & \text{is optimal if} & 0 \le \theta \le \theta^{(1)} \\ \mathbf{x}^{(2)} & \text{is optimal if} & \theta^{(1)} \le \theta \le \theta^{(2)} \\ \vdots & & \\ \mathbf{x}^{(p)} & \text{is optimal if} & \theta^{(p-1)} \le \theta \le \theta^{(p)} \end{array}$$

and

$$\mathbf{x}^{(p+1)} \quad \text{is optimal if} \quad \theta^{(p)} \le \theta$$

Notice that although the optimal value of any particular variable x_j may change discontinuously as θ is increased (due to x_j being pivoted into or out of the basis), the optimal objective value z_{opt} will be a continuous piecewise linear function of θ. This can be demonstrated by "tracking" θ as it passes through any one of its critical values:

$$\begin{array}{l} \text{For} \quad \theta^{(i-1)} < \theta < \theta^{(i)}, \quad z_{\text{opt}} = (\mathbf{c} + \theta\mathbf{s})^T\mathbf{x}^{(i)} \\ \text{for} \quad \theta = \theta^{(i)}, \quad z_{\text{opt}} = (\mathbf{c} + \theta^{(i)}\mathbf{s})^T\mathbf{x}^{(i)} = (\mathbf{c} + \theta^{(i)}\mathbf{s})^T\mathbf{x}^{(i+1)} \\ \text{and for} \quad \theta^{(i)} < \theta < \theta^{(i+1)}, \quad z_{\text{opt}} = (\mathbf{c} + \theta\mathbf{s})^T\mathbf{x}^{(i+1)} \end{array}$$

As a simple illustration of parametric programming, let us consider the following problem:

$$\begin{array}{lr} \text{Max } z = & -x_1 + x_2 \\ \text{subject to} & x_1 + 2x_2 \le 10 \\ & 2x_1 + x_2 \le 11 \\ & x_1 - 2x_2 \le 3 \\ \text{and} & x_1, x_2 \ge 0 \end{array}$$

The feasible region is diagramed in Figure 7-1. After the slack variables x_3, x_4, and x_5 have been added in that order to the three constraints, the problem

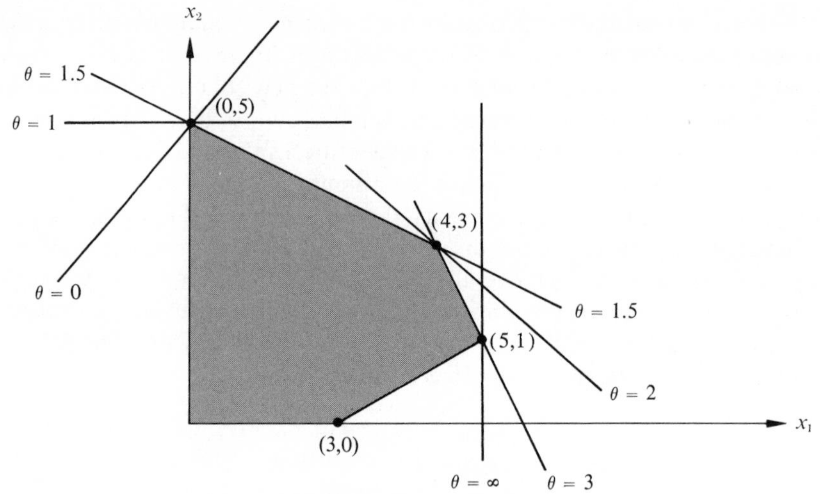

Figure 7-1

can be solved by the primal simplex method, using standard pivots. The optimal solution is found to be $(x_1,x_2) = (0,5)$, with the optimal tableau as follows:

	x_1	x_2	x_3	x_4	x_5
$x_{\mathbf{B}1} = x_2 = 5$	0.5	1	0.5	0	0
$x_{\mathbf{B}2} = x_4 = 6$	1.5	0	−0.5	1	0
$x_{\mathbf{B}3} = x_5 = 13$	2.0	0	1.0	0	1
Max $z = 5$	1.5	0	0.5	0	0

Suppose now that, although we have accurate estimates for all the other parameters, we are uncertain as to the true value of the cost coefficient c_1, knowing only that it is greater than or equal to -1. We may want to determine the optimal solution to the linear program for *all* possible values of c_1. This can be done by perturbing the cost vector $\mathbf{c}^T = [-1, 1, 0, 0, 0]$ in the direction $\mathbf{s} = [1, 0, 0, 0, 0]$, so that the general cost vector is given by

$$\mathbf{c}^* = \mathbf{c} + \theta\mathbf{s} = \begin{bmatrix} -1 \\ 1 \\ 0 \\ 0 \\ 0 \end{bmatrix} + \theta \begin{bmatrix} 1 \\ 0 \\ 0 \\ 0 \\ 0 \end{bmatrix} = \begin{bmatrix} \theta - 1 \\ 1 \\ 0 \\ 0 \\ 0 \end{bmatrix}$$

(Note that although the feasible region can be displayed geometrically in two dimensions, the algebraic space of the problem is E^5—slack variables receive the same algebraic treatment as nonslacks.) For the original problem $\theta = 0$, and the optimal objective hyperplane

$$z_{\text{opt}} = \mathbf{c}^T\mathbf{x} = -x_1 + x_2 = 5$$

is so labeled in Figure 7-1.

Our initial sensitivity analysis requires the following computations:

$$\mathbf{s_B}^T = [s_2, s_4, s_5] = [0, 0, 0]$$

$$\mathbf{s_B}^T\mathbf{y}_1 - s_1 = 0 - 1 = -1$$

and $$\mathbf{s_B}^T\mathbf{y}_3 - s_3 = 0 - 0 = 0$$

where $(\mathbf{s_B}^T\mathbf{y}_j - s_j)$ is identically zero for all basic variables x_j. The first critical value of θ is given by (7-20):

$$\theta^{(1)} = \min_j \left(\frac{-(z_j - c_j)}{\mathbf{s_B}^T\mathbf{y}_j - s_j}, \quad \mathbf{s_B}^T\mathbf{y}_j - s_j < 0\right) = \frac{-(z_1 - c_1)}{\mathbf{s_B}^T\mathbf{y}_1 - s_1} = \frac{-1.5}{-1} = 1.5$$

As long as θ does not exceed 1.5, the current solution will remain optimal; for example, when $\theta = 1$ the objective function is

$$\text{Max } z = \mathbf{c}^{*T}\mathbf{x} = [0, 1, 0, 0, 0]\cdot\mathbf{x} = x_2$$

and, as can be seen in Figure 7-1, the point $(x_1,x_2) = (0,5)$ is still the unique optimum.

When $\theta = \theta^{(1)} = 1.5$, however, an alternate optimum is introduced. In order to find it we employ the postoptimal technique of Section 7.6. The adjusted values of the reduced costs are given by (7-18):

$$z_1^* - c_1^* = z_1 - c_1 + \theta(\mathbf{s_B}^T\mathbf{y}_1 - s_1) = 1.5 + 1.5(-1) = 0$$

and $$z_3^* - c_3^* = z_3 - c_3 + \theta(\mathbf{s_B}^T\mathbf{y}_3 - s_3) = 0.5$$

Note that the same values would have been obtained from the transformation formula

$$z_j^* - c_j^* = (z_j - c_j) + \sum_{i=1}^{m} (\Delta c_{\mathbf{B}i})y_{ij} - \Delta c_j \tag{7-12}$$

because, in fact, $\Delta c_j \equiv \theta s_j$ for all j. The reduced costs of the basic variables remain at zero and the new value of the objective

$$z^* = \mathbf{c}^{*T}\mathbf{x} = [0.5, 1, 0, 0, 0]\cdot(0, 5, 0, 6, 13) = 5$$

is found to be the same as the old (in general, this need not happen). Let these new values be substituted into the optimal tableau of the original problem. Then, because the nonbasic variable x_1 has a reduced cost of zero, it can be pivoted into the basis to produce an alternate optimal solution:

	x_1	x_2	x_3	x_4	x_5
$x_{\mathbf{B}1} = x_2 = 3$	0	1	$\frac{2}{3}$	$-\frac{1}{3}$	0
$x_{\mathbf{B}2} = x_1 = 4$	1	0	$-\frac{1}{3}$	$\frac{2}{3}$	0
$x_{\mathbf{B}3} = x_5 = 5$	0	0	$\frac{5}{3}$	$-\frac{4}{3}$	1
Max $z = 5$	0	0	$\frac{1}{2}$	0	0

The existence of alternate optima at $\theta = 1.5$ is due to the fact that the objective function

$$\text{Max } z = 0.5x_1 + x_2$$

is now parallel to the first constraint; its optimal hyperplane (see Figure 7-1) passes through the extreme points (0,5) and (4,3).

When θ is increased above $\theta^{(1)} = 1.5$, the new solution (4,3) becomes the unique optimum; this is true, for example, at $\theta = 2.0$, where the objective function is

$$\text{Max } z = x_1 + x_2$$

In order to determine the value at which (4,3) ceases to be optimal, we perform a second sensitivity analysis *based on the latest optimal tableau*:

$$\mathbf{s_B}^T = [s_2, s_1, s_5] = [0, 1, 0]$$

$$\mathbf{s_B}^T\mathbf{y}_3 - s_3 = -\tfrac{1}{3} - 0 = -\tfrac{1}{3}$$

$$\mathbf{s_B}^T\mathbf{y}_4 - s_4 = \tfrac{2}{3} - 0 = \tfrac{2}{3}$$

and $$\theta_2 = \min_j \left(\frac{-(z_j - c_j)}{\mathbf{s_B}^T\mathbf{y}_j - s_j},\ \mathbf{s_B}^T\mathbf{y}_j - s_j < 0\right) = \frac{-(z_3 - c_3)}{\mathbf{s_B}^T\mathbf{y}_3 - s_3} = \frac{-\frac{1}{2}}{-\frac{1}{3}} = 1.5$$

Notice that we have represented the new critical value as θ_2, not $\theta^{(2)}$. The above tableau contains an optimal solution to the adjusted linear program

$$\text{Max } z = (\mathbf{c} + \theta^{(1)}\mathbf{s})^T\mathbf{x}$$

subject to the original constraints

Let us denote the original cost vector $\mathbf{c}$ by $\mathbf{c}_{\text{orig}}$ and the adjusted, or current, cost vector by

$$\mathbf{c}_{\text{curr}} = \mathbf{c}_{\text{orig}} + \theta^{(1)}\mathbf{s} \tag{7-22}$$

The sensitivity analysis just performed tells us that $(x_1,x_2) = (4,3)$ will continue to be the unique optimal solution to the linear program

$$\text{Max } z = (\mathbf{c}_{\text{curr}} + \theta\mathbf{s})^T\mathbf{x}$$

subject to the original constraints

until θ reaches $\theta_2 = 1.5$. At that point the cost vector will be

$$\mathbf{c}^* = \mathbf{c}_{\text{curr}} + \theta_2\mathbf{s}$$

or, from (7-22),

$$\mathbf{c}^* = \mathbf{c}_{\text{orig}} + \theta^{(1)}\mathbf{s} + \theta_2\mathbf{s} = \mathbf{c}_{\text{orig}} + \theta^{(2)}\mathbf{s}$$

where $\theta^{(2)} \equiv \theta^{(1)} + \theta_2 = 1.5 + 1.5 = 3$.

Returning to the above sensitivity analysis, let us determine the new optimal solution that is introduced when θ reaches a "relative" value of 1.5 with respect to the adjusted problem or an "actual" value of 3.0 with respect to the original problem. Using (7-18) and bearing in mind that the current values of the reduced costs are based on $\mathbf{c}_{\text{curr}}$, not on $\mathbf{c}_{\text{orig}}$, we have

$$z_3^* - c_3^* = z_3 - c_3 + \theta_2(\mathbf{s_B}^T\mathbf{y}_3 - s_3) = \tfrac{1}{2} + 1.5(-\tfrac{1}{3}) = 0$$

and

$$z_4^* - c_4^* = z_4 - c_4 + \theta_2(\mathbf{s_B}^T\mathbf{y}_4 - s_4) = 0 + 1.5(\tfrac{2}{3}) = 1$$

Because the new cost vector is

$$\mathbf{c}^* = \mathbf{c}_{\text{curr}} + \theta_2\mathbf{s} = \mathbf{c}_{\text{orig}} + \theta^{(2)}\mathbf{s} = (2, 1, 0, 0, 0)$$

the adjusted value of the objective function is given by

$$z^* = \mathbf{c}^{*T}\mathbf{x} = [2, 1, 0, 0, 0]\cdot(4, 3, 0, 0, 5) = 11$$

After these new values have been substituted into the above tableau, the variable x_3 can be pivoted into the basis to produce the following alternative optimal solution:

	x_1	x_2	x_3	x_4	x_5
$x_{B1} = x_2 = 1$	0	1	0	0.2	−0.4
$x_{B2} = x_1 = 5$	1	0	0	0.4	0.2
$x_{B3} = x_3 = 3$	0	0	1	−0.8	0.6
Max $z = 11$	0	0	0	1.0	0

Note that the objective function

$$\text{Max } z = 2x_1 + x_2$$

is now parallel to the second constraint. Again the optimal hyperplane is shown in Figure 7-1, labeled with its "actual" θ value, $\theta = \theta^{(2)} = 3$.

When θ is increased beyond $\theta^{(2)}$, the solution $(x_1,x_2) = (5,1)$ becomes the unique optimum. In order to determine the next critical value, we try to perform another sensitivity analysis:

$$\mathbf{s_B}^T = [s_2, s_1, s_3] = [0, 1, 0]$$

$$\mathbf{s_B}^T\mathbf{y}_4 - s_4 = 0.4 - 0 = 0.4$$

and

$$\mathbf{s_B}^T\mathbf{y}_5 - s_5 = 0.2 - 0 = 0.2$$

This time, however, all of the $(\mathbf{s_B}^T\mathbf{y}_j - s_j)$ turn out to be nonnegative, and it follows that the current solution must remain optimal as θ approaches infinity. Our conclusion can be verified by observing that for any positive value of θ the general objective function

$$\text{Max } z = \mathbf{c}^{*T}\mathbf{x} = (\theta - 1)x_1 + x_2$$

is equivalent to

$$\text{Max } z = \left(1 - \frac{1}{\theta}\right)x_1 + \frac{1}{\theta}x_2$$

Thus, as θ approaches infinity, the objective function becomes, in the limit,

$$\text{Max } z = x_1$$

for which the optimal extreme point is still (5,1).

The results of our parametric programming analysis may be summarized as follows. Let the objective function of the given linear program be replaced by

$$\text{Max } z = (\theta - 1)x_1 + x_2$$

where θ is a nonnegative scalar parameter. Then the optimal solution to the resulting problem is

$$\mathbf{x}^{(1)} \equiv (0,5) \qquad \text{if } 0 \leq \theta \leq 1.5$$

$$\mathbf{x}^{(2)} \equiv (4,3) \qquad \text{if } 1.5 \leq \theta \leq 3.0$$

and

$$\mathbf{x}^{(3)} \equiv (5,1) \qquad \text{if } 3.0 \leq \theta$$

This section concludes with a brief discussion of a second, less important class of perturbation problems that also require postoptimal sensitivity analysis. As usual, we assume that the optimal solution $\mathbf{x_B}$, with basis matrix $\mathbf{B}$, already has been obtained for the linear programming problem P:

$$\begin{aligned} &\text{Max } z = \mathbf{c}^T\mathbf{x} \\ &\text{subject to} \quad \mathbf{Ax} = \mathbf{b} \\ &\text{and} \qquad\qquad \mathbf{x} \geq \mathbf{0} \end{aligned}$$

Suppose now that the right-hand-side vector $\mathbf{b}$ is replaced by

$$\mathbf{b}^* = \mathbf{b} + \phi\mathbf{t} \tag{7-23}$$

where $\mathbf{t}$ is a specified m-component vector and ϕ is a nonnegative scalar parameter. We want to determine the maximum value of ϕ for which the variables of $\mathbf{x_B}$ continue to form an optimal basis. Because the cost coefficients and the columns $\mathbf{y}_j \equiv \mathbf{B}^{-1}\mathbf{a}_j$ are not affected by the substitution (7-23), neither are the reduced costs, and the current basic solution will remain dual feasible regardless of how large ϕ becomes. As ϕ increases, however, the values of the basic variables change,

$$\mathbf{x}^*_\mathbf{B} = \mathbf{B}^{-1}\mathbf{b}^* = \mathbf{B}^{-1}\mathbf{b} + \phi\mathbf{B}^{-1}\mathbf{t} = \mathbf{x_B} + \phi\mathbf{u} \tag{7-24}$$

where $\mathbf{u} \equiv \mathbf{B}^{-1}\mathbf{t}$, and the current solution will remain optimal only as long as $\mathbf{x}^*_\mathbf{B}$ is nonnegative. Clearly, if $\mathbf{u} \geq \mathbf{0}$, the solution $\mathbf{x}^*_\mathbf{B}$ will be primal feasible and therefore optimal for all positive values of ϕ. But if some u_i is negative, then $x^*_{\mathbf{B}i} \geq 0$ if and only if

$$\phi \leq \frac{-x_{\mathbf{B}i}}{u_i}$$

and it follows that the current solution ceases to be optimal as soon as ϕ exceeds

$$\phi^{(1)} = \min_i \left(\frac{-x_{\mathbf{B}i}}{u_i}, \quad u_i < 0 \right) \tag{7-25}$$

Notice that $\phi^{(1)}$ can be zero if the original solution $\mathbf{x_B}$ was degenerate.

As far as any further sensitivity analysis or parametric programming is concerned, perturbations of the right-hand-side vector are treated in essentially the same way as perturbations of the cost vector. The major difference is that when ϕ reaches one of its critical values, the new alternate optimal solution is obtained via a *dual* rather than a primal simplex pivot.

7.8 PROBLEMS INVOLVING STRUCTURAL CHANGES

Postoptimality problems in this category arise when variables or constraints are added to or deleted from the linear program P after its optimal solution $\mathbf{x_B}$ has been obtained. We shall discuss four simple structural changes:

(1) The addition of a single nonnegative variable
(2) The deletion of a single variable
(3) The addition of a single constraint
(4) The deletion of a single constraint

Problems involving multiple additions or deletions then can be handled by repeated application of the procedures we shall develop in this section.

The first of these four postoptimal alterations is the easiest to deal with. In order to add to problem P a nonnegative variable x_{n+1}, having activity column $\mathbf{a}_{n+1}$ and cost coefficient c_{n+1}, it is necessary only to compute

$$\mathbf{y}_{n+1} = \mathbf{B}^{-1}\mathbf{a}_{n+1} \qquad \text{and} \qquad z_{n+1} - c_{n+1} = \mathbf{c_B}^T\mathbf{y}_{n+1} - c_{n+1}$$

where $\mathbf{B}^{-1}$ and $\mathbf{c_B}$ are the basis inverse and basic cost vector associated with the optimal solution $\mathbf{x_B}$. If $(z_{n+1} - c_{n+1}) \geq 0$, then $\mathbf{x_B}$ remains optimal for the new problem; if not, pivoting proceeds via the primal simplex method, with x_{n+1} entering the basis at the next iteration (if standard pivots are being used, a new tableau column must be created). Note that the same overall sequence of pivots would have been generated if we had included x_{n+1} in the formulation originally, but had stubbornly refused to consider it as a candidate for basis entry until all the other reduced costs had finally become nonnegative.

The deletion of a variable x_q from problem P, together with subsequent reoptimization, is also quite straightforward. First we should remark that "deleting" a variable is the same as forcing it to take on a value of zero, so that it cannot form part of the optimal solution. Thus the deletion of a nonbasic variable would be a totally superfluous operation and need not concern us at all. To delete a basic variable $x_{\mathbf{B}r} = x_q > 0$ from the problem, we simply pivot it out of the basis using the dual simplex entry criterion (7-9); this results in a new basic solution that is dual feasible but, in general, not primal feasible. Then, after scratching out or otherwise ignoring the tableau column of x_q, we can reoptimize via the dual simplex method.

Turning to the third type of structural change, suppose that the constraint

$$\boldsymbol{\alpha}\mathbf{x} \leq \alpha_0 \tag{7-26}$$

where $\boldsymbol{\alpha}$ is an n-component row vector of coefficients and α_0 is a scalar constant, is added postoptimally to problem P. The first step is to determine whether the optimal solution $\mathbf{x_B}$ satisfies the new constraint; if it does, then it must also be an optimal solution to the adjusted problem. This follows from the fact that an extra constraint cannot enlarge the feasible region of a linear program; it can only reduce it or leave it unchanged. Thus, no solution better than $\mathbf{x_B}$ can possibly spring into existence when a new constraint is added; if $\mathbf{x_B}$ remains feasible, it remains optimal.

Assume, however, that $\mathbf{x_B}$ violates (7-26), so that another optimal solution must be sought. Let the new constraint be converted to standard form by means of a slack variable x_s. The adjusted problem then has the following set of $m + 1$ constraints:

$$\begin{aligned} \mathbf{Ax} \qquad &= \mathbf{b} \\ \boldsymbol{\alpha}\mathbf{x} + x_s &= \alpha_0 \end{aligned} \tag{7-27}$$

with $\mathbf{x}$ and x_s required to be nonnegative, as usual. We now assert that the variables $x_{\mathbf{B}1}, \ldots, x_{\mathbf{B}m}, x_s$ form a basis for the adjusted problem. This is verified easily by collecting their activity columns into an $(m + 1) \times (m + 1)$ matrix

$$\mathbf{B}^* = \begin{bmatrix} \mathbf{B} & \mathbf{0} \\ \boldsymbol{\beta} & 1 \end{bmatrix}$$

where $\boldsymbol{\beta}$ is an m-component row vector containing the new-constraint coefficients of the variables in $\mathbf{x_B}$; that is, if $x_{\mathbf{B}p} = x_q$, then $\beta_p = \alpha_q$. Thus $\mathbf{B}^*$ must be a basis matrix because it has an inverse, namely,

$$(\mathbf{B}^*)^{-1} = \begin{bmatrix} \mathbf{B}^{-1} & \mathbf{0} \\ -\boldsymbol{\beta}\mathbf{B}^{-1} & 1 \end{bmatrix} \tag{7-28}$$

The reader can check that $\mathbf{B}^*$ and $(\mathbf{B}^*)^{-1}$ are partitioned conformably for multiplication and that $\mathbf{B}^*(\mathbf{B}^*)^{-1} = \mathbf{I}_{m+1}$.

Using (7-28), we can compute all the values required to fill in a standard simplex tableau for the adjusted problem (the computations are somewhat different when revised simplex pivots are being used). The values of the basic variables are given by

$$\begin{bmatrix} \mathbf{x_B} \\ x_s \end{bmatrix} = (\mathbf{B}^*)^{-1}\mathbf{b}^* = \begin{bmatrix} \mathbf{B}^{-1} & \mathbf{0} \\ -\boldsymbol{\beta}\mathbf{B}^{-1} & 1 \end{bmatrix} \begin{bmatrix} \mathbf{b} \\ \alpha_0 \end{bmatrix} = \begin{bmatrix} \mathbf{B}^{-1}\mathbf{b} \\ \alpha_0 - \boldsymbol{\beta}\mathbf{B}^{-1}\mathbf{b} \end{bmatrix} \tag{7-29}$$

where $\mathbf{b}^*$ denotes the right-hand-side vector of the constraint set (7-27); note that, by assumption, the value of x_s is negative. Because each column of the activity matrix is now $\mathbf{a}_j^* = (\mathbf{a}_j, \alpha_j)$, the $\mathbf{y}$ columns must be

$$\mathbf{y}_j^* \equiv (\mathbf{B}^*)^{-1}\mathbf{a}_j^* = \begin{bmatrix} \mathbf{B}^{-1} & \mathbf{0} \\ -\boldsymbol{\beta}\mathbf{B}^{-1} & 1 \end{bmatrix} \begin{bmatrix} \mathbf{a}_j \\ \alpha_j \end{bmatrix} = \begin{bmatrix} \mathbf{y}_j \\ \alpha_j - \boldsymbol{\beta}\mathbf{B}^{-1}\mathbf{a}_j \end{bmatrix} \tag{7-30}$$

The reduced cost of any nonbasic variable x_j is

$$z_j^* - c_j \equiv \mathbf{c}_\mathbf{B}^{*T}\mathbf{y}_j^* - c_j = [\mathbf{c}_\mathbf{B}^T, 0] \cdot (\mathbf{y}_j, \alpha_j - \boldsymbol{\beta}\mathbf{B}^{-1}\mathbf{a}_j) - c_j = z_j - c_j$$

the same as before. This implies that the solution $(\mathbf{x_B}, x_s)$ is dual feasible. Finally, the value of the objective function also is unchanged:

$$z^* = [\mathbf{c}_\mathbf{B}^T, 0] \cdot (\mathbf{x_B}, x_s) = z$$

Examining the new values, we see that none of the entries in the optimal tableau of problem P need to be changed. It is necessary only to add a new row for the $(m + 1)$th basic variable x_s, whose value and y elements are given by (7-29) and (7-30); a new tableau column also must be added for x_s. The adjusted problem then can be solved by the dual simplex method.

Notice that the sign of α_0 in (7-26) was not specified. Thus, a constraint of the opposite inequality form can be multiplied through by -1 and added to problem P exactly as above. The question of adding an *equality* constraint will be reserved for Exercise 7-20.

The fourth type of structural change, the deletion of a constraint, gives rise to a postoptimality problem only if the constraint in question is binding on the optimal solution $\mathbf{x_B}$ (an equality constraint is automatically binding). The deletion of a nonbinding constraint does enlarge the feasible region, but the new solutions are all inferior to $\mathbf{x_B}$; this proposition, if it is not obvious already, can easily be verified graphically. It follows that when a constraint is to be deleted postoptimally from problem P, the first step should be to see whether or not it has a positive-valued slack or surplus variable in the optimal basis $\mathbf{x_B}$; if it has, then it cannot be binding and $\mathbf{x_B}$ must be an optimal solution to the adjusted problem.

Suppose, however, that the constraint to be deleted is found to be *binding* on $\mathbf{x_B}$. In this case the simplest way to proceed is via the addition of one or two new variables. Let the constraint in question be

$$a_{r1}x_1 + \cdots + a_{rn}x_n = b_r \tag{7-31}$$

which is the rth constraint of problem P, and consider the implications of adding to it both a slack and a surplus variable, as follows:

$$a_{r1}x_1 + \cdots + a_{rn}x_n + x_{n+1} - x_{n+2} = b_r \tag{7-32}$$

where $$x_{n+1}, x_{n+2} \geq 0$$

Now instead of requiring the left side of (7-31) to equal b_r, we are permitting it to be greater than or less than b_r as well. For example, $(a_{r1}x_1 + \cdots + a_{rn}x_n) < b_r$ can be accommodated by letting x_{n+1} take on a positive value; inasmuch as x_{n+1} is a slack variable, this has no effect on the other constraints nor on the value of the objective function. Similarly, it is quite feasible for $(a_{r1}x_1 + \cdots + a_{rn}x_n)$ to exceed b_r, because the surplus can be absorbed by a positive value of x_{n+2}. Thus, by introducing x_{n+1} and x_{n+2} we have, in effect, deleted the constraint (7-31). The procedures for adding new variables to problem P were described earlier in this section; note that $\mathbf{a}_{n+1}$ is the m-component unit vector $\mathbf{e}_r$, while $\mathbf{a}_{n+2} = -\mathbf{e}_r$ and $c_{n+1} = c_{n+2} = 0$.

As a final remark, it may well happen that the constraint (7-31) began as an inequality and therefore already includes either a slack or a surplus variable. If so, its deletion can be accomplished by adding only a single new variable of the opposite type.

EXERCISES

Section 7.2

7-1. Turn to the example in Section 5.5 that illustrates the application of the two-phase simplex method, and write the dual of the phase-1 problem Q. Then, for each of the three basic solutions obtained during phase 1, compute the associated dual solution and verify that:

(a) The primal solution and the associated dual solution have the same objective value.

(b) Negative reduced costs correspond to unsatisfied dual constraints.

Now trace what happens to the dual problem, the associated dual solution, and the primal reduced costs in the conversion from phase 1 to phase 2, and verify (a) and (b) for the initial phase-2 solution.

Section 7.4

7-2. Solve by the dual simplex method:

$$
\begin{aligned}
\text{Min } z &= 3x_1 + x_2 + x_3 \\
\text{subject to} \quad & x_1 + 2x_2 \geq 8 \\
& 3x_1 - 2x_2 - x_3 \geq 6 \\
& -x_1 - x_2 + 4x_3 \geq 2 \\
\text{and} \quad & x_1, x_2, x_3 \geq 0
\end{aligned}
$$

7-3. Solve by the dual simplex method:

$$
\begin{aligned}
\text{Min } z &= 2x_1 + x_2 + 2x_3 \\
\text{subject to} \quad & x_1 + x_2 + x_3 \geq 2 \\
& x_1 - 3x_2 - 2x_3 \geq 0 \\
& -3x_1 + x_2 \geq -3 \\
\text{and} \quad & x_1, x_2, x_3 \geq 0
\end{aligned}
$$

7-4. Solve the following problem *without* using artificial variables:

$$
\begin{aligned}
\text{Min } z &= 2x_1 + 3x_2 + x_3 \\
\text{subject to} \quad & x_1 + x_2 + x_3 = 10 \\
& 2x_1 - x_2 \geq 3 \\
& x_1 + 3x_2 \leq 12 \\
\text{and} \quad & x_1, x_2, x_3 \geq 0
\end{aligned}
$$

7-5. In solving a linear program by the dual simplex method, can a variable that has just entered the basis be removed on the next pivot? Can a variable that has just been removed from the basis reenter on the next pivot?

7-6. (a) Is it possible for cycling to occur when a linear program is solved by the dual simplex method? If so, how?

(b) Can alternative optimal extreme points and basic feasible solutions—if they exist—be obtained by means of dual simplex pivots?

7-7. Outline the steps by which a minimization problem would be solved by the dual simplex method using *revised* simplex pivots. How does the operation count for a revised dual simplex pivot compare with that of a standard dual simplex pivot? Does the revised method have the same computational advantages over the standard approach as it has in the case of primal simplex pivoting?

Section 7.6

7-8. Consider the linear programming problem

$$\begin{aligned}
\text{Max } z &= 2x_1 + x_2 + 4x_3 - x_4 \\
\text{subject to } \quad & x_1 + 2x_2 + x_3 - 3x_4 + x_5 = 8 \\
& -x_2 + x_3 + 2x_4 + x_6 = 0 \\
& 2x_1 + 7x_2 - 5x_3 - 10x_4 + x_7 = 21 \\
\text{and} \quad & x_i \geq 0 \quad i = 1, \ldots, 7
\end{aligned}$$

where x_5, x_6, and x_7 are slack variables. The optimal solution to this problem is contained in the following tableau:

	x_1	x_2	x_3	x_4	x_5	x_6	x_7
$x_{B1} = x_1 = 8$	1	0	3	1	1	2	0
$x_{B2} = x_2 = 0$	0	1	−1	−2	0	−1	0
$x_{B3} = x_7 = 5$	0	0	−4	2	−2	3	1
Max $z = 16$	0	0	1	1	2	3	0

For each of the discrete parameter changes listed below, make the necessary corrections to the optimal tableau and solve the adjusted problem. (The algebraic symbols a_{ij}, b_i, etc., are used in the usual way to refer to elements of the above problem.)

(a) Change c_1 to 1.
(b) Change $\mathbf{c}^T$ to [1, 2, 3, 4].
(c) Change b_3 to 11.
(d) Change $\mathbf{b}$ to (3, −2, 4).
(e) Change a_{21} to −1.
(f) Change a_{32} to 5.

7-9. It was recommended in the text that, in dealing with the problem that arises when a *basic* activity column $\mathbf{a}_q$ is altered postoptimally, one should attempt to pivot x_q out of the basis before substituting $\mathbf{a}_q^*$ for $\mathbf{a}_q$. When revised simplex pivots are being used, however, it is better to recalculate the auxiliary matrix inverse (5-32) *directly*, without first removing x_q (although the adjusted solution that results from this procedure stands an excellent chance of being neither primal nor dual feasible). Show how Lemma 5.1 can be used to accomplish this postoptimal alteration. Why was it not recommended that the optimal *standard* simplex tableau be adjusted directly when a basic activity column $\mathbf{a}_q$ is changed?

Section 7.7

7-10. In the problem of Exercise 7-8, by how much can the cost coefficient c_2 be increased before the current solution $\mathbf{x_B} = (x_1, x_2, x_7) = (8, 0, 5)$ ceases to be optimal? Answer the same question for c_1. How far can the cost vector $\mathbf{c}$ be perturbed in the direction $\mathbf{s} = [1, 1, 0, 0, 0, 0, 0]$ before optimality is disrupted?

7-11. In Exercise 7-8, how far can the right-hand-side vector $\mathbf{b}$ be perturbed in the direction $\mathbf{t} = (-1, 0, 0)$ before the basic solution associated with x_1, x_2, and x_7 becomes infeasible? Repeat the question for the direction $\mathbf{t} = (3, 1, 1)$.

7-12. Why would sensitivity analysis of the cost vector **c** be more useful to a real-world decision-maker than sensitivity analysis of the right-hand-side vector **b**?

7-13. Let the cost coefficient c_4 in the problem of Exercise 7-8 be increased continuously from its initial value of -1, so that

$$c_4^* = c_4 + \theta = \theta - 1$$

and perform a complete parametric programming analysis of this perturbation. Identify all critical values of the parameter θ and all optimal basic solutions, and sketch the optimal objective value as a function of θ.

7-14. The optimal solution to the linear program

$$\text{Max } z = x_1 - 3x_2$$

$$\text{subject to} \quad \begin{aligned} x_1 - 2x_2 + x_3 \qquad\qquad &= 2 \\ x_1 - x_2 \qquad + x_4 \qquad &= 3 \\ -2x_1 + x_2 \qquad\qquad + x_5 &= 2 \end{aligned}$$

$$\text{and} \qquad x_i \geq 0 \qquad i = 1, \ldots, 5$$

is contained in the following tableau:

	x_1	x_2	x_3	x_4	x_5
$x_{B1} = x_1 = 2$	1	-2	1	0	0
$x_{B2} = x_4 = 1$	0	1	-1	1	0
$x_{B3} = x_5 = 6$	0	-3	2	0	1
Max $z = 2$	0	1	1	0	0

Let the cost coefficient $c_2 = -3$ be increased continuously from its initial value and perform a parametric programming analysis. Graph the feasible region (in two dimensions) and sketch in the optimal objective hyperplane for each critical value of the parameter θ.

7-15. Obtain the optimal solution to the following linear program for all real values of b_3:

$$\text{Max } z = 2x_1 + x_2 - x_3$$

$$\text{subject to} \quad \begin{aligned} 3x_1 + 2x_2 + x_3 &\leq 20 \\ -x_1 - x_2 + x_3 &\leq 0 \\ x_1 + x_2 + x_3 &\leq b_3 \end{aligned}$$

$$\text{and} \qquad x_1, x_2, x_3 \geq 0$$

Plot the optimal value of z as a function of b_3.

7-16. In the text we did not discuss the postoptimal perturbation of an activity column of problem P, which would take the form

$$\mathbf{a}_j^* = \mathbf{a}_j + \phi\mathbf{t}$$

What overall strategy would be employed in seeking the first critical value of ϕ, that is, the maximum value of ϕ for which the variables of $\mathbf{x_B}$ continue to form an optimal basis? What computations would have to be performed and what difficulties might arise?

Section 7.8

7-17. Solve the problem that results when each of the following structural changes is made in the linear program of Exercise 7-8.

(a) Add a new variable x_8, with activity column $\mathbf{a}_8 = (1, -2, 1)$ and cost coefficient $c_8 = 4$.

(b) Delete the variable x_7.

(c) Delete the variable x_1.

(d) Delete the variable x_2.

(e) Add the constraint $x_1 + x_2 \leq 6$.

7-18. Solve the postoptimality problems that result when each of the three constraints is deleted (singly, not successively!) from the linear program of Exercise 7-8.

7-19. Use the dual simplex method as necessary to determine the new optimal solution when each of the following constraints is added to the final tableau in Section 7.4:

(a) $x_2 \geq 1$

(b) $x_1 + x_2 + x_3 \geq 20$

(c) $x_3 \geq x_1$

(d) $2x_2 - x_3 \geq 8$

(e) $2x_1 - 6x_2 - x_3 \leq 13$

7-20. What computational steps are required for the postoptimal addition of an equality constraint

$$\boldsymbol{\alpha}\mathbf{x} = \alpha_0$$

to problem P? Note that at the current optimal solution the left-hand side $\boldsymbol{\alpha}\mathbf{x}$ may be less than, equal to, or greater than α_0; be sure to consider all three possibilities in your answer.

7-21. Suppose a constraint has a zero-valued slack variable in the (degenerate) optimal basis of some linear program. If that constraint is deleted, will the optimal solution change or remain the same?

7-22. Suppose that after solving problem P the analyst discovers that he used a completely wrong constraint; that is, instead of

$$a_{r1}x_1 + \cdots + a_{rn}x_n = b_r$$

he should have used

$$a_{r1}^* x_1 + \cdots + a_{rn}^* x_n = b_r^*$$

How should he proceed in order to rectify his error and solve the correct problem?

7-23. Consider the following optimization problem:

$$\text{Max } z = \mathbf{c}^T\mathbf{x}$$

$$\text{subject to} \quad \mathbf{A}\mathbf{x} = \mathbf{b}$$

$$\mathbf{x} \geq \mathbf{0}$$

$$\text{and subject to } \textit{either } \boldsymbol{\alpha}\mathbf{x} \leq \alpha_0 \textit{ or } \boldsymbol{\beta}\mathbf{x} \leq \beta_0$$

where A is an $m \times n$ matrix, $\boldsymbol{\alpha}$ and $\boldsymbol{\beta}$ are n-component row vectors, and α_0 and β_0 are scalars. What we mean here is that $\hat{\mathbf{x}}$ is feasible provided that $\mathbf{A}\hat{\mathbf{x}} = \mathbf{b}$ and $\hat{\mathbf{x}} \geq \mathbf{0}$ *and* provided that either $\boldsymbol{\alpha}\hat{\mathbf{x}} \leq \alpha_0$ or $\boldsymbol{\beta}\hat{\mathbf{x}} \leq \beta_0$ (or both). Is the feasible region necessarily convex? How can this problem be solved?

7-24. (a) Show that a constraint having a slack variable can be deleted postoptimally by pivoting that slack into the basis and then eliminating an entire row of the tableau.

(b) Use this approach to delete the first constraint from the linear program of Exercise 7-8.

(c) Contrast this procedure for deleting a constraint with the addition-of-variables method discussed in the text.

7-25. Suppose that after a linear program has been solved it is belatedly discovered that a constant value $Q > 0$ should have been assigned to one of its variables x_q (i.e., x_q was not really a variable at all); x_q may or may not appear in the optimal basis. Describe two methods for dealing with this postoptimality problem and comment on the relative advantages of each.

8

THE TRANSPORTATION PROBLEM

8.1 THE PROBLEM IN ITS BASIC CONTEXT

By now the reader should be fairly familiar with the transportation problem, which was introduced in Example 1.2 and discussed in some detail in Section 6.5. For convenience we shall begin this chapter by outlining it again in its classic form. Various amounts of some homogeneous commodity that are initially stored at each of m different *origins*, or *sources*, must be shipped to n *destinations*, or *sinks*, where they will be sold, processed industrially, or used for some other purpose. Typically, the origins might be warehouses and the destinations retail stores. The amount of commodity available at the ith origin is $a_i > 0$, $i = 1, \ldots, m$, and the amount required at the jth destination is $b_j > 0$, $j = 1, \ldots, n$, where it is assumed that the total supply equals total demand:

$$\sum_{i=1}^{m} a_i = \sum_{j=1}^{n} b_j \tag{8-1}$$

If the cost of shipping one unit of commodity from the ith origin to the jth destination is c_{ij} for all i and j, what shipping plan will satisfy the demands at minimum cost?

This problem could be illustrated with a network flow diagram of the type shown in Figures 1-2 and 6-2. To formulate it, let x_{ij} be the amount to be shipped from origin i to destination j. The optimal shipping plan is found by solving the following linear program in standard form:

$$\text{Min } z = \sum_{i=1}^{m} \sum_{j=1}^{n} c_{ij} x_{ij}$$

$$\text{subject to} \quad \sum_{j=1}^{n} x_{ij} = a_i \qquad \text{where } a_i > 0, \; i = 1, \ldots, m \tag{8-2}$$

$$\sum_{i=1}^{m} x_{ij} = b_j \qquad \text{where } b_j > 0, \; j = 1, \ldots, n$$

$$\text{and} \qquad x_{ij} \geq 0 \qquad \text{for all } i \text{ and } j$$

The constraints must be satisfied as equalities because of (8-1); that is, no extra commodity is available at the origins.

The problem (8-2), subject to the side condition (8-1), is known as the *transportation problem.* Notice first that we do not really restrict anything by requiring the *supplies* a_i and the *demands* b_j to be positive. Because all the variables are nonnegative in (8-2), it follows immediately that each a_i and b_j must be greater than or equal to zero if a feasible solution is to exist. Moreover, a supply a_p could be zero only if $x_{pj} \equiv 0$ for all j, in which case the pth origin, along with those variables x_{pj}, might as well be eliminated from the problem (an analogous remark would apply if any $b_q = 0$). We do permit the costs c_{ij} to be negative, however. Thus our transportation format (8-2) also can represent maximization problems, because maximizing

$$\sum_{i,j} c_{ij} x_{ij}$$

is equivalent to minimizing

$$\sum_{i,j} (-c_{ij}) x_{ij}$$

A negative cost would not occur in the simple shipping context described above, but would be perfectly reasonable under other circumstances—for example, if c_{ij} were defined in a maximization problem as the net profit or loss associated with producing a unit in the ith plant and transporting it to the jth city for sale there at the prevailing price.

Notice that the transportation problem always has a feasible solution; for example, the reader can easily verify that one feasible solution is

$$x_{ij} = \frac{a_i b_j}{T} \qquad \text{for all } i \text{ and } j$$

where

$$T = \sum_{i=1}^{m} a_i = \sum_{j=1}^{n} b_j$$

Furthermore, because of nonnegativity, each variable is bounded above and below:

$$0 \leq x_{ij} \leq \min\{a_i, b_j\} \qquad \text{for all } i \text{ and } j$$

Hence the problem cannot be unbounded and an optimal solution must exist.

8.2 ALTERNATIVE FORMATS AND VARIATIONS

The transportation problem actually arises in several different but, as we shall see, equivalent forms. Frequently, for example, more supply is available than is needed to meet all the demands; if the demands must be satisfied as equalities, the constraints become

$$\sum_{j=1}^{n} x_{ij} \leq a_i \qquad i = 1, \ldots, m$$

and

$$\sum_{i=1}^{m} x_{ij} = b_j \qquad j = 1, \ldots, n \tag{8-3}$$

The ith supply constraint can be converted to standard form by the addition of a slack variable $x_{i,n+1}$:

$$\sum_{j=1}^{n} x_{ij} + x_{i,n+1} = a_i$$

Supposing we sum this equation over all origins, $i = 1, \ldots, m$, producing

$$\sum_{i=1}^{m} \sum_{j=1}^{n} x_{ij} + \sum_{i=1}^{m} x_{i,n+1} = \sum_{i=1}^{m} a_i \tag{8-4}$$

Then, if the destination constraints are summed over j and the result subtracted from (8-4), we find

$$\sum_{i=1}^{m} x_{i,n+1} = \sum_{i=1}^{m} a_i - \sum_{j=1}^{n} b_j \equiv b_{n+1} \tag{8-5}$$

where b_{n+1} is defined as the amount of *excess commodity*, that is, the amount by which total supply exceeds total demand. Equation (8-5) thus expresses the obvious fact that the "total slack" for any feasible shipping plan must equal the excess commodity b_{n+1}. In fact, we may think of the unused b_{n+1} units in any feasible solution as being shipped to a *dummy destination*, the $(n + 1)$th, at a per-unit shipping cost of zero. This suggests adding (8-5) to the constraint set in standard form,[1] thereby producing a transportation problem with m origins and $n + 1$ destinations that satisfies the format (8-2):

$$\text{Min } z = \sum_{i=1}^{m} \sum_{j=1}^{n+1} c_{ij} x_{ij}$$

$$\text{subject to} \quad \sum_{j=1}^{n+1} x_{ij} = a_i \qquad i = 1, \ldots, m$$

$$\sum_{i=1}^{m} x_{ij} = b_j \qquad j = 1, \ldots, n + 1$$

$$\text{and} \quad x_{ij} \geq 0 \qquad \text{for all } i \text{ and } j$$

where $c_{i,n+1} \equiv 0$, $i = 1, \ldots, m$ (and total supply equals total demand). Incidentally, this problem can be varied slightly by requiring that all commodity at the ith origin that is *not* shipped to a real destination must be stored at a per-unit cost of $c_{i,n+1}$; the formulation is the same as above, except that the slack costs $c_{i,n+1}$ are, in general, different from zero.

When excess commodity is available, as in the case just discussed, the oversatisfying of demands is usually not *explicitly* forbidden, so that the constraints may be written as

$$\sum_{j=1}^{n} x_{ij} \leq a_i \qquad i = 1, \ldots, m$$

$$\text{and} \quad \sum_{i=1}^{m} x_{ij} \geq b_j \qquad j = 1, \ldots, n$$

[1] Because (8-5) was derived from the other constraints, the enlarged constraint set obviously will be redundant, and it is now clear that redundancy also was present in the transportation problem as originally stated. This point will be discussed in Section 8.3.

If some of the shipping costs c_{ij} are negative, the optimal solution may well call for more units to be sent to certain destinations than are demanded there. However, when all costs are nonnegative—as is true of most transportation problems arising in the real world—it can never be optimal to ship more units than necessary (except when one or more $c_{ij} = 0$, in which case there must still exist at least one optimal shipping plan that does not overfulfill any of the demands). Therefore the demand constraints might as well be represented as

$$\sum_{i=1}^{m} x_{ij} = b_j \qquad j = 1, \ldots, n$$

which places the problem within the domain of the case discussed previously.

Consider now another variation in which total supply exceeds total demand, but in which every unit must be shipped to some destination (i.e., some demands must be oversatisfied):

$$\begin{aligned} &\text{Min } z = \sum_{i=1}^{m} \sum_{j=1}^{n} c_{ij} x_{ij} \\ &\text{subject to} \quad \sum_{j=1}^{n} x_{ij} = a_i \qquad i = 1, \ldots, m \\ &\qquad\qquad\quad \sum_{i=1}^{m} x_{ij} \geq b_j \qquad j = 1, \ldots, n \\ &\text{and} \qquad\qquad x_{ij} \geq 0 \qquad \text{for all } i \text{ and } j \end{aligned} \tag{8-6}$$

This is the "mirror image," in a sense, of the variation (8-3) discussed earlier, but there is an important difference: Conversion to standard form introduces surplus variables instead of slacks, and the problem simply cannot be fitted to our transportation format (8-2). Nevertheless, it will turn out that the standard transportation algorithm, which we shall present below, can be modified slightly to solve problems of the type (8-6); the student will have the opportunity of discovering this for himself in Exercise 8-31.

Incidentally, in some problems it may not be possible to ship from every origin to every destination. For example, there might not be any highway or railway line linking warehouse p with retail outlet q. In such a case the variable x_{pq} is identically zero and need not be included in the formulation. Because transportation problems are so easy to solve, however, it is better to preserve the format (8-2) by retaining x_{pq} and assigning to c_{pq} an extremely large positive value. This maneuver insures that x_{pq} will be zero in the optimal solution, as desired (a positive value of x_{pq} would indicate that the problem had no feasible solution).

8.3 MATHEMATICAL PROPERTIES OF THE TRANSPORTATION PROBLEM

We have observed that the transportation problem (8-2) is a linear program in standard form. Its constraints can be displayed as follows:

$$
\begin{array}{llllllll}
x_{11}+x_{12}+\cdots+x_{1n} & & & & & & & =a_1 \\
 & & x_{21}+x_{22}+\cdots+x_{2n} & & & & & =a_2 \\
 & & & & \cdots & & & \vdots \\
 & & & & & x_{m1}+x_{m2}+\cdots+x_{mn} & & =a_m \\
x_{11} & & +x_{21} & & +\cdots+x_{m1} & & & =b_1 \\
 & x_{12} & & +x_{22} & +\cdots & +x_{m2} & & =b_2 \\
 & \cdots & & \cdots & & \cdots & & \vdots \\
 & x_{1n} & & +x_{2n}+\cdots & & & +x_{mn} & =b_n
\end{array}
\tag{8-7}
$$

Note that a problem with m origins and n destinations has $m + n$ constraints and mn nonnegative variables. Each variable x_{ij} appears in exactly two constraints, one associated with the ith origin and the other with the jth destination.

If we switch to the usual matrix/vector notation, the transportation problem becomes

$$
\begin{aligned}
&\text{Min } z = \mathbf{c}^T\mathbf{x} \\
&\text{subject to} \quad \mathbf{Ax} = \mathbf{b} \\
&\text{and} \qquad\qquad \mathbf{x} \geq \mathbf{0}
\end{aligned}
\tag{8-8}
$$

where $\mathbf{x} \equiv (x_{11}, x_{12}, \ldots, x_{1n}, x_{21}, x_{22}, \ldots, x_{2n}, \ldots, x_{m1}, x_{m2}, \ldots, x_{mn})$, $\mathbf{b} \equiv (a_1, a_2, \ldots, a_m, b_1, b_2, \ldots, b_n)$, $\mathbf{c}$ is analogous to $\mathbf{x}$, and $\mathbf{A}$ is an $(m + n) \times (mn)$ matrix containing the coefficients of the constraints (8-7). The above ordering of the constraints—origins first, then destinations—gives the $\mathbf{A}$ matrix a very special structure; for example, in a problem having three origins and four destinations ($m = 3$, $n = 4$) it would be

$$
\mathbf{A} = \begin{bmatrix}
1 & 1 & 1 & 1 & 0 & 0 & 0 & 0 & 0 & 0 & 0 & 0 \\
0 & 0 & 0 & 0 & 1 & 1 & 1 & 1 & 0 & 0 & 0 & 0 \\
0 & 0 & 0 & 0 & 0 & 0 & 0 & 0 & 1 & 1 & 1 & 1 \\
1 & 0 & 0 & 0 & 1 & 0 & 0 & 0 & 1 & 0 & 0 & 0 \\
0 & 1 & 0 & 0 & 0 & 1 & 0 & 0 & 0 & 1 & 0 & 0 \\
0 & 0 & 1 & 0 & 0 & 0 & 1 & 0 & 0 & 0 & 1 & 0 \\
0 & 0 & 0 & 1 & 0 & 0 & 0 & 1 & 0 & 0 & 0 & 1
\end{bmatrix}
$$

In general, the constraint matrix for an m-origin, n-destination problem can be written as

$$\mathbf{A} = \begin{bmatrix} \mathbf{J}_{mn}^1 & \mathbf{J}_{mn}^2 & \cdots & \mathbf{J}_{mn}^m \\ \mathbf{I}_n & \mathbf{I}_n & \cdots & \mathbf{I}_n \end{bmatrix} \tag{8-9}$$

where $\mathbf{J}_{mn}{}^k$ is an $m \times n$ submatrix having all 1's in the kth row and 0's everywhere else, and $\mathbf{I}_n$ is an $n \times n$ identity matrix. The column of the $\mathbf{A}$ matrix associated with any variable x_{ij} is evidently of the form

$$\mathbf{a}_{ij} \equiv \mathbf{e}_i + \mathbf{e}_{m+j} \tag{8-10}$$

where $\mathbf{e}_i$ and $\mathbf{e}_{m+j}$ are unit vectors in E^{m+n}; this is easily seen from (8-7). Note that we are now using two subscripts to identify a variable and its column vector.

Referring to (8-7), it is not difficult to show that the constraints of any feasible transportation problem are redundant. Subtracting the sum of the demand constraints from the sum of the supply constraints produces

$$\sum_{i=1}^{m} \sum_{j=1}^{n} x_{ij} - \sum_{j=1}^{n} \sum_{i=1}^{m} x_{ij} = \sum_{i=1}^{m} \sum_{j=1}^{n} 0 \cdot x_{ij} = \sum_{i=1}^{m} a_i - \sum_{j=1}^{n} b_j = 0$$

Because this equation identifies a linear combination of the left-hand sides of the constraints that adds up to the null vector,[1] we know from Section 2.3 that the constraints are *redundant* (assuming that total supply equals total demand; if not, the problem has no feasible solution). Moreover, every single constraint in the problem can be said to be redundant in the sense that the scalar coefficient of every row in the above linear combination is nonzero.

To investigate the "amount" of redundancy in the transportation problem we arbitrarily remove the last destination constraint from the set (8-7), leaving a total of $m + n - 1$ rows. Consider the columns of the reduced $\mathbf{A}$ matrix that are associated with the variables

$$x_{1n}, x_{2n}, \ldots, x_{mn}, x_{11}, x_{12}, \ldots, x_{1,n-1}$$

Collecting them in that order into an $(m + n - 1) \times (m + n - 1)$ matrix produces

$$\begin{bmatrix} \mathbf{I}_m & \mathbf{J}_{m,n-1}^1 \\ \mathbf{0} & \mathbf{I}_{n-1} \end{bmatrix}$$

[1] Strictly speaking, the null vector is equal to a linear combination of the rows of the $\mathbf{A}$ matrix (8-9), with the scalar coefficients being $+1$ for the first m rows and -1 for the last n.

where the submatrix $\mathbf{0}$ is $(n - 1) \times m$ and where $\mathbf{J}^1_{m,n-1}$ is defined as earlier in this section. This matrix has an inverse, namely

$$\begin{bmatrix} \mathbf{I}_m & -\mathbf{J}^1_{m,n-1} \\ \mathbf{0} & \mathbf{I}_{n-1} \end{bmatrix}$$

so its rows and, therefore, the $m + n - 1$ rows remaining in the $\mathbf{A}$ matrix are linearly independent. A similar invertible submatrix could be found if any other constraint were removed from the set (8-7). Thus, we have proved that *the transportation problem has a "total" of only one redundant constraint*, in the sense that redundancy vanishes if just one constraint is deleted.

Let us imagine solving the transportation problem (8-8) via the standard two-phase simplex method, without bothering to remove the redundant constraint (one artificial variable therefore will remain in the basis with a value of zero throughout phase 2). Since phase 1 begins with an identity matrix consisting of the $m + n$ artificial columns, the nonbasic $\mathbf{y}$ columns must initially be

$$\mathbf{y}_{ij} \equiv \mathbf{a}_{ij} = \mathbf{e}_i + \mathbf{e}_{m+j}$$

Suppose that some real variable x_{pq} is chosen to enter the basis. The column $\mathbf{y}_{pq}$ contains only 0's and 1's, so the pivot element y_{rpq} must necessarily equal 1. (The triply-subscripted notation is regrettable, but if two subscripts are used to identify the column vector $\mathbf{a}_{ij}$, then presumably two are needed for the vector $\mathbf{y}_{ij}$ and three for its rth component y_{rij}. In general, then, the first subscript of any y element y_{rij} will refer to a row and the second and third will jointly identify a column.) From (4-75) it can be seen that, except for its last component, the $\mathbf{\Phi}$ vector for the first pivot is made up only of 0's and (-1)'s. Consider now the simplex transformation of the tableau column $\mathbf{y}_{ij}$:

$$\text{(New column } ij) = (\text{old column } ij) + y_{rij}\mathbf{\Phi} \tag{4-76}$$

where we have altered the subscripts to conform to our new notation. We know that y_{rij} is either 0 or $+1$. If $y_{rij} = 0$ the new vector $\hat{\mathbf{y}}_{ij}$ is the same as before, and if $y_{rij} = 1$ the new vector is obtained by adding 0's and (-1)'s to 0's and $(+1)$'s in various combinations. Thus, in either case, after the first pivot, $\hat{\mathbf{y}}_{ij}$ consists entirely of (-1)'s, 0's, and $(+1)$'s.

It turns out that when a transportation problem is solved via the simplex method, *every y element in every tableau throughout both phases is* $-1, 0$, or $+1$, although, of course, this is not true of the basic variables or of the reduced costs. The student saw an illustration of this remarkable property in the example of Section 6.5. We shall postpone its proof for a few pages, but it is convenient to discuss some of the implications now. Notice, for example, that the pivot element will be $+1$ at every iteration, so that each component of the $\mathbf{\Phi}$ vector except the last (the reduced-cost component) will be -1, 0, or $+1$. It follows from (4-76) that no nontrivial multiplications or divisions are ever involved in the updating of $\mathbf{x_B}$ and the $\mathbf{y}_{ij}$; the reader also may check this statement by referring to the simplex transformation formulas listed together in Section 4.8. Similarly, because y_{rij} in (4-76) will always be -1, 0, or $+1$ (except for the $\mathbf{x_B}$ column), the reduced costs $(z_{ij} - c_{ij})$ also may be updated without multiplying or dividing. Thus the only arithmetic operations required during an entire simplex pivot are addition and subtraction,[1] a computational property that allows the transportation problem to be solved with great speed and accuracy either by hand or on a digital computer. In fact, the reader will discover that the transportation algorithm presented below is merely a disguised version of the simplex method, designed specifically to exploit that freedom from multiplication and division.

The fact that the y elements are always ± 1 or 0 leads to other interesting consequences:

(1) Whenever all supplies a_i and demands b_j are integers, so are the values of all basic variables at every iteration and in the eventual optimal solution.
(2) Whenever all costs c_{ij} are integers, so are the reduced costs $(z_{ij} - c_{ij})$ at every iteration.

These two statements can be verified from the transformation formulas. In the case of (1), for example, the basic variables are updated via

$$\hat{x}_{\mathbf{B}r} = \frac{x_{\mathbf{B}r}}{y_{rpq}} \tag{4-60}$$

and

$$\hat{x}_{\mathbf{B}i} = x_{\mathbf{B}i} - \frac{x_{\mathbf{B}r}}{y_{rpq}} y_{ipq} \qquad \text{for all } i \neq r \tag{4-61}$$

where x_{pq} is replacing $x_{\mathbf{B}r}$ in the basis and the subscripts have been modified as usual. When the a_i and b_j are integral, the basic variables begin as integers in phase 1, and because y_{rpq} is always 1 they remain as integers throughout. This result seems intuitively persuasive: If it is optimal to ship a fraction of a unit of commodity from the ith origin to the jth destination, then it should be

[1] If the value of the objective function is updated at every iteration via

$$\hat{z} = z - \frac{x_{\mathbf{B}r}}{y_{rpq}}(z_{pq} - c_{pq}) \tag{4-74}$$

where x_{pq} is replacing $x_{\mathbf{B}r}$ in the basis, then a single nontrivial multiplication per pivot is required. This procedure is not necessary, however, inasmuch as the current value of z is not used in pivoting; it is quite possible simply to ignore the objective until after the optimal solution $\mathbf{x}_\mathbf{B}^*$ has been obtained, at which time $z_{\text{opt}} = \mathbf{c}_\mathbf{B}^{*T}\mathbf{x}_\mathbf{B}^*$ can be calculated.

optimal, in general, to ship as much as possible, which (until other shipments have been planned) would be min $\{a_i,b_j\}$, an integer. Incidentally, it should be noted that fractional a_i and b_j can be converted to integers by multiplying them all by their least common denominator or by a large constant; the same is true of fractional c_{ij}. These maneuvers have the effect of changing the units and permit transportation problems to be solved by means of computer algorithms that use integer variables only. Integers require fewer memory locations than real numbers, so larger problems can be handled within a given core size without recourse to secondary memory.

8.4 THE TRANSPORTATION TABLEAU

The transportation problem, which is known also as the Hitchcock problem, was first formulated in 1941 by Frank L. Hitchcock, who proceeded to sketch out a procedure for obtaining a solution. Investigators in several countries studied the problem and published articles about it during the 1940's, and by the end of that decade an efficient procedure for solving it had been developed that remains the basis of all present-day methods. Principal credit for this success belongs to George Dantzig [3]. His approach consisted essentially of a simple and elegant adaptation of the simplex method, and, although a few frills and computational devices have since been proposed, no significant improvement upon it has appeared. It is to this *transportation algorithm* that we now turn our attention.

Instead of using the simplex tableau as the major computational vehicle, we shall work with a different sort of diagram known as a *transportation tableau*, an example of which is shown in Figure 8-1. This tableau is designed to display all essential elements, both constant and variable, of any given transportation problem (8-8). Each row except the last is associated with an origin and each column except the last with a destination. The per-unit cost c_{ij} of shipping from the ith origin to the jth destination is entered in the upper left portion of the square, or *cell*, which lies in row i and column j; we shall refer to this cell as (i,j). Any given feasible solution to the transportation problem can be represented by writing the value of each variable x_{ij} in the lower right portion of the cell (i,j). We shall adopt the convention that when a basic feasible solution is being displayed in a transportation tableau, the value of any particular x_{ij} is actually entered *if and only if x_{ij} is in the basis*; note that because the values of the nonbasic variables (known to be zero) are not shown in the tableau, the appearance of one or more zeros indicates that the basis is degenerate. Finally, the various supplies and demands are listed in the right-most column and bottom row, respectively; thus one can readily determine whether a given basic solution satisfies all the constraints by summing the x_{ij} entries across the rows and down the columns. (The elements $\hat{u}_i$ and $\hat{v}_j$ will be discussed later.)

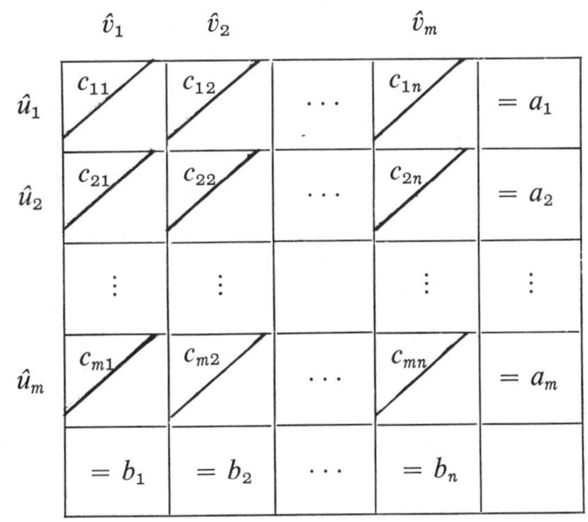

Figure 8-1

In order to establish the transportation algorithm, we shall need to prove two important theorems, which describe how linear dependence and independence manifest themselves graphically on the transportation tableau. Before doing so, however, we must use the rest of this section to state and discuss a single crucial definition on which those theorems depend.

DEFINITION 8.1. An ordered sequence of four or more different cells in a transportation tableau will be called a *loop* if it possesses the following properties:

(a) any two consecutive cells must lie either in the same row or in the same column; and
(b) no three consecutive cells may lie in the same row nor in the same column,

where the first cell is considered to follow the last in the sequence.

In other words, the loop is to be treated as an unending circular sequence, so that, for example, the second-to-last, the last, and the first cell would be "three consecutive cells" and would have to satisfy property (b). It is clear that if, say, the $(k-1)$th and kth cells are in the same row, then the kth and $(k+1)$th must be in the same column but in different rows. Thus, a typical loop might be

$$\{(p,q), (p,r), (s,r), (s,t), \ldots, (v,w), (v,q)\}$$

Because the definition of a loop does not specify a starting point or a direction, however, the same loop might be represented also as

$$\{(s,r), (p,r), (p,q), (v,q), (v,w), \ldots, (s,t)\}$$

Some specific examples are diagramed schematically in Figure 8-2, using "skeleton tableaus" with four origins and five destinations. The loop {(2,1), (2,4), (4,4), (4,1)} shown in Figure 8-2(a) includes only four cells and is therefore of the simplest possible type; a somewhat more complicated loop,

$$\{(1,1), (1,2), (2,2), (2,3), (4,3), (4,5), (3,5), (3,1)\}$$

appears in 8-2(b). The sequences of cells indicated in 8-2(c) and (d) fail to satisfy the above definition and are not loops.

The reader may have noticed that *every loop must have an even number of cells.* This is easily demonstrated. Given any loop composed of p cells, arbitrarily select a starting point and a direction and number the cells in their proper sequence from 1 to p. Suppose that cells 1 and 2 are, say, in the same row, so that the step from 1 to 2 involves a column change. Then the step from 2 to 3 is a row change, from 3 to 4 is a column change, and so on; in general,

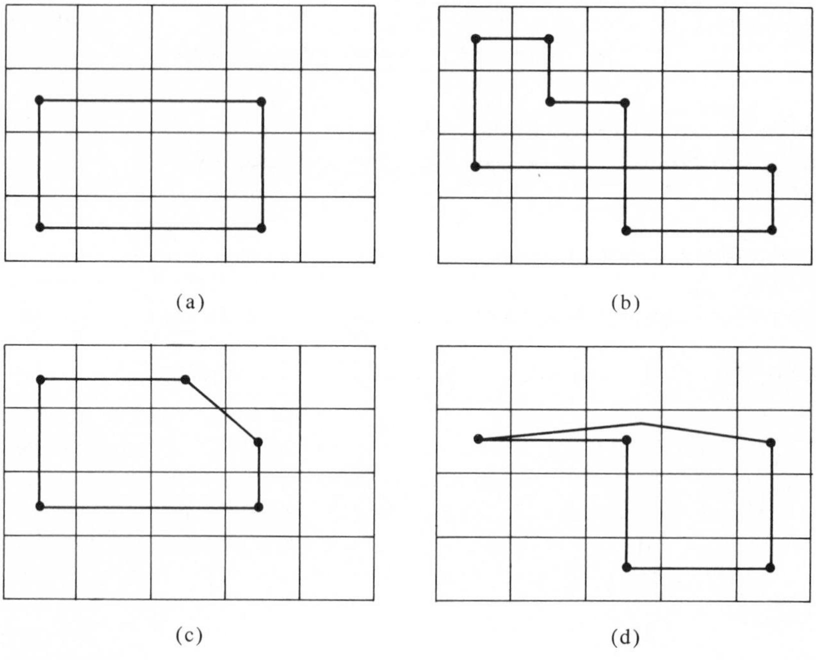

Figure 8-2

the step to cell k involves a column change if and only if k is even. Because the step to cell 2 is a column change, however, the step from p to 1 must be a row change and the step from $p - 1$ to p, a column change; hence p is even.

8.5 LOOPS AND LINEAR DEPENDENCE

In this section we shall establish the theoretical foundation for the development of the transportation algorithm. As will become evident later on, this algorithm is essentially an adaptation of the simplex method that utilizes the notion of a loop to perform simplex pivots directly on the transportation tableau. We remark first that a set S of cells of a transportation tableau is said to *contain a loop* if the cells of S or any subset of them can be sequenced so as to form a loop. Then we have the following theorem.

THEOREM 8.1. Let S denote any set of columns of the **A** matrix in the transportation problem (8-8). Then the columns of S are linearly dependent if and only if their corresponding cells in the transportation tableau contain a loop.

The proof depends on the fact that the column of the **A** matrix associated with the cell (i,j) can be represented as

$$\mathbf{a}_{ij} = \mathbf{e}_i + \mathbf{e}_{m+j} \tag{8-10}$$

where $\mathbf{e}_i$ and $\mathbf{e}_{m+j}$ are unit vectors in E^{m+n}. Suppose the cells associated with the columns of S contain a loop

$$\{(p,q), (p,r), (s,r), (s,t), \ldots, (v,w), (v,q)\}$$

Then S includes the columns

$$\begin{aligned}
\mathbf{a}_{pq} &= \mathbf{e}_p + \mathbf{e}_{m+q} \\
\mathbf{a}_{pr} &= \mathbf{e}_p + \mathbf{e}_{m+r} \\
\mathbf{a}_{sr} &= \mathbf{e}_s + \mathbf{e}_{m+r} \\
\mathbf{a}_{st} &= \mathbf{e}_s + \mathbf{e}_{m+t} \\
&\vdots \\
\mathbf{a}_{vw} &= \mathbf{e}_v + \mathbf{e}_{m+w} \\
\mathbf{a}_{vq} &= \mathbf{e}_v + \mathbf{e}_{m+q}
\end{aligned}$$

But the linear combination

$$\mathbf{a}_{pq} - \mathbf{a}_{pr} + \mathbf{a}_{sr} - \mathbf{a}_{st} + \cdots + \mathbf{a}_{vw} - \mathbf{a}_{vq} \tag{8-11}$$

equals the null vector, as the reader can easily verify, using the fact that a loop contains an even number of cells. Therefore this particular group of columns is linearly dependent, and so is the set S.

Going in the other direction, suppose the columns of S are linearly dependent; then there exist scalars λ_{ij} not all zero such that

$$\sum_{\mathbf{a}_{ij} \in S} \lambda_{ij}\mathbf{a}_{ij} = \mathbf{0} \tag{8-12}$$

To simplify matters, let us remove from S all columns $\mathbf{a}_{ij}$ for which $\lambda_{ij} = 0$, so that now all $\lambda_{ij} \neq 0$ in (8-12). Take any column

$$\mathbf{a}_{pq} = \mathbf{e}_p + \mathbf{e}_{m+q} \quad \text{in } S$$

Because $\lambda_{pq} \neq 0$, the set S must contain at least one other column associated with the qth destination, that is, one other column whose second subscript is q; if this were not true, then the $(m + q)$th component of the vector equation (8-12) would reduce to $\lambda_{pq} \cdot 1 = 0$, which is impossible. Accordingly, suppose that

$$\mathbf{a}_{rq} = \mathbf{e}_r + \mathbf{e}_{m+q} \quad \text{is in } S$$

Then by a similar line of reasoning there must be at least one other column in S whose first subscript is r:

$$\mathbf{a}_{rs} = \mathbf{e}_r + \mathbf{e}_{m+s} \quad \text{is in } S$$

Using the same argument once more, there must be another column whose second subscript is s. If it is $\mathbf{a}_{ps}$, then we have found four columns in S, namely, $\mathbf{a}_{pq}$, $\mathbf{a}_{rq}$, $\mathbf{a}_{rs}$, and $\mathbf{a}_{ps}$, whose cells in the transportation tableau form a loop. Therefore, either the proof is complete or else S contains some other column whose second subscript is s:

$$\mathbf{a}_{ts} = \mathbf{e}_t + \mathbf{e}_{m+s} \quad \text{is in } S$$

As before, there must be another column whose first subscript is t. If it is $\mathbf{a}_{tq}$, a four-cell loop is completed; if not, then

$$\mathbf{a}_{tu} = \mathbf{e}_t + \mathbf{e}_{m+u} \quad \text{is in } S$$

Again, S contains another column whose second subscript is u. If its first subscript is *one that has been identified* already (i.e., either r or p), then a loop is completed (containing either four or six cells); if not, etc., etc. By extending the reasoning the reader can see that because the number of columns in S is finite, a loop must be formed eventually, Q.E.D.

The implications of Theorem 8.1 are extremely important. For example, we saw in Section 8.3 that a total of one constraint in the transportation problem (8-8) is redundant; that is, when (8-8) is solved via the simplex method, each basis in phase 2 contains exactly $m + n - 1$ real variables x_{ij} and one zero-valued artificial variable. It follows from Theorem 8.1 that any feasible solution involving $m + n - 1$ real variables is a *basic* feasible solution[1] if and only if its corresponding cells in the transportation tableau do *not* contain a loop. Theorem 8.3 will make use of this fact to establish a set of rules for generating basic feasible solutions directly on the tableau; thus, in solving transportation problems we will be able to begin at phase 2 and will not need to be concerned with phase 1 at all.

Theorem 8.1 also allows us to prove most of an assertion that was made in Section 8.3 concerning the y elements associated with the transportation problem (we restrict our attention to phase 2 only):

THEOREM 8.2. Let the columns $\mathbf{b}_1, \ldots, \mathbf{b}_{m+n-1}$ constitute a feasible basis for the transportation problem (8-8). Then, in the representation of any nonbasic column $\mathbf{a}_{pq}$ as a linear combination of basis vectors,

$$\mathbf{a}_{pq} = \sum_{k=1}^{m+n-1} y_{kpq}\mathbf{b}_k \tag{8-13}$$

every y element is -1, 0, or $+1$.

In the statement of the theorem the symbol $\mathbf{b}_k$ denotes the kth column in the basis, which is, of course, some column $\mathbf{a}_{ij}$ from the constraint matrix. It is evident from (8-13) that the columns $\mathbf{a}_{pq}, \mathbf{b}_1, \mathbf{b}_2, \ldots, \mathbf{b}_{m+n-1}$ together constitute a linearly dependent set. Therefore their corresponding cells in the transportation tableau contain a loop that must include the cell (p,q) (because a loop involving only the linearly independent basis vectors is not possible). Suppose the loop is

$$\{(p,q), (p,r), (s,r), (s,t), \ldots, (v,w), (v,q)\} \tag{8-14}$$

where $\mathbf{a}_{pr}, \mathbf{a}_{sr}, \ldots,$ and $\mathbf{a}_{vq}$ are basis vectors. Then, as we saw in the proof of Theorem 8.1, the linear combination (8-11) is equal to the null vector, so that

$$\mathbf{a}_{pq} = \mathbf{a}_{pr} - \mathbf{a}_{sr} + \mathbf{a}_{st} - \cdots - \mathbf{a}_{vw} + \mathbf{a}_{vq} \tag{8-15}$$

[1] It will be natural and convenient for us to deal with the transportation problem (8-8) in its intact form, that is, preserving all $m + n$ constraints. However, we shall be using the terms "basis" and "basic feasible solution" to refer to sets of $m + n - 1$ linearly independent columns $\mathbf{a}_{ij}$ or to their associated variables x_{ij}; therefore the reader should be aware that in this usage the artificial column that is formally required to complete the $(m + n) \times (m + n)$ basis matrix is simply ignored.

This is the unique representation of $\mathbf{a}_{pq}$ as a linear combination of basis vectors; therefore the y elements associated with the basis vectors in (8-15) are $+1$, -1, $+1, \ldots, -1$, and $+1$, and all other y elements are zero, Q.E.D.

Incidentally, the loop (8-14) could have been represented also as

$$\{(p,q), (v,q), (v,w), \ldots, (s,t), (s,r), (p,r)\}$$

that is, with the cells listed in opposite order. In that case (8-15) would have been replaced by

$$\mathbf{a}_{pq} = \mathbf{a}_{vq} - \mathbf{a}_{vw} + \ldots + \mathbf{a}_{st} - \mathbf{a}_{sr} + \mathbf{a}_{pr}$$

and the respective y elements would have been exactly the same.

8.6 THE INITIAL BASIC FEASIBLE SOLUTION

At last the stage has been set for the presentation of an overall approach for solving transportation problems, an approach that will depend almost completely upon Theorems 8.1 and 8.2. We begin with a discussion of the various methods for finding an initial basic feasible solution; this will occupy our attention throughout the present section. Then in Sections 8.7 and 8.8 we shall develop an algorithm—*the* transportation algorithm—that will allow us to compute reduced costs and pivot from one BFS to another until the optimum is reached. Finally, after the method has been summarized and a sample problem solved, the last sections will consist of miscellaneous remarks on computational matters and on one or two other topics.

It was mentioned earlier that there exist certain direct procedures for constructing initial basic feasible solutions to transportation problems, "direct" in the sense that iterative search schemes (such as phase 1 of ordinary linear programming) are not needed. The simplest of these procedures is known as the *northwest-corner rule*, and the simplest way to explain it is to work out an example; later on, the validity of the northwest-corner rule and of several similar methods for generating basic feasible solutions will be formally established by the proof of Theorem 8.3. Consider the transportation problem whose associated tableau, with values assigned to some of the variables, is shown in Figure 8-3. Because

$$\sum_i a_i = \sum_j b_j = 24.7$$

we know at once that feasible solutions must exist. Begin by imagining that the tableau is blank, that is, that no values have been entered for any of the variables,[1]

[1] The reader may want to draw a 5×7 rectangular grid on a piece of scrap paper, copy the supplies and demands from Figure 8-3, and work through the example step by step.

1 / 4.0	4 / 1.6	4 /	6 /	2 /	3 /	= 5.6
2 /	1 / 0.8	5 / 3.6	1 / 5.6	1 /	4 /	= 10.0
4 /	5 /	6 /	4 / 0.0	2 / 2.1	1 /	= 2.1
6 /	4 /	5 /	3 /	2 / 2.9	1 / 4.1	= 7.0
= 4.0	= 2.4	= 3.6	= 5.6	= 5.0	= 4.1	

Figure 8-3

so that initially all $x_{ij} = 0$. Set $x_{11} = \min(a_1,b_1) = \min(5.6,4.0) = 4.0$ and enter this value in the cell (1,1), which lies in the northwest corner of the tableau. Note that we have chosen the largest possible *feasible* value for x_{11}: Four units of commodity is the maximum amount that can be accepted at destination 1. Because the first destination constraint has been satisfied exactly, no other variables in the first column will be assigned positive values, and we can mentally "cross out," or eliminate, column 1 from the tableau. On the other hand, the first origin constraint is still *undersatisfied*, so at least one other variable in the first row will have to take on a positive value.

We now locate the northwest corner of that portion of the tableau that has not been crossed out and set the variable there equal to *its* largest possible feasible value:

$$x_{12} = \min(a_1 - x_{11}, b_2) = \min(1.6,2.4) = 1.6$$

Again we have satisfied one constraint exactly (the first origin constraint) while undersatisfying another; this time the first row is crossed out, so that the cell (2,2) becomes the new northwest corner. Continuing in the same manner through two more iterations, we set $x_{22} = 0.8$ and then $x_{23} = 3.6$.

At this point, row 1 and columns 1, 2, and 3 have been crossed out, and the northwest corner is the cell (2,4). When we compute the value to be assigned to x_{24}, however, we find that a tie occurs:

$$x_{24} = \min\left(a_2 - \sum_j x_{2j}, b_4\right) = \min(5.6,5.6) = 5.6$$

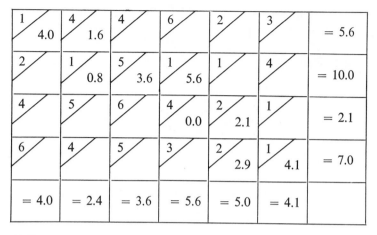

This situation actually requires *no* special handling, except that an arbitrary tie-breaking choice must be made: When the value 5.6 is entered in the cell (2,4), we have the option of crossing out either row 2 or column 4, *but not both*. Suppose we choose the former. The new northwest corner is then (3,4), and its maximum feasible value is

$$x_{34} = \min\left(a_3, b_4 - \sum_i x_{i4}\right) = \min(2.1,0) = 0$$

We proceed as usual, entering a value of zero in (3,4) and, because the fourth destination constraint now is satisfied, crossing out column 4. In practice, these steps can be compressed as follows: Enter the value 5.6 in the cell (2,4), then enter a zero either in the cell below or in the cell to the right of it, and finally cross out both row 2 and column 4. The presence of a zero means, of course, that the basic feasible solution will be degenerate.

The northwest corner is now (3,5), and we continue as before, setting $x_{35} = 2.1$ and $x_{45} = 2.9$. At the final stage every cell but (4,6) has been crossed out; the last value to be entered in the tableau is

$$x_{46} = \min\left(a_4 - \sum_j x_{4j}, b_6 - \sum_i x_{i6}\right) = \min(4.1,4.1) = 4.1$$

This satisfies the fourth origin and sixth destination constraints simultaneously (given that total supply equals total demand, would it have been possible for a tie *not* to have occurred here?), thereby producing a feasible solution to the transportation problem involving exactly $m + n - 1$ variables. In constructing the solution we moved only downward and to the right in the tableau, never "circling back"; thus we could not possibly have formed a loop, and it follows from Theorem 8.1 that our feasible solution must be basic as well.

Although the student may already be convinced that the above method will "work" for any transportation problem, we still must demonstrate it formally. Instead of devising a proof specifically for the northwest-corner rule, however, we prefer to treat it as a member of a more general family of methods, all of which can be established with a single theorem, as follows.

THEOREM 8.3. The following generalized procedure must yield a basic feasible solution to any transportation problem having m origins and n destinations:

(a) Beginning with a blank transportation tableau, so that all $x_{ij} = 0$ initially, choose any variable x_{pq} and set $x_{pq} = \min(a_p, b_q)$.

(b) If $a_p < b_q$, so that the pth origin constraint is satisfied, eliminate or cross out the pth row of the tableau. If $a_p > b_q$, eliminate the qth column. If

$a_p = b_q$, set some other variable in either row p or column q equal to zero and eliminate both row p and column q.

(c) Choose any variable x_{rs} that is not in an eliminated row or column and set

$$x_{rs} = \min \left(a_r - \sum_{j=1}^{n} x_{rj},\ b_s - \sum_{i=1}^{m} x_{is} \right)$$

eliminating row r and/or column s in accordance with the logic of step (b).

(d) Repeat step (c) until values have been assigned to exactly $m + n - 1$ variables; these values then constitute a (possibly degenerate) basic feasible solution.

Notice that the theorem does not specify a criterion for determining which of the eligible (i.e., not eliminated) variables is to be added to the solution at each iteration. In fact, the various procedures for generating basic feasible solutions, two of which will be outlined below, differ only in having different rules for selecting the variables x_{pq} and x_{rs}; in the northwest-corner rule, for example, one always chooses the "upper left-most" of the eligible variables.

The student who has read through the northwest-corner example carefully should find the proof of Theorem 8.3 quite easy to follow. Observe first that in order for any x_{rs} to be selected in step (c) both the rth origin and the sth destination constraints currently must be *undersatisfied.* Therefore each x_{rs} is assigned a positive value and, although some zero values may be introduced when ties occur, the solution that eventually is constructed will be entirely nonnegative. Next, it is clear that exactly one constraint is satisfied for each variable added to the solution, so that a total of $m + n - 1$ constraints have been *explicitly* satisfied when the procedure ends. This *implies* that the remaining constraint, which is redundant and can be expressed as a linear combination of the others, must be satisfied too; therefore the solution is at least *feasible.*

In order to show that it is also *basic* we shall proceed by contradiction. Assume that when a value is assigned to the variable x_{rs} at some stage, a loop

$$\{(r,s), (r,t), (u,t), \ldots, (v,w), (v,s)\}$$

is formed on the tableau, where (r,t), (u,t), . . . , and (v,s) are all cells in which x_{ij} values have been entered *previously*. Because x_{rs} was selected last, it was selected after x_{rt}, in particular, so the assignment of a value to x_{rt} must have satisfied the tth destination constraint; if it had satisfied the rth origin constraint, row r would have been crossed out and x_{rs} subsequently could not have been chosen. The reader should verify that this conclusion is valid even if x_{rs} or x_{rt} was assigned a value of zero [recall that the addition of two variables to the solution when a tie occurs in step (c) can be considered as two successive single additions]. Now, because the tth destination constraint was satisfied by the assignment of a value to x_{rt}, the selection of x_{ut} could only have satisfied the uth origin constraint; working on through the loop, we eventually conclude that the choices of x_{vw} and x_{vs} must have satisfied the vth origin and sth destination constraints, respectively. This means that column s must have been crossed out *before* x_{rs} was selected, which is impossible. Therefore, our assumption was false; the procedure we are considering can never lead to the formation of a loop, and any solution it generates must be basic, Q.E.D.

Although the northwest-corner rule will produce a perfectly legitimate basic feasible solution to any transportation problem, it is very seldom used in practice. In solving a problem via the transportation algorithm, as we shall see later on, the initial BFS must serve as a starting point for a sequence of pivots that ultimately leads to the optimal solution. Because these pivots are time-consuming, it is clearly desirable to begin with a good initial solution—"good" in the sense of being close to the optimum—provided that a prohibitive amount of time does not have to be spent in searching for it. From this point of view, the northwest-corner rule, which completely ignores the shipping costs c_{ij}, can hardly be expected to produce a very good starting solution, except by sheer chance. It seems much more sensible to devote a little extra effort to the determination of a better initial BFS, in the hope that overall solution time will be reduced.

One obvious and simple rule for adding variables to the solution is to choose at any stage the eligible variable whose cost c_{ij} is smallest; that is, the criterion for selecting x_{rs} in step (c) of Theorem 8.3 is that c_{rs} must be less than or equal to c_{ij} for every other eligible variable x_{ij}—similarly, in step (a) c_{pq} must be the smallest cost in the entire tableau. This "matrix minimum" approach is, in fact, quite a reasonable one to use: It usually generates a low-cost initial BFS and does not require a great deal of auxiliary computation. Incidentally, whenever the minimum cost is not unique, the tie is broken arbitrarily; in many computer codes the variable selected is simply the first one scanned.

A more sophisticated criterion for choosing each variable x_{rs} would be one that takes into account not only the cost c_{rs}, but also the costs of the other variables that might have to be included in the basis if x_{rs} is not. For example, suppose that the smallest cost in the tableau is c_{pq}, but that several other costs in row p and column q are almost as small. Then clearly it would not be as important to include x_{pq} in the initial solution as it would be to include some other variable x_{uv} whose cost c_{uv}, although perhaps fairly high, is still substantially lower than all other costs in row u. This is the approach taken by *Vogel's method* [23], which proceeds as follows, beginning with the first iteration. For each row i in the tableau, $i = 1, \ldots, m$, find the lowest and second-lowest costs in the row and compute their difference. Similarly, compute this difference for each column, so that a total of $m + n$ differences are eventually obtained.

1 / 4.0	4	4	6	2 / 1.6	3	= 5.6
2	1 / 2.4	5	1 / 5.6	1 / 2.0	4	= 10.0
4	5	6	4	2	1 / 2.1	= 2.1
6	4	5 / 3.6	3	2 / 1.4	1 / 2.0	= 7.0
= 4.0	= 2.4	= 3.6	= 5.6	= 5.0	= 4.1	

Figure 8-4

Then the first variable to be included in the solution will be the one with the smallest cost in whichever row or column has the greatest difference. Suppose it is x_{pq}; then set $x_{pq} = \min(a_p,b_q)$ and eliminate row p and/or column q in the usual manner. Now the entire process is repeated until a basic solution has been generated, with the differences at each iteration being calculated only for the surviving rows and columns (and with a cost c_{ij} being eligible for consideration as lowest or second-lowest only if both row i and column j have survived). The procedure is terminated when either the number of surviving rows or the number of surviving columns has been reduced to one, because at that point the values of all remaining variables are completely determined.

Both Vogel's method and the matrix minimum approach were suggested in the 1950's and both are still widely used today. Of the two, the former tends to produce better solutions, but requires more time; extra logical testing is involved in locating both the lowest *and* the second-lowest cost in each row and column. For an illustration of Vogel's method, let us return to the tableau of Figure 8-3, imagining now that it is blank.[1] The differences for the rows are 1, 0, 1, and 1, and those for the columns are 1, 3, 1, 2, 1, and 0. The largest of these is associated with column 2, whose smallest cost is $c_{22} = 1$. Accordingly, set $x_{22} = \min(a_2,b_2) = \min(10.0,2.4) = 2.4$ and eliminate column 2. Now the row differences are 1, 0, 1, and 1 and the column differences are 1, 1, 2, 1 and 0, so the next variable to be selected is x_{24}. Continuing in this manner, we eventually derive the basic feasible solution shown in Figure 8-4, provided that when the

[1] Incidentally, the costs c_{ij} in the tableau have all been given positive integer values strictly for simplicity and convenience. None of the theorems or algorithms in this chapter place any restrictions on the c_{ij} (although the supplies a_i and demands b_j have been assumed throughout to be *positive* real numbers).

maximum difference at any stage is not unique, the following order of preference is used for breaking the tie: row 1 first, then row 2, ..., row m, column 1, ..., and column n. The above solution has a total cost of 42.1 units, as compared to 48.9 for the northwest-corner solution; the minimum cost for this problem will turn out to be 40.5.

8.7 THE TRANSPORTATION ALGORITHM: THE REDUCED COSTS

In the next two sections we shall develop Dantzig's algorithm for performing simplex pivots directly on the transportation tableau. The overall plan of attack will be exactly the same as in the revised simplex method of linear programming: At each iteration determine the reduced costs, test for optimality, choose the entering and exiting variables, and compute the new basic feasible solution. Inasmuch as our procedures for pivoting will rely heavily on the properties of loops, the reader may want to review the appropriate material in Sections 8.4 and 8.5.

We first consider the question of how to evaluate the reduced costs $z_{ij} - c_{ij}$. One possibility is to take the direct approach: For each nonbasic variable x_{ij}, determine $\mathbf{y}_{ij}$ and then calculate

$$z_{ij} - c_{ij} = \mathbf{c_B}^T\mathbf{y}_{ij} - c_{ij}$$

The vectors $\mathbf{y}_{ij}$ cannot be obtained without some difficulty, however, and it turns out to be far more efficient to derive the reduced costs indirectly, by making use of the dual problem. Let $u_1, \ldots, u_m$ and $v_1, \ldots, v_n$ be the dual variables associated with the origin and destination constraints, respectively; then the dual of the transportation problem (8-8) has the following set of constraints:

$$\left.\begin{cases} u_i + v_j \le c_{ij} \\ \text{with } u_i \text{ and } v_j \text{ unrestricted} \end{cases}\right\} \text{for all } i \text{ and } j$$

These can easily be verified by referring to the primal constraints (8-7). We now recall a definition from Section 7.2: Given any basic feasible solution to a primal linear program in standard form, with basis matrix $\mathbf{B}$ and cost vector $\mathbf{c_B}$, the set of values

$$\hat{\mathbf{u}}^T = \mathbf{c_B}^T\mathbf{B}^{-1} \tag{7-1}$$

constitutes the *associated solution* to the dual problem. It was shown that this definition led to an alternative expression for the reduced cost of each primal variable:

$$z_j - c_j = \mathbf{c_B}^T\mathbf{B}^{-1}\mathbf{a}_j - c_j = \hat{\mathbf{u}}^T\mathbf{a}_j - c_j \qquad \text{for all } j \tag{7-3}$$

where $\mathbf{a}_j$ is the jth column of the primal constraint matrix $\mathbf{A}$. In the case of the transportation problem, the associated dual solution can be represented as the row vector

$$[\hat{\mathbf{u}},\hat{\mathbf{v}}] = [\hat{u}_1, \ldots, \hat{u}_m, \hat{v}_1, \ldots, \hat{v}_n]$$

and (7-3) becomes

$$z_{ij} - c_{ij} = [\hat{\mathbf{u}},\hat{\mathbf{v}}] \cdot \mathbf{a}_{ij} - c_{ij} = \hat{u}_i + \hat{v}_j - c_{ij} \qquad \text{for all } i \text{ and } j \qquad (8\text{-}16)$$

using the fact that $\mathbf{a}_{ij}$ is the sum of two unit vectors. The reader is reminded again that the redundant constraint has not been deleted, so that the basis includes one zero-valued artificial variable.

Given any basic feasible solution to the transportation problem, equation (8-16) provides the key to evaluating the reduced costs. Because $z_{ij} - c_{ij} = 0$ for basic variables, we may write

$$\hat{u}_i + \hat{v}_j = c_{ij} \qquad \text{for all } x_{ij} \text{ in the basis} \qquad (8\text{-}17)$$

producing a system of $m + n - 1$ linear equations in $m + n$ unknowns. Had we deleted the redundant constraint from the transportation problem, this system would have had $m + n - 1$ unknowns—one for each primal constraint—and the values of the $\hat{u}_i$ and $\hat{v}_j$ would be completely determined. Instead, by retaining the redundant constraint, we generated an extra dual variable and introduced a single "degree of freedom" into the system (8-17); thus we may assign an arbitrary value, say, zero, to any one of the $\hat{u}_i$ or $\hat{v}_j$ and then solve the $m + n - 1$ simultaneous equations for the remaining $m + n - 1$ unknowns.

For a more rigorous explanation of why the system (8-17) should be indeterminate, let us pretend that we are solving the transportation problem (8-8) by ordinary linear programming methods. We add an artificial variable to each of the $m + n$ constraints and discover at the end of phase 1 that one of them will have to remain in the basis with a value of zero throughout phase 2. Suppose it is the artificial variable w_1 associated with the first origin constraint; then, in effect, w_1 continues to be part of the problem, and the unit column vector $\mathbf{e}_1$ must be adjoined to the constraint matrix $\mathbf{A}$. Following the model of (8-16), the reduced cost of this artificial variable for any phase-2 basis can be expressed as $[\hat{\mathbf{u}},\hat{\mathbf{v}}] \cdot \mathbf{e}_1 - c^* = \hat{u}_1 - c^*$, where c^* is whatever arbitrary cost was assigned to w_1 (remember that the "artificial row" in the simplex tableau consists entirely of zeros, so that the choice of a value for c^* has no effect on anything that happens in phase 2). But because w_1 is basic, its reduced cost is known to be zero, which implies that

$$\hat{u}_1 = c^* \qquad (8\text{-}18)$$

This combines with (8-17) to form a system of $m + n$ equations in $m + n$ unknowns, containing one arbitrary parameter. Moreover, every constraint in the transportation problem participates in the redundancy, so any one of the artificial variables might have been carried into phase 2. Thus $\hat{u}_1$ in (8-18) might just as well be replaced by any of the other unknowns.

The actual process of solving the system (8-17) is, of course, trivial. After a single arbitrary assignment, typically $\hat{u}_1 = 0$, the rest of the values are determined by simple addition and subtraction. When this has been done, the reduced costs of all nonbasic variables can be calculated from (8-16). Consider, for example, the basic feasible solution in Figure 8-3, which was generated via the northwest-corner rule. The system (8-17) includes the following equations:

$$\begin{aligned} \hat{u}_1 + \hat{v}_1 &= c_{11} = 1 & \qquad \hat{u}_3 + \hat{v}_4 &= c_{34} = 4 \\ \hat{u}_1 + \hat{v}_2 &= c_{12} = 4 & \qquad \hat{u}_3 + \hat{v}_5 &= c_{35} = 2 \\ \hat{u}_2 + \hat{v}_2 &= c_{22} = 1 & \qquad \hat{u}_4 + \hat{v}_5 &= c_{45} = 2 \\ \hat{u}_2 + \hat{v}_3 &= c_{23} = 5 & \qquad \hat{u}_4 + \hat{v}_6 &= c_{46} = 1 \\ \hat{u}_2 + \hat{v}_4 &= c_{24} = 1 \end{aligned}$$

Note that an equation has been written for the basic variable x_{34} even though its value is zero. We arbitrarily set $\hat{u}_1 = 0$; solving sequentially,

$$\begin{aligned} \hat{v}_1 &= 1 - \hat{u}_1 = 1 - 0 = 1 \\ \hat{v}_2 &= 4 - \hat{u}_1 = 4 - 0 = 4 \\ \hat{u}_2 &= 1 - \hat{v}_2 = 1 - 4 = -3 \end{aligned}$$

and so on

The values of the associated dual solution are entered just above and to the left of the transportation tableau, as shown in Figure 8-5; this, of course, accounts for the labels $\hat{u}_i$ and $\hat{v}_j$ in the generalized tableau of Figure 8-1.

Now we can use (8-16) to obtain the reduced cost of every nonbasic variable:

$$\begin{aligned} z_{13} - c_{13} &= \hat{u}_1 + \hat{v}_3 - c_{13} = 0 + 8 - 4 = +4 \\ z_{14} - c_{14} &= \hat{u}_1 + \hat{v}_4 - c_{14} = 0 + 4 - 6 = -2 \\ z_{15} - c_{15} &= \hat{u}_1 + \hat{v}_5 - c_{15} = 0 + 2 - 2 = 0 \end{aligned}$$

and so on

Instead of crowding all these values into the tableau, we enter the reduced cost

	1	4	8	4	2	1	
0	1 / 4.0	4 / 1.6	4 / (4)	6	2	3	= 5.6
−3	2	1 / 0.8	5 / 3.6	1 / 5.6	1	4	= 10.0
0	4	5	6 / (2)	4 / 0.0	2 / 2.1	1	= 2.1
0	6	4	5 / (3)	3 / (1)	2 / 2.9	1 / 4.1	= 7.0
	= 4.0	= 2.4	= 3.6	= 5.6	= 5.0	= 4.1	

Figure 8-5

$z_{ij} - c_{ij}$, enclosed in parentheses, in the cell (i,j) if and only if $z_{ij} - c_{ij} > 0$, which implies that x_{ij} is a candidate for basis entry. This has been done in Figure 8-5. We then observe that because four of the reduced costs are positive (and we are minimizing), the current basic feasible solution is not optimal. Applying the usual simplex criterion, the variable with the largest reduced cost, here x_{13}, will be chosen to enter the basis at the next iteration.

8.8 THE TRANSPORTATION ALGORITHM: PIVOTING TO A NEW BFS

So far we have developed a general procedure for evaluating the reduced costs associated with any basic feasible solution to the transportation problem. If all the $z_{ij} - c_{ij}$ are nonpositive, the current BFS is known to be optimal; if not, the variable x_{pq} is chosen to enter the basis, where

$$z_{pq} - c_{pq} = \max_{i,j} (z_{ij} - c_{ij}) \qquad \text{(8-19)}$$

In order to proceed further, we must next obtain the vector $\mathbf{y}_{pq}$, which will be used to select the exiting variable and then to determine the new values of all basic variables. But, in fact, $\mathbf{y}_{pq}$ has been identified already; in the proof of Theorem 8.2 it was established that there must exist on the transportation tableau a loop of the form

$$\{(p,q), (p,r), (s,r), (s,t), \ldots, (v,w), (v,q)\} \qquad \text{(8-14)}$$

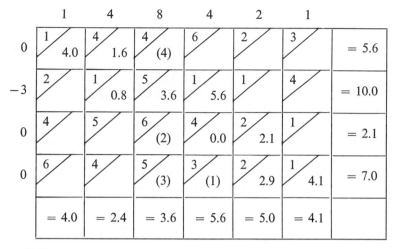

where $\mathbf{a}_{pr}$, $\mathbf{a}_{sr}$, . . ., and $\mathbf{a}_{vq}$ are basic columns, and that

$$\mathbf{a}_{pq} = \mathbf{a}_{pr} - \mathbf{a}_{sr} + \mathbf{a}_{st} - \cdots - \mathbf{a}_{vw} + \mathbf{a}_{vq} \tag{8-15}$$

This representation of $\mathbf{a}_{pq}$ as a linear combination of basis vectors serves to define $\mathbf{y}_{pq}$: The y elements associated with $\mathbf{a}_{pr}$, $\mathbf{a}_{sr}$, . . ., and $\mathbf{a}_{vq}$ are $+1$, -1, . . ., and $+1$, while all other y elements are zero.

Provided that we can find the above loop—when solving by hand this is done simply by trial and error—equation (8-15) contains all the information we need to pivot x_{pq} into the basis. Let us first rewrite the simplex exit criterion in accordance with the notation of the transportation problem: If x_{pq} is to enter the basis, then the basic variable $x_{\mathbf{B}r}$ must leave, where

$$\frac{x_{\mathbf{B}r}}{y_{rpq}} = \min_i \left(\frac{x_{\mathbf{B}i}}{y_{ipq}}, \quad y_{ipq} > 0\right) \tag{4-59}$$

Inasmuch as every positive y_{ipq} equals $+1$, the exiting variable is selected via

$$x_{\mathbf{B}r} = \min_i (x_{\mathbf{B}i}, \quad y_{ipq} > 0) \tag{8-20}$$

The criterion (8-20) is quite easy to apply because the basic variables whose y elements are positive correspond to *every other cell* in the loop (8-14); referring to (8-15), the variables with positive y elements would be x_{pr}, x_{st}, . . ., and x_{vq}. As usual, if the minimum in (8-20) is not unique, the tie may be broken arbitrarily.

All that remains now is to update the values of the basic variables. Given that x_{pq} is to replace $x_{\mathbf{B}r}$ in the basis, the appropriate simplex transformation formulas, with notation revised as before, are

$$\hat{x}_{\mathbf{B}r} = \frac{x_{\mathbf{B}r}}{y_{rpq}} \tag{4-60}$$

and

$$\hat{x}_{\mathbf{B}i} = x_{\mathbf{B}i} - \frac{x_{\mathbf{B}r}}{y_{rpq}} y_{ipq} \qquad \text{for all } i \neq r \tag{4-61}$$

The rth variable in the new basis, whose postpivot value is given by (4-60), is, of course, x_{pq}. Because the pivot element y_{rpq} equals 1, these formulas immediately reduce to

$$\left.\begin{aligned} &\hat{x}_{\mathbf{B}r} = x_{\mathbf{B}r} \\ &\hat{x}_{\mathbf{B}i} = x_{\mathbf{B}i} - y_{ipq}x_{\mathbf{B}r} = x_{\mathbf{B}i}\,(\pm)\,x_{\mathbf{B}r} && \text{for all other } x_{\mathbf{B}i} \text{ in } L \\ \text{and} \quad &\hat{x}_{\mathbf{B}i} = x_{\mathbf{B}i} && \text{for all } x_{\mathbf{B}i} \text{ not in } L \end{aligned}\right\} \tag{8-21}$$

where L is the set of basic variables whose cells are included in the loop (8-14). Notice that, in transforming a basic variable $x_{\mathbf{B}i}$ in L, the choice of whether to add or subtract $x_{\mathbf{B}r}$ is governed by whether the cell of $x_{\mathbf{B}i}$ occupies an "even" or "odd" position in the loop. The formulas (8-21) are actually quite straightforward and simple to use—several examples appear below.

One additional remark is in order at this point. As might have been expected, the phenomenon of degeneracy arises and persists in the transportation tableau exactly as in ordinary linear programming. When the current BFS is degenerate, the exiting variable $x_{\mathbf{B}r}$ chosen by (8-20) may equal zero. If it does, a zero-for-zero pivot will result in which the value of every variable remains the same, as can be seen from (8-21); because the entering variable will take on the value $\hat{x}_{\mathbf{B}r} = x_{\mathbf{B}r} = 0$, the new BFS will be degenerate also. The condition of degeneracy can arise originally when a tie occurs in choosing the exiting variable via (8-20). In that case one or more of the basic variables will take on the new value

$$\hat{x}_{\mathbf{B}i} = x_{\mathbf{B}i} - x_{\mathbf{B}r} = 0$$

and a degenerate BFS will be produced.

To illustrate the pivoting operation let us consider Figure 8-6(a), which represents the upper left-hand segment of some large transportation tableau (with costs c_{ij} omitted). The current values of five basic variables are shown in the diagram—$x_{11} = 3.4$, $x_{12} = 1.1$, and so on—while the other four variables are assumed to be nonbasic. Suppose first that the variable x_{21} has been chosen to enter the basis. The loop (8-14) is immediately seen to be

$$\{(2,1), (1,1), (1,2), (2,2)\}$$

Thus the entering column $\mathbf{a}_{21}$ can be expressed in terms of basic columns as

$$\mathbf{a}_{21} = \mathbf{a}_{11} - \mathbf{a}_{12} + \mathbf{a}_{22}$$

and the y elements for $\mathbf{a}_{11}$, $\mathbf{a}_{12}$, and $\mathbf{a}_{22}$ are $+1$, -1, and $+1$, respectively. At this point the reader should pause to verify that if we had written the loop in its opposite sense, that is, as

$$\{(2,1), (2,2), (1,2), (1,1)\}$$

we would have been led to the same y elements. Applying the exit criterion (8-20),

$$x_{\mathbf{B}r} = \min_i (x_{\mathbf{B}i}, \quad y_{ipq} > 0) = \min (x_{11}, x_{22}) = \min (3.4, 2.0) = 2.0$$

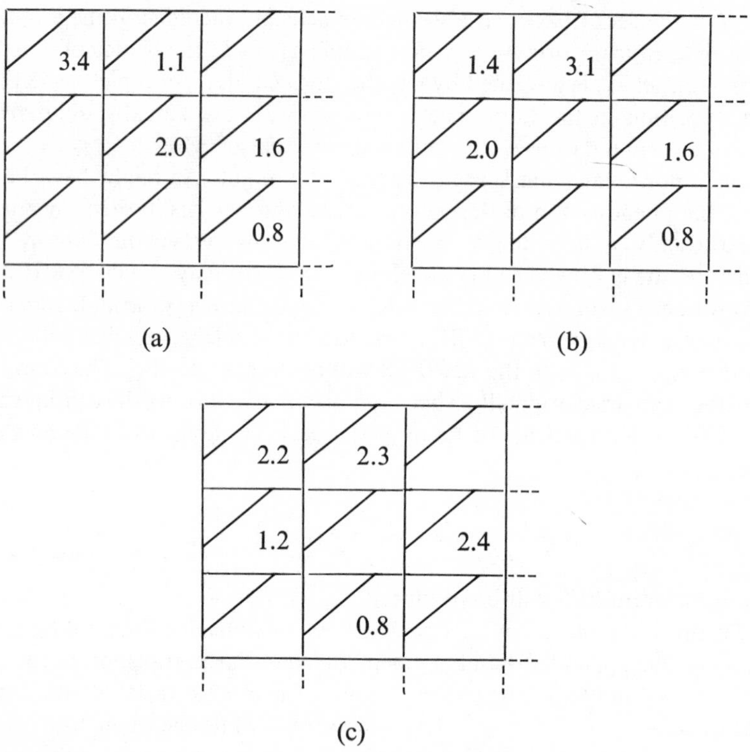

Figure 8-6

so x_{22} will leave the basis and will play the role of $x_{\mathbf{B}r}$ in the transformation formulas (8-21). The new values are

$$\hat{x}_{21} = x_{\mathbf{B}r} = 2.0 \quad \text{(the entering variable)}$$

$$\hat{x}_{11} = x_{11} - x_{\mathbf{B}r} = 3.4 - 2.0 = 1.4$$

$$\hat{x}_{12} = x_{12} + x_{\mathbf{B}r} = 1.1 + 2.0 = 3.1$$

and

$$\hat{x}_{22} = x_{22} - x_{\mathbf{B}r} = 2.0 - 2.0 = 0 \quad \text{(nonbasic)}$$

with the other values remaining unchanged; the new solution is diagramed in Figure 8-6(b).

In performing this pivot, what we did, in effect, was to assign 2.0 units to be shipped from origin 2 to destination 1 and then proceed around the loop making a sequence of adjustments in which each change in the value of a variable exactly compensated for the previous change. Thus, when x_{21} was increased from 0 to 2.0, the value of x_{11} had to be decreased by 2 units *in order that the amount being shipped to destination 1 would not oversatisfy the demand there.* When x_{11} was decreased, however, x_{12} had to be increased to "soak up" the excess supply at origin 1; by the same token, the increase in x_{12} required a decrease in x_{22} in order to avoid shipping too much to destination 2. And this final change completed the loop: The decrease in x_{22} matched the original increase in x_{21}, insuring that the supply constraint at origin 2 would remain exactly satisfied.

With a little practice the student will find that the pivoting procedure is extremely simple. Once the entering variable has been selected and the correct loop identified, it is necessary only to let the eye travel around the loop twice. On the first circuit the values in every other cell (i.e., the first basic cell, the third, the fifth, etc.) are checked and the smallest of them becomes $x_{\mathbf{B}r}$; the loop is then traversed a second time and $x_{\mathbf{B}r}$ is alternately subtracted from and added to the $x_{\mathbf{B}i}$ values in the various cells. To illustrate with another example, suppose that the variable x_{32} is to be brought into the basis which is diagramed (in part) in Figure 8-6(b). After some trial and error we discover that the loop is

$$\{(3,2), (1,2), (1,1), (2,1), (2,3), (3,3)\}$$

The values in the "odd" cells, counting the nonbasic cell (3,2) as cell number zero, are 3.1, 2.0, and 0.8, so $x_{\mathbf{B}r} = x_{33} = 0.8$. We now move around the loop a second time to calculate the new values, which are shown in Figure 8-6(c):

$$\hat{x}_{32} = x_{\mathbf{B}r} = 0.8 \quad \text{(the entering variable)}$$

$$\hat{x}_{12} = x_{12} - x_{\mathbf{B}r} = 3.1 - 0.8 = 2.3$$

$$\hat{x}_{11} = x_{11} + x_{\mathbf{B}r} = 1.4 + 0.8 = 2.2$$

$$\hat{x}_{21} = x_{21} - x_{\mathbf{B}r} = 2.0 - 0.8 = 1.2$$

$$\hat{x}_{23} = x_{23} + x_{\mathbf{B}r} = 1.6 + 0.8 = 2.4$$

and

$$\hat{x}_{33} = x_{33} - x_{\mathbf{B}r} = 0.8 - 0.8 = 0 \quad \text{(nonbasic)}$$

8.9 SUMMARY

The theoretical development of the transportation algorithm is now complete. Given any basic feasible solution, we have seen how to compute the reduced costs, choose an entering variable, determine its **y** vector, identify the exiting variable, and generate a new BFS. The cycle can be repeated until an optimal solution is obtained. Note that we do not bother transforming the objective value z at every iteration; its final value may be computed directly, via $z = \mathbf{c_B}^T\mathbf{x_B}$, after the optimal $\mathbf{x_B}$ has been found.

The algorithm presented in this chapter for solving the transportation problem (8-8) may be summarized as follows:

(1) Construct a transportation tableau[1] and enter the m supplies a_i, the n demands b_j, and the mn costs c_{ij}, as shown in Figure 8-1.

(2) Use any method that conforms to Theorem 8.3 to generate an initial basic feasible solution.

(3) Set $\hat{u}_1 = 0$ and solve the system of equations

$$\hat{u}_i + \hat{v}_j = c_{ij} \qquad \text{for all } x_{ij} \text{ in the basis} \tag{8-17}$$

entering the values $\hat{u}_i$ and $\hat{v}_j$ in the positions shown in Figure 8-1. Determine all reduced costs $z_{ij} - c_{ij} = \hat{u}_i + \hat{v}_j - c_{ij}$. If all $z_{ij} - c_{ij} \leq 0$, the current solution is optimal; if not, go on to step 4.

(4) Select the variable x_{pq} with the largest positive reduced cost to enter the basis. Identify the loop

$$\{(p,q), (p,r), (s,r), (s,t), \ldots, (v,w), (v,q)\} \tag{8-14}$$

check the values in the odd cells to find $x_{\mathbf{B}r}$, and then shift $x_{\mathbf{B}r}$ units of commodity "around the loop" to obtain a new BFS. Return to step 3.

Notice that if total supply does not equal total demand, so that the problem has no feasible solution (presumably due to an error or misprint), that condition will be discovered in step 2.

As a final and more comprehensive illustration of the transportation algorithm, we shall finish solving the problem of Figure 8-5. Recall that the basic feasible solution depicted there was originally generated by the northwest-corner rule; the values $\hat{u}_i$ and $\hat{v}_j$ of the associated dual solution were subsequently obtained, the reduced costs calculated, and each positive $z_{ij} - c_{ij}$ entered in parentheses in its appropriate cell. The largest reduced cost is $z_{13} - c_{13} = 4$, so the variable x_{13} now enters the basis. The loop is

$$\{(1,3), (1,2), (2,2), (2,3)\}$$

with the values in the odd cells being $x_{12} = 1.6$ and $x_{23} = 3.6$; therefore $x_{\mathbf{B}r} = x_{12} = 1.6$ and 1.6 units must be shifted around the loop. The new solution appears in Figure 8-7.

[1] The labor involved in solving a transportation problem by hand can be minimized by laying out a single tableau, including supplies, demands, and costs, with pen and ink, and then using pencil to write in and erase the successive basic feasible solutions, associated dual solutions, and reduced costs.

To obtain the dual solution associated with this BFS we arbitrarily set $\hat{u}_1 = 0$ and then solve the system of equations (8-17):

$$\hat{u}_1 = 0$$
$$\hat{v}_1 = c_{11} - \hat{u}_1 = 1$$
$$\hat{v}_3 = c_{13} - \hat{u}_1 = 4$$
$$\hat{u}_2 = c_{23} - \hat{v}_3 = 1$$
$$\cdots$$

and finally
$$\hat{v}_6 = c_{46} - \hat{u}_4 = -3$$

These values are displayed above and to the left of the tableau in Figure 8-7. When we compute the reduced costs of the nonbasic variables we find that four of them are positive:

$$z_{31} - c_{31} = \hat{u}_3 + \hat{v}_1 - c_{31} = 1$$
$$z_{33} - c_{33} = \hat{u}_3 + \hat{v}_3 - c_{33} = 2$$
$$z_{43} - c_{43} = \hat{u}_4 + \hat{v}_3 - c_{43} = 3$$
$$z_{44} - c_{44} = \hat{u}_4 + \hat{v}_4 - c_{44} = 1$$

These values also appear in the tableau; evidently, the present solution is not optimal and x_{43} will enter the basis at the next iteration.

	1	0	4	0	−2	−3	
0	1 / 4.0	4	4 / 1.6	6	2	3	= 5.6
1	2	1 / 2.4	5 / 2.0	1 / 5.6	1	4	= 10.0
4	4 / (1)	5	6 / (2)	4 / 0.0	2 / 2.1	1	= 2.1
4	6	4	5 / (3)	3 / (1)	2 / 2.9	1 / 4.1	= 7.0
	= 4.0	= 2.4	= 3.6	= 5.6	= 5.0	= 4.1	

Figure 8-7

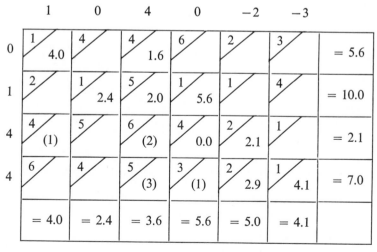

	1	0	4	0	1	0	
0	1 / 4.0	4	4 / 1.6	6	2	3	= 5.6
1	2	1 / 2.4	5 / 2.0	1 / 5.6	1 / (1)	4	= 10.0
1	4	5	6	4	2 / 2.1	1	= 2.1
1	6	4	5 / 0.0	3	2 / 2.9	1 / 4.1	= 7.0
	= 4.0	= 2.4	= 3.6	= 5.6	= 5.0	= 4.1	

Figure 8-8

Two more pivots are required to reach the optimum, and the tableaus they produce are shown in Figures 8-8 and 8-9. The first of these pivots is of the zero-for-zero variety, with x_{43} replacing x_{34} in the basis; note that the associated dual solution changes (compare Figures 8-7 and 8-8) while all x_{ij} values remain the same. The tableau in Figure 8-9 is optimal, but because the nonbasic variable x_{36} has a reduced cost of zero, we know that an alternate optimum would be generated by pivoting x_{36} into the basis. The minimum total cost is $z_{\text{opt}} = \mathbf{c_B}^T\mathbf{x_B} = 40.5$.

	1	1	4	1	1	0	
0	1 / 4.0	4	4 / 1.6	6	2	3	= 5.6
0	2	1 / 2.4	5	1 / 5.6	1 / 2.0	4	= 10.0
1	4	5	6	4	2 / 2.1	1 / (0)	= 2.1
1	6	4	5 / 2.0	3	2 / 0.9	1 / 4.1	= 7.0
	= 4.0	= 2.4	= 3.6	= 5.6	= 5.0	= 4.1	

Figure 8-9

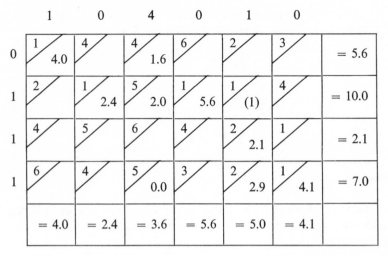

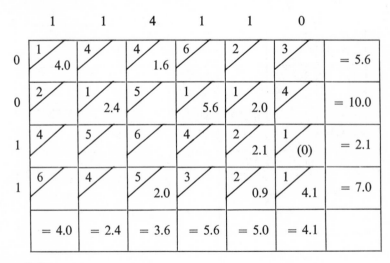

8.10 COMPUTATIONAL REMARKS

The algorithm presented in this chapter was designed, of course, especially to exploit the structure of the transportation problem. As a result, it is far more efficient in solving problems of that type than the simplex method would be. The transportation algorithm has two major advantages over the simplex method: It is faster and it is able to handle much larger problems within a given amount of primary computer memory. The first of these derives mainly from the fact that it uses no arithmetic operations other than addition and subtraction; by comparison, the multiplications and divisions needed in ordinary linear programming are almost an order of magnitude slower, whether by hand or on a digital computer. The second advantage is even more decisive: A problem with m origins and n destinations, which could be solved on a transportation tableau having mn cells, would require a standard simplex tableau with $m + n - 1$ rows and mn columns, for a total of $m + n - 1$ times as many entries. For example, using the transportation algorithm, a problem having 50 origins and 1000 destinations can be solved entirely within the core memory of a moderately large computer, whereas a direct application of the standard simplex method would require sophisticated peripheral equipment and vast quantities of secondary storage.[1]

In addition to its great speed and low storage requirement, the transportation algorithm also has the highly useful capability of allowing problems to be solved entirely in the integer mode. Recall that, if all supplies a_i and demands b_j are integers, so will be the values of the basic variables at every iteration; similarly, integer costs c_{ij} guarantee that the associated dual variables and reduced costs will have integer values throughout. Because integers require fewer memory locations than real numbers, these properties can lead to further savings in storage space. Moreover, integer arithmetic is not subject to round-off error. For these reasons, when the costs or the supplies and demands of a transportation problem are given originally in decimal form, it is frequently desirable to convert them to integers—this can be done usually by multiplying them all by a sufficiently large constant—so that an integer-only computer code can be used.

Our earlier remarks about degeneracy and cycling in the simplex method apply to the transportation algorithm as well. Theoretically, it might be possible to stumble into a sequence of degenerate pivots on the transportation tableau that would lead to a repeated basic feasible solution and, therefore, to infinite cycling. In practice, however, this has never occurred (to our knowledge), and the extra logical testing required to eliminate the remote possibility of cycling is, in the long run, a waste of time.

[1] If the revised simplex method were used, the basis inverse would require about $(m + n)^2 = m^2 + 2mn + n^2$ entries; the columns $\mathbf{a}_{ij}$ would not have to be stored explicitly, however, because $\mathbf{a}_{ij}$ is known to equal $\mathbf{e}_i + \mathbf{e}_{m+j}$. Still, for $m = 50$ and $n = 1000$ the revised simplex approach would require about 22 times as many memory locations as would the transportation tableau.

8.11 NETWORK FLOW THEORY

The transportation problem has many variations that are of great interest in management science and engineering. For example, upper bounds, or *capacities*, may be imposed on the amounts of commodity shipped from certain origins to certain destinations; these lead to additional constraints of the form

$$x_{ij} \le d_{ij} \tag{8-22}$$

Another possibility is that there may be *intermediate* locations that are neither origins nor destinations, but through which a commodity can be trans-shipped. In a typical problem, a company might be required to send manufactured goods from its factories to warehouses and then on to individual stores or customers. Goods neither originate nor are demanded at the kth warehouse, so the constraint on the flow of commodity into and out of it would be

$$\sum_i x_{ik} - \sum_j x_{kj} = 0 \tag{8-23}$$

where x_{ik} and x_{kj} are the amounts shipped from factory i to warehouse k and from warehouse k to customer j, respectively. Because it guarantees that every unit that enters the warehouse also will leave it, (8-23) is known as a *conservation-of-flow constraint*.

The inclusion of constraints such as (8-22) or (8-23) disrupts the precise structure of the transportation problem; the columns of the $\mathbf{A}$ matrix are no longer all of the form

$$\mathbf{a}_{ij} = \mathbf{e}_i + \mathbf{e}_{m+j} \tag{8-10}$$

and Dantzig's algorithm cannot be applied. This does not mean, however, that we must fall back on the simplex method (observe that the extra constraints mentioned above are linear) to solve such problems. They still possess a great deal of special structure and are, in fact, part of the subject matter of another branch of applied mathematics known as *network flow theory* or *graph theory*.

This is not the place for a detailed discussion of network flow problems.[1] Briefly, their distinguishing characteristic is that they can be modeled on spatial diagrams of the type shown in Figure 8-10. This diagram is called a *network* or *graph* and consists of a set of points, or *nodes*, that are linked in various ways by directed lines known as *arcs*. In the typical context the nodes represent geographic locations such as cities, ports, factories, warehouses, and retail stores,

[1] The subject of network flow theory is best presented by itself in a one-semester course or seminar. When this subject is given partial coverage at the end of a linear programming course—as is sometimes done—it is usually at the expense of the material in our Chapter 6 and parts of Chapters 3 and 7.

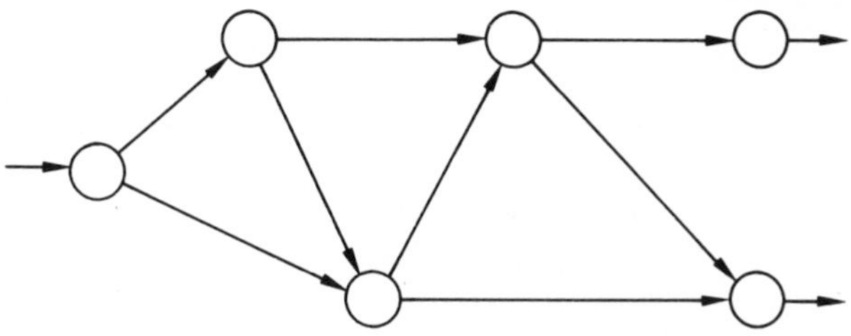

Figure 8-10

and the arcs represent shipping or travel routes between them (e.g., roads, rail lines, commercial air routes). Flow or travel along any given arc normally is restricted to a single direction, which is indicated by an arrowhead.

Depending upon the variety of network flow problem, there may be associated with each of the various arcs a per-unit shipping cost, a capacity, a physical distance, and/or a travel time. Possible objectives include shipping a commodity from origins to destinations at minimum total cost, sending the maximum amount of commodity through a capacitated network, identifying the shortest path (in time or distance) from node P to node Q, and so on. Each of these problem types can be solved by means of one or more algorithms that are designed especially to take advantage of whatever structure is available. In general, the algorithms use only the simplest arithmetic, and most of them are based upon the principles of duality and complementary slackness, just as the transportation algorithm was; in fact, the transportation problem is itself a species of network flow problem.

Our purpose in introducing network flow problems in this last section was twofold: first, to notify the reader of a large and important subclass of linear programs and, second, to encourage him to go ahead and read about them in some text such as Ford and Fulkerson [10]. Generally speaking, once linear programming has been thoroughly mastered, network flow theory is relatively easy to understand; its algorithms are much easier to apply than the various simplex methods and depend upon very few additional theoretical results. We can be confident in saying that network flow problems will pose no serious difficulties to any student who has read our text through to the very end.

EXERCISES

The diagrams used in Exercises 8-11, 8-13, 8-17, et al., to display transportation problem data all follow the scheme of the transportation tableau: The costs c_{ij} are arranged in a rectangular array, with an additional column of supplies a_i attached at the right and an additional row of demands b_j at the bottom.

Section 8.2

8-1. Formulate the following as a transportation problem in the standard format (8-2):

> A company wishes to send various quantities of some commodity from its four manufacturing plants to its four storage warehouses. A total of $10i$ units are available at the ith plant, $i = 1, \ldots, 4$, and $8j$ units are demanded at the jth warehouse, $j = 1, \ldots, 4$. No units are permitted to be shipped from an odd-numbered plant to an odd-numbered warehouse; otherwise the per-unit cost of shipping from plant i to warehouse j is $c_{ij} = ij$. What shipping plan yields the lowest possible total cost?

8-2. Under what circumstances would the (nonbasic) solution

$$x_{ij} = \frac{a_i b_j}{T} \qquad \text{for all } i \text{ and } j,$$

where

$$T = \sum_{i=1}^{m} a_i = \sum_{j=1}^{n} b_j$$

be an optimal solution to the transportation problem (8-2)?

8-3. Would the optimal solution to a transportation problem in the standard form (8-2) be changed if each cost coefficient c_{ij} were multiplied by a positive constant h? What if a real constant k were *added* to each c_{ij}? Answer the same two questions when the constraints are presented in the form (8-3).

Section 8.3

8-4. Suppose that the first origin constraint were deleted from (8-7). Identify a set of $m + n - 1$ columns of the reduced $\mathbf{A}$ matrix that can be arranged into an $(m + n - 1) \times (m + n - 1)$ invertible submatrix.

8-5. *This exercise is only for students who have had a thorough course in linear algebra.*

(a) Let $\mathbf{A}_k$ be a $k \times k$ matrix formed by taking any k rows and any k columns of the transportation problem's constraint matrix $\mathbf{A}$, where $k \leq m + n$. Show *either* that the determinant of $\mathbf{A}_k$ is zero *or* that at least one column of $\mathbf{A}_k$ contains exactly one 1 and no column of $\mathbf{A}_k$ is composed entirely of 0's.

(b) Prove that the determinant of $\mathbf{A}_k$ must be -1, 0, or $+1$ (this is called the *unimodularity* property of the matrix $\mathbf{A}$).

(c) Use the above results to prove that the $\mathbf{y}$ columns associated with any basic feasible solution to the transportation problem consist entirely of integers.

(d) Show that, in fact, the $\mathbf{y}$ columns consist entirely of -1's, 0's, and $+1$'s.

Section 8.4

8-6. (a) Prove that no loop can contain fewer than four cells.

(b) Prove that any loop containing exactly four cells is a rectangle.

8-7. Identify *all* loops contained in each of the following sets of cells:

(a) (1,2), (1,3), (2,1), (2,3), (3,1)

(b) (1,2), (1,3), (2,1), (2,3), (3,1), (3,2)

(c) (1,1), (1,2), (2,1), (2,2), (2,3), (3,2), (3,3)

(d) (1,1), (1,2), (1,3), (1,4), (2,4), (3,4), (4,4)

(e) (1,2), (1,4), (2,1), (2,2), (3,3), (4,1), (4,3), (4,4)

(f) (1,3), (1,4), (2,2), (2,3), (3,1), (3,4), (4,1), (4,2)

Section 8.5

8-8. Is it possible for a transportation problem with $m > 1$ and $n > 1$ to have a basic feasible solution that includes x_{rs}, but no other variable in either row r (i.e., associated with origin r) or column s? Prove or disprove.

8-9. The nine cells (1,2), (1,5), (2,1), (2,3), (2,4), (3,3), (4,4), (4,5), and (5,2) form a basic set on a five-origin, five-destination transportation tableau. What loop is formed when each of the following cells is added to the basic set?

(a) (1,1) (c) (5,1) (e) (5,4)

(b) (4,1) (d) (5,3) (f) (5,5)

8-10. Consider the ordered sequence of cells

{(2,2), (1,2), (1,1), (2,1), (2,2), (3,2), (3,3), (2,3)}

on a transportation tableau having several rows and columns. Does this "figure 8" constitute a loop according to our definition in Section 8.4? Prove that no basic feasible solution could possibly contain as many as six of the seven different variables that correspond to its member cells. Might it ever be possible to encounter this figure-8 pattern in solving a transportation problem?

Section 8.6

8-11. Find an initial basic feasible solution to the following transportation problem, using (a) the northwest-corner rule, (b) the matrix-minimum approach, and (c) Vogel's method. Contrast the three methods and comment on the results obtained.

1	4	2	2	= 3.1
3	4	2	4	= 5.0
4	3	2	3	= 7.5
2	4	4	4	= 12.8
= 10.4	= 2.2	= 4.0	= 11.8	

8-12. Suppose we are given a set of variables that constitutes a basis for a transportation problem having m origins and n destinations. We can determine the values of these basic variables by setting all nonbasic variables equal to zero, deleting any one constraint, and then solving the resulting system of $m + n - 1$ constraint equations in $m + n - 1$ unknowns.

(a) Prove that setting all nonbasic variables equal to zero reduces at least one constraint equation to the form $x_{pq} = a_p$ or $x_{pq} = b_q$, where x_{pq} is a basic variable (i.e., no other unknowns appear on the left-hand side).

(b) Suppose, in accordance with part (a), that some equation of the form $x_{pq} = a_p$ or $x_{pq} = b_q$ has been found. Let the value a_p or b_q be substituted for x_{pq} throughout the above system of equations, and show that the *reduced system* that results must again include at least one equation having only one unknown on the left-hand side.

(c) By repeated applications of part (b), show that the original system of equations is *triangular*. A set of k linear equations in k unknowns forms a triangular system if, for some numbering of the unknowns x_1 through x_k, the equations can be arranged in the form

$$\begin{aligned} a_{11}x_1 + a_{12}x_2 + \cdots + a_{1k}x_k &= b_1 \\ a_{22}x_2 + \cdots + a_{2k}x_k &= b_2 \\ \cdots \\ a_{kk}x_k &= b_k \end{aligned}$$

Section 8.7

8-13. Find a basic feasible solution to the following transportation problem via the northwest-corner rule. Is your solution immediately optimal?

4	2	4	6	= 30
2	0	2	4	= 40
3	1	3	5	= 20
5	3	5	7	= 10
= 30	= 40	= 20	= 10	

8-14. In solving a transportation problem, how can the value of each reduced cost $z_{ij} - c_{ij}$ be obtained directly, that is, without forming the dual problem? Might it ever be reasonable to adopt this direct approach?

8-15. (a) In solving a transportation problem with m origins and n destinations, how many additions and subtractions are required at each iteration to obtain the values $\hat{u}_i$ and $\hat{v}_j$ and then to calculate all the reduced costs?

(b) How many multiplications would be required at each iteration to update the value of the objective z? If z is not updated, how many multiplications are required to calculate its final value after the optimal basic solution has been found? Would you ever recommend updating z at each pivot?

Section 8.9

8-16. Outline an automatic logical procedure for identifying, at each iteration of the transportation algorithm, the unique loop that connects the entering cell with the set of basic cells. Such a procedure must be specified when the algorithm is programmed on a digital computer.

8-17. Solve the following transportation problem three times via Dantzig's algorithm, using (a) the northwest-corner rule, (b) the matrix-minimum approach, and (c) Vogel's method to find an initial basic feasible solution. Compare the overall efficiencies of the three approaches.

2.3	1.3	2.9	1.9	= 5.2
−0.6	0.0	3.9	2.1	= 0.5
0.4	−2.1	0.5	0.0	= 3.1
1.2	0.0	3.0	1.8	= 3.6
= 3.2	= 1.0	= 2.8	= 5.4	

Exercises 8-18 through 8-20. Solve each of the following transportation problems, using any procedure to obtain the initial BFS. Find all distinct optimal solutions (i.e., distinct in having different values for some of the variables x_{ij}).

8-18.

6	6	4	4	7	6	= 16.5
5	5	4	5	4	7	= 12.2
4	4	5	1	3	2	= 8.4
4	5	5	4	6	6	= 10.0
= 7.4	= 4.6	= 6.0	= 9.2	= 8.8	= 11.1	

8-19.

2	1	2	1	2	= 20
3	2	2	2	1	= 23
2	2	3	1	2	= 30
= 21	= 14	= 17	= 9	= 12	

8-20.

2	−3	1	4	−2	2	0	= 125
3	−1	2	0	3	1	−2	= 50
−4	2	3	1	0	0	−1	= 80
−2	0	2	3	0	2	−2	= 70
−4	1	2	2	1	1	−3	= 100
= 60	= 95	= 40	= 60	= 85	= 55	= 30	

8-21. Solve Exercise 8-18 as a *maximization* rather than a minimization problem; that is, use the objective function

$$\text{Max } z = \sum_i \sum_j c_{ij} x_{ij}$$

8-22. Solve Exercise 8-19 as a maximization problem.

8-23. Solve Exercise 8-20 as a maximization problem.

General

8-24. Prove that the problem

$$\begin{aligned}
\text{Min } z &= \sum_{i=1}^{n} \sum_{j=1}^{n} c_{ij} x_{ij} \\
\text{subject to } &\sum_{j=1}^{n} x_{ij} = 1 \qquad i = 1, \ldots, n \\
&\sum_{i=1}^{n} x_{ij} = 1 \qquad j = 1, \ldots, n \\
\text{and all } &x_{ij} \geq 0
\end{aligned} \tag{8-24}$$

where the c_{ij} are any real numbers, would have exactly the same optimal solution (or solutions) if the nonnegativity constraints were replaced by the requirement that

$$\text{All } x_{ij} = 0 \text{ or } 1$$

That is, prove that when (8-24) is solved, the optimal value of every variable will turn out to be 0 or 1. Why is (8-24) known as the *matching*, or *assignment, problem*?

Use the result of Exercise 8-24 to formulate and solve the following problem.

8-25. A certain manufacturing plant has contracted for five major production jobs; one job is to be performed on each of its five assembly lines. The production cost, in hundreds of dollars, of running each job on each line is given in the table below. What assignment of jobs to lines will minimize total production cost?

Job	*Lines*				
	1	*2*	*3*	*4*	*5*
A	15	18	14	17	20
B	14	10	9	13	18
C	6	9	5	10	16
D	8	11	6	10	16
E	12	8	9	13	13

8-26. Repeat Exercise 8-25 under the additional constraint that neither job 3 nor job 4 can be performed on line 1.

8-27. Repeat Exercise 8-25 under the additional constraint that job 1 must be performed on either line 2 or line 3.

8-28. Suppose that the entries in the table of Exercise 8-25 are the *profits* associated with performing each job on each assembly line. Find the assignment that maximizes total profit.

8-29. Uwanimus University is reviewing bids for the construction of a complex of four new science laboratories. Five different construction companies have submitted bids, in thousands of dollars, for each lab, as shown in the table below. The university has decided to award each of the four jobs to a different company, with one company necessarily receiving no contract. How should the contracts be awarded in order to minimize the total cost to the university (i.e., to minimize the sum of the four bids accepted)?

Company	*Laboratories*			
	1	*2*	*3*	*4*
A	70	80	65	65
B	65	80	80	85
C	75	80	75	80
D	70	75	70	80
E	60	70	80	70

8-30. Four bachelors, who have fortunes of 1, 2, 3, and 4 million dollars, and four spinsters, who also have fortunes of 1, 2, 3, and 4 million, would like to get married. The local matchmaker, whom they have all secretly consulted, charges a fee of $100hw$ dollars for arranging a marriage, where h and w are the fortunes of the husband and wife, respectively, in millions of dollars. How should the matchmaker pair off his eight clients in order to maximize his total fees?

8-31. In this exercise we shall develop a method for solving the problem

$$\text{Min } z = \sum_{i=1}^{m-1} \sum_{j=1}^{n} c_{ij}x_{ij}$$

$$\text{subject to} \quad \sum_{j=1}^{n} x_{ij} = a_i \qquad i = 1, \ldots, m-1$$

$$\sum_{i=1}^{m-1} x_{ij} \geq b_j \qquad j = 1, \ldots, n$$

$$\text{and} \quad x_{ij} \geq 0 \qquad \text{for all } i \text{ and } j$$

This is a variation of the transportation problem in which all supply must be shipped and some demands accordingly must be oversatisfied.

(a) Use the nonnegative surplus variables x_{mj}, $j = 1, \ldots, n$, to transform the above problem into the following form:

$$
\begin{aligned}
\text{Min } z &= \sum_{i=1}^{m} \sum_{j=1}^{n} c_{ij} x_{ij} \\
\text{subject to } \quad \sum_{j=1}^{n} x_{ij} &= a_i \qquad i = 1, \ldots, m-1 \\
-\sum_{j=1}^{n} x_{mj} &= a_m \\
\sum_{i=1}^{m-1} x_{ij} - x_{mj} &= b_j \qquad j = 1, \ldots, n \\
\text{and} \qquad x_{ij} &\geq 0 \qquad \text{for all } i \text{ and } j
\end{aligned}
\tag{8-25}
$$

where $c_{mj} = 0$, $j = 1, \ldots, n$, and

$$a_m = \sum_{j=1}^{n} b_j - \sum_{i=1}^{m-1} a_i \leq 0$$

In effect, we have created a dummy origin, which can be thought of as having "negative supply."

(b) The new problem can be solved on a transportation tableau having m origins and n destinations, with the supply a_m at the mth origin being the nonpositive number defined above. Letting the constraint coefficients in (8-25) be collected into an $(m + n) \times (mn)$ matrix $\mathbf{A}$, show that if the ordered set of tableau cells

$$\{(p,q), (p,r), (s,r), (s,t), \ldots, (v,w), (v,q)\}$$

constitutes a loop, then

$$\delta_{pq}\mathbf{a}_{pq} - \delta_{pr}\mathbf{a}_{pr} + \delta_{sr}\mathbf{a}_{sr} - \delta_{st}\mathbf{a}_{st} + \cdots + \delta_{vw}\mathbf{a}_{vw} - \delta_{vq}\mathbf{a}_{vq} = \mathbf{0}$$

where $\mathbf{a}_{ij}$ is the column of $\mathbf{A}$ associated with x_{ij} and

$$\delta_{ij} = \begin{cases} -1 & \text{if } i = m \\ +1 & \text{otherwise} \end{cases}$$

(c) Write the dual of (8-25). Given a primal basic feasible solution, what set of equations must be solved to obtain the values $\hat{u}_i$ and $\hat{v}_j$ of the associated dual solution? Prove that the reduced costs of the nonbasic variables are then given by

$$z_{ij} - c_{ij} = \begin{cases} -\hat{u}_i - \hat{v}_j - c_{ij} & \text{if } i = m \\ +\hat{u}_i + \hat{v}_j - c_{ij} & \text{otherwise} \end{cases}$$

(d) Use the results of parts (b) and (c) to state a step-wise algorithm for solving problems of the form (8-25).

8-32. Let 40, 35, 60, and 50 units of commodity be available at origins 1, 2, 3, and 4, respectively, and let 25, 50, 30, 10, and 45 be the minimum numbers of units required at destinations 1 through 5. Shipping costs c_{ij} are given in the table below. Assuming that all units available must be shipped to the various destinations, find the minimum-cost shipping plan by means of the algorithm developed in Exercise 8-31. Use the northwest-corner rule (suitably modified, of course) to obtain the initial BFS.

3	2	4	1	6
2	2	4	6	3
3	1	2	1	3
4	5	3	2	2

REFERENCES

1. G. Birkhoff and S. MacLane, *A Survey of Modern Algebra*, Third Edition, Macmillan Co., New York, 1965.
2. A. Charnes, W. W. Cooper, and A. Henderson, *An Introduction to Linear Programming*, John Wiley & Sons, New York, 1953.
3. G. Dantzig, "Application of the Simplex Method to a Transportation Problem," Chapter 23 in Koopmans [17], 1951.
4. G. Dantzig, "Upper Bounds, Secondary Constraints, and Block Triangularity in Linear Programming," *Econometrica*, **23**:2, April 1955, pp. 174–183.
5. G. Dantzig, *Linear Programming and Extensions*, Princeton University Press, Princeton, N.J., 1963.
6. G. Dantzig, A. Orden, and P. Wolfe, "The Generalized Simplex Method for Minimizing a Linear Form Under Linear Inequality Restraints," *Pacific Journal of Mathematics*, **5**:2, June 1955, pp. 183–195.
7. G. Dantzig and P. Wolfe, "Decomposition Principle for Linear Programs," *Operations Research*, **8**:1, January–February 1960, pp. 101–111.
8. R. Dorfman, P. Samuelson, and R. Solow, *Linear Programming and Economic Analysis*, McGraw-Hill, New York, 1958.
9. S. Dreyfus and M. Freimer, "A New Approach to the Duality Theory of Mathematical Programming," Appendix II of *Applied Dynamic Programming*, R. Bellman and S. Dreyfus, Princeton University Press, Princeton, N.J., 1962.

10. L. Ford and D. Fulkerson, *Flows in Networks*, Princeton University Press, Princeton, N.J., 1962.
11. D. Gale, H. W. Kuhn, and A. W. Tucker, "Linear Programming and the Theory of Games," Chapter 19 in Koopmans [17], 1951.
12. S. I. Gass, *Linear Programming: Methods and Applications*, Third Edition, McGraw-Hill, New York, 1969.
13. A. J. Goldman and A. W. Tucker, "Theory of Linear Programming," article 4 in Kuhn and Tucker [18], 1956.
14. G. Hadley, *Linear Algebra*, Addison-Wesley, Reading, Mass., 1961.
15. G. Hadley, *Linear Programming*, Addison-Wesley, Reading, Mass., 1962.
16. F. B. Hildebrand, *Methods of Applied Mathematics*, Second Edition, Prentice-Hall, Englewood Cliffs, N.J., 1965.
17. T. C. Koopmans, Ed., *Activity Analysis of Production and Allocation*, John Wiley & Sons, New York, 1951.
18. H. W. Kuhn and A. W. Tucker, Eds., *Linear Inequalities and Related Systems*, Princeton University Press, Princeton, N.J., 1956.
19. C. E. Lemke, "The Dual Method of Solving the Linear Programming Problem," *Naval Research Logistics Quarterly*, **1**:1, March 1954, pp. 36–47.
20. B. Noble, *Applied Linear Algebra*, Prentice-Hall, Englewood Cliffs, N.J., 1969.
21. W. Orchard-Hays, "Background, Development, and Extensions of the Revised Simplex Method," Report RM-1433, The RAND Corporation, April 1954.
22. W. Orchard-Hays, "Evolution of Linear Programming Computing Techniques," *Management Science*, **4**:2, January 1958, pp. 183–190.
23. N. Reinfeld and W. Vogel, *Mathematical Programming*, Prentice-Hall, Englewood Cliffs, N.J., 1958.
24. M. Simonnard, *Linear Programming*, Prentice-Hall, Englewood Cliffs, N.J., 1966.
25. H. Wagner, "The Dual Simplex Algorithm for Bounded Variables," *Naval Research Logistics Quarterly*, **5**:3, September 1958, pp. 257–261.
26. P. Wolfe, "An Extended Composite Algorithm for Linear Programming," Paper P-2373, The RAND Corporation, July 1961.
27. P. Wolfe, "A Technique for Resolving Degeneracy in Linear Programming," *Journal of the Society of Industrial and Applied Mathematics*, **11**:2, June 1963, pp. 205–211.

INDEX

activity coefficients, 51
activity matrix, 51
alternative optima, 120–121
artificial variables, 133, 140–141, 145–147, 148
assignment problem, 278
associated dual solution, 199–202
automatic algorithm, 203

basic column, 36, 99
basic feasible solution, 90–98, 99–100
  definition of, 90
  degenerate, 108
  existence of, 95–98
  as an extreme point, 92–93
  optimality of, 103
basic solution, 36
basic variable, 36, 99
basis or basis matrix, 27, 30, 99
basis inverse, 100
binary variable, 172, 178
binding constraint, 71, 79
block pivot, 129–130

canonical form of linear program, 50–54
caterer's problem, 195
Charnes, A., 119, 147
closed half-space, 39
complementary slackness, 77–79
computation times of simplex method, 157–158
cone, 46
constraint(s), 2, 9, 51, 52–53
  binding and nonbinding, 71
  relaxation of, 71
convergence of simplex method, 118, 119
convex combination, 40
convex set, 39–41
  adjacent extreme points on, 43
  edge of, 43
  extreme point, 39
  extreme ray, 43
cost coefficients, 51
cost vector, 51
cycling in simplex method, 119

Dantzig, G. B., 18, 147, 248, 260

degeneracy in simplex method, 108, 118–120
degenerate basic solution, 37, 108
diet problem, 168–169
Dreyfus, S., 64
dual constraints, 56, 74
dual feasibility, 204
dual problem *or* dual, 55–62
  of canonical-form primal, 55–56
  definition of, 55–56
  economic interpretation of, 69–75
  equality constraints, 58–59
  resource-allocation context, 47–50
  of standard-form primal, 58–59
  unrestricted variables, 59–60
dual simplex method, 203–210
  dual feasibility, 204
  entry criterion for maximization, 210
  entry criterion for minimization, 208
  exit criterion, 209
  infeasibility theorem, 207
  initial solution, 205
  versus primal simplex method, 203–204
  strategy, 204
  summary, 209
dual variables, 56
  economic interpretation, 70–73
duality lemmas, 63–64
duality theorem, 64–68
dynamic programming, 8, 183

edge of a convex set, 43, 127
efficient algorithm, 203
euclidean space, 6, 25
existence theorem of linear programming, 75–76
extreme point, 39, 91–94
  adjacent, 43, 127
extreme ray, 43, 125

Farkas' lemma, 64
feasible region, 6
feasible solution, 2, 6
  basic, 90–98, 99–100
  initial basic, 133–137
formulations of linear programs, 10–13, 163–194
Freimer, M., 64

game theory, 87–88, 185
gaussian elimination, 161–162
geometry of linear programming, 14–17, 37–43
graph theory (*see* network flow theory)

Hitchcock, F., 248
Hitchcock problem (*see* transportation problem)
hyperplane, 38–39, 40
  bounding *or* supporting, 42

identity matrix, 27
incomplete information, 197
indifference curve, 14
infeasible problem, 7, 16
infeasible solution, 2, 6
initial basic solution, 116, 133–137
input-output analysis (*see* resource allocation)
integer variable, 172, 178
inverse of a matrix, 28, 30
investment scheduling, 181–183

job scheduling (*see* production scheduling)
job training, 178–181

Lemke, C. E., 204
Leontief model, 86
line, 38
linear algebra, 24–31
linear combination, 26
linear dependence, 26, 29
  on transportation tableau, 251–254
linear equations, 31–37
  inconsistent, 32
  indeterminate, 32
  redundant, 32, 33
  solution of, 32
  system of, 32
linear independence, 26, 30
linear program, 8–10
  canonical form, 50–54
  case problems, 163–194
  definition of, 8–9
  feasible region a convex set, 41
  geometry of, 14–17, 37–43
  as a mathematical program, 7
  sample formulations, 10–13
  standard form, 54–55
linking constraints, 189
loop on transportation tableau, 249, 250–254

machine scheduling (*see* production scheduling)
marginal value, 70–73
(*see also* dual variables)
matching problem (*see* assignment problem)
mathematical program, 6–7
constraint, 6
feasible region, 6
feasible solution, 6
infeasible problem, 7
objective function, 6
optimal solution, 6
unbounded problem, 7
unbounded solution, 7
matrix, 24–29
conformable, 27
definition of, 24
equality, 25
identity, 27
inverse, 28, 30
multiplication, 27
partitioned, 28–29
scalar multiple of, 25
singular and nonsingular, 28
submatrix, 28
sum, 26
transpose of, 25
measure of effectiveness, 2
minimax problem, 184–185, 188
model, 3, 5
multiple optima in linear programming, 120–121

network flow theory, 272–273
nonbasic variable, 36, 99
nonlinear program, 7–8
nonnegative variable, 9, 51, 53
nonnegativity restriction, 9
nonpositive variable, 61
northwest-corner rule, 254–256, 258

objective *or* objective function, 6, 51–52
one-phase simplex method (*see* penalties method)
operation count, 155–156, 161–162
operations research, 3–5
optimality theorem of simplex method, 103
optimization problem, 2, 3
deterministic and stochastic, 5
optimum *or* optimal solution, 2, 6
Orden, A., 147

parametric programming, 213, 223–229
partitioning of a matrix, 28–29
penalties method, 147–149
phase 1 of two-phase simplex method, 133–137, 145
phase 2 of two-phase simplex method, 139–141, 146–147
piecewise linear objective function, 167–168, 170–171, 182–183, 191, 192–193
pivot (*see* simplex pivot)
postoptimality problems, 197–199, 212–233
addition of a constraint, 231–232, 237
addition of a variable, 230
deletion of a constraint, 232–233, 238
deletion of a variable, 230–231
discrete parameter changes, 213–217
parametric programming, 213, 221, 223–229
sensitivity analysis, 198, 213, 220–222
structural changes, 230–233
primal and dual problems, 55–62, 77–79
complementary slackness, 77–79
duality lemmas, 63–64
duality theorem, 64–68
existence theorem, 75–76
primal problem, 56
primal simplex method (*see* simplex method)
production scheduling, 169–173
setup time, 171
start-up time, 171
production/inventory problems, 178–181, 185–194
back orders, 185, 190–191
employment levels, 191–193
inventory carrying cost, 185
limited storage capacity, 187–188
manpower constraints, 179, 192
material balance equation, 180
multiple plants, 187
multiple products, 188–189
multistage manufacturing, 193–194
overtime costs, 186
safety stock, 190
semifinished products, 193
stock-out, 190

program (*see* linear program, mathematical program)
program-planning, 4

ray, 41
(*see also* extreme ray)
real and artificial variables, 133
recursive constraints, 180, 182, 186
reduced cost, 100, 102–103
redundant constraints in two-phase simplex method, 136–137
redundant linear equations, 32, 33
requirements vector, 51
resource allocation, 10–11, 47–50, 164–168
restricted basis entry, 146
revised simplex method, 149–152
auxiliary matrix, 151
operation count, 155–156
Φ vector, 151
pivot, 150–151
versus standard simplex method, 155–157
summary, 152

scheduling problems
investment scheduling, 181–183
job training, 178–181
production scheduling, 169–173
production/inventory problems, 178–181, 185–194
sensitivity analysis, 198, 213, 220
of cost vector, 221–222
critical value in, 222
of right-hand-side vector, 229–230
shadow price, 70
(*see also* dual variables)
simplex method, 89–90, 94–95, 99–124
computation times, 157–158
computational procedures, 110–114
convergence of, 118, 119
cycling, 119
degeneracy, 108, 118–120
dual (*see* dual simplex method)
economic interpretation, 122–124
entry criterion, 103
exit criterion, 108
initial basic feasible solution, 116, 133–137
minimization versus maximization, 121–122
multiple optima, 120–121
operation count, 155

simplex method (*cont'd*):
optimality theorem, 103
Φ vector, 113
pivot, 106–114
(*see also* simplex pivot)
pivot element, 114
reduced costs in, 103
revised (*see* revised simplex method)
tableau, 112–114
timing, 157–158
transformation formulas, 113, 114
unboundedness theorem, 107
uniqueness of solution, 120–121
simplex pivot, 94, 106–114
degeneracy in, 108
dual (*see* dual simplex method)
entering variable, 103
exiting variable, 108
Φ vector, 113
pivot element, 114
revised (*see* revised simplex method)
tableau, 112–114
transformation formulas, 113, 114
unboundedness in, 107
simplex tableau, 112–114
simplex transformation formulas, 113, 114
simultaneous equations (*see* linear equations)
slack variable, 54–55
solution, 6
(*see also* feasible solution, basic feasible solution)
spanning set, 27, 30
special structure, 177
standard form of linear program, 54–55
standard simplex method (*see* simplex method)
straight line *or* line segment, 38
submatrix, 28
surplus variable, 55
system of equations (*see* linear equations)
systems analysis, 4

transportation algorithm, 260–265
dual variables in, 260–262
entry criterion, 263
exit criterion, 264
loop, 249, 250–254
pivot, 263–265
reduced costs, 260–262
summary, 268
transformation formulas, 264

transportation problem, 239–271
activity matrix, 244–245
alternative formulations, 241–243
assignment *or* matching problem, 278
capacitated arcs, 87, 177
case problem, 173–178
computational properties, 246–248, 271
definition of, 239–240, 244
demands, 240
destination, 239
dual problem, 260
dummy destination, 242
fixed charge, 178
initial solution, 254–260
integrality properties, 246–248, 271
northwest-corner rule, 254–256, 258
origin, 239
redundancy of constraints, 245–246
supplies, 240
tableau, 248–249
transportation algorithm, 260–265
triangularity of constraints, 276
unimodularity, 274
Vogel's method, 258–259
transportation tableau, 248–249
Tucker tableau, 56
two-phase simplex method, 133–142, 145–147
artificial variables, 133, 140–141
phase 1, 133–137, 145
phase 2, 139–141, 146–147
redundant constraints, 136–137
summary, 141–142

unbounded solution, 7, 15–16, 107, 125
unboundedness theorem of simplex method, 107
unrestricted variable, 53
dual of, 59–60
upper-bound constraints, 166

variable, 6
artificial, 133
basic, 36
binary, 172
dual, 56
integer, 172
real, 133
slack, 54
surplus, 55
vector
basis, 27
definition, 24
in euclidean space, 25
inequality relations, 26
linear combination, 26
linear dependence and independence, 26, 29, 30
null, 26
row and column, 24
scalar or dot product, 27
spanning set, 27
unit, 26
(*see also* matrix)
Vogel, W., 258, 259
Vogel's method, 258–259
von Neumann, J., 64

Wolfe, P., 147

y vector, 100

zero-for-zero pivot, 108, 118–120